Feeding the World

Also by Vaclav Smil

China's Energy

Energy in the Developing World

Energy Analysis in Agriculture

Biomass Energies

The Bad Earth

Carbon Nitrogen Sulfur

Energy Food Environment

Energy in China's Modernization

General Energetics

China's Environmental Crisis

Global Ecology

Energy in World History

Cycles of Life

Energies

Feeding the World

A Challenge for the Twenty-First Century

Vaclav Smil

The MIT Press
Cambridge, Massachusetts
London, England

This book was set in Sabon by Best-set Typesetter Ltd., Hong Kong, printed on recycled paper, and bound in the United States of America.

Library of Congress Cataloging-in-Publication Data

Smil, Vaclav.
Feeding the world: A challenge for the twenty-first century / Vaclav Smil.
p. cm.
Includes bibliographical references and index.
ISBN 0-262-19432-5 (hc: alk. paper)
1. Agricultural productivity. 2. Food supply. I. Title.
S494.5.P75S55 2000
338. 1′9—dc21 99-41747
CIP

Contents

Acknowledgments

Most of this book was written during a leave of absence paid for by a grant from the Canadian Donner Foundation, which also covered the cost of research trips to Rome and Beijing.

My debt to hundreds of experts on whose work I relied in writing this book is obvious. The following individuals provided particularly useful advice, help or inspiration: Nicholas Alexandratos (FAO), Dennis Avery (Hudson Institute), Ken Cassman (University of Nebraska), Giuseppe Celestini (FAO), Tim Dyson (London School of Economics), Jacques de Guerney (FAO), Walter Burgos Leon (FAO), Patrick Luciani (Canadian Donner Foundation), François Mazaud (FAO), Per Pinstrup-Andersen (International Food Policy Research Institute), Amitava Roy (International Fertilizer Development Center), Albert Tacon (FAO), and Mao Yushi (Unirule Institute of Economics).

Marjorie Halmarson and Theresa Nichols prepared the illustrations; Greg McNamee and Deborah Cantor-Adams edited the book, and Ori Kometani designed the cover.

Introduction

"To produce a mighty book, you must choose a mighty theme," wrote Herman Melville, the author of my favorite American novel, *Moby-Dick* (Melville 1851). Naturally, I do not interpret Melville's admonition as a matter of simple *sequitur*. After more than two decades of writing books I do not have any illusions about the ease with which one can produce a mighty opus. I am, however, certain that the theme of this volume—how can we best feed some ten billion people who will likely inhabit the Earth by the middle of the twenty-first century—is sufficiently mighty.

Why am I not trying to answer, in the first place, a grander and more fascinating question of how many people the Earth can feed? Because this question has no single answer today, and it will not have one tomorrow (Smil 1994a; Cohen 1995). No complex models are needed in order to demonstrate a large range of plausible outcomes. We would not even have to increase the existing agricultural inputs in order to feed many more than ten billion people in a global economy guided by concerns about consumption equity and offering everybody frugal, largely vegetarian but nutritionally adequate diets. On the other hand, even today's six billion people could not be fed if North America's current average per capita food supply (of which about 40 percent is wasted!) were to become the global norm in a world that would be using much higher agricultural inputs with no better efficiencies than we do today.

The question of how many people the Earth could eventually feed will be asked again and again, but answering it in a definite manner is futile and counterproductive. Futile because it takes only a very short time before many requisite assumptions of detailed scenarios become either completely indefensible or highly questionable. Counterproductive because human population will cease growing for reasons unrelated to food supply. There is a very high probability that humanity will not double in number again, and that its 2050 total of around ten billion people may be very close to (or perhaps even a bit above) its long-term maximum. A more practical, and a more meaningful, inquiry is thus to look into the best means of securing the requisite nutrition for ten billion people. Or, more precisely, it is to ask: Can human ingenuity produce enough food to support healthy and vigorous life for all those people without irreparably damaging the integrity of the biosphere?

A systematic attempt to answer this question should be guided by three important principles: it requires an appreciation of complex realities whose adequate understanding is essential to tackle these questions; it demands a consideration of the entire food chain, from the fundamentals of field cropping to food intakes; and, if rational solutions are to prevail, it must concentrate on improving chainwide efficiencies of production and consumption. Without appreciating this complexity of food supply and demand it is easy to succumb either to catastrophist fears or to cornucopian dismissal—in short, to Ehrlichian doom or to Simonian giddiness.

Catastrophists and Cornucopians

During the early 1960s Paul Ehrlich's specialty was butterflies. Then came a well-publicized vasectomy and *The Population Bomb*, which refashioned him into a principal guru of global catastrophism (Ehrlich 1968). He concluded that "the battle to feed all humanity is over" and that "at this late date nothing can prevent a substantial increase in the world death rate." Not content with this prophecy, Ehrlich darkened his vision a year later by forecasting the end of all important animal life in the ocean by the summer of 1979. This event was to bring "almost instantaneous starvation in Japan and China" and force the Chinese armies to attack Russia "on a broad front on 13 October." Ehrlich's bottom line was suitably apocalyptic: "Most of the people who are going to die in the greatest cataclysm in the history of man have already been born" (Ehrlich 1969).

Obviously, Ehrlich's global population maximum would have to be well below the 1970 total of about 3.7 billion people. The only way to achieve this optimum would be by eliminating about half of humanity. The other leading catastrophist, Lester Brown of the Worldwatch Institute in Washington, has been wholesaling doom almost as long as Ehrlich. Since the mid-1970s he has been promising steadily rising food prices and a new era of permanent global food crises, if not one of a global famine.

To dismiss these extreme claims is easy: realities have been quite different. Nearly three decades after Ehrlich declared the global food fight over and prophesied Chinese armies attacking Russia, Russian military missions are busily selling advanced fighter jets to the Chinese, whose per capita food production is higher than at any time in the country's long history, and whose average food availability is almost as high as the comfortable Japanese mean. There has been no substantial increase in the world's death rate, but just the opposite: life expectancy at birth in the world's most populous countries rose impressively, with China's mean, now at sixty-nine years, surprisingly close to the Western level.

And, contrary to Brown's undeviating forecasts of ever more costly food, the world prices of every staple foodstuff are, in inflation-adjusted terms, lower than a generation ago, many of them at their lowest level during this century or even since the beginning of systematic recordkeeping began a few hundred years ago. Food self-sufficiency in the world's most populous nations and gains of five to seven years of life expectancy per decade would seem to me to be very convincing proofs that the world food situation is not unraveling even in places where one might have expected the worst. Ehrlich does not see it that way, supplanting *The Population Bomb* with *The Population Explosion* (Ehrlich and Ehrlich 1990), and Brown has raised new fears about the world's future food supply with his *Who Will Feed China?*, a work that is, as we shall see in chapter 9, based on wrong data and dubious interpretations (Brown 1995a).

Curiously, catastrophists have no closer allies in their denial of realities than their antagonists, the ebullient cornucopians. These techno-optimists revel in large population increases as the source of endless human inventiveness and they consider food scarcities or environmental decay as merely temporary aberrations. Moreover, their principal protagonist, Julian Simon, claimed that food has no long-run, physical limit (Simon 1981, 1996). Dismissing these visions is also easy, inasmuch as Simon's claim is obvious nonsense. If the global grain output were to continue growing only as fast as it has done during the 1980s (almost 2 percent a year), the annual harvest of cereals would surpass the Earth's mass in less than 1,500 years, roughly an equivalent of the time elapsed since the dissolution of the Western Roman Empire.

But dismissal of extreme claims should not mean the total rejection of either message. A close reading reveals a wealth of irrefutable facts and valuable arguments in the writings of both of these extreme camps—but these patches of sensibility are surrounded by the flood of a true-believer bias precluding recognition of vastly more complex, largely unpredictable, and repeatedly counterintuitive realities. Catastrophists believe that the Earth's true carrying capacity is merely a fraction of existing global population, and that the planet has been overpopulated for several generations; they also maintain that its food production has been an increasingly unsustainable exercise underpinned by the extraction of nonrenewable resources and resulting in a grave damage to the biosphere.

I find much to agree with in this analysis. To begin with, I am convinced that ever larger populations are not of any benefit to the biosphere—or to themselves. That does not mean I have an ideal total for humanity in mind. As a scientist I know that a multitude of natural, technical and social factors constrains the human choices about the planet's carrying capacity, that our understanding of these limits constantly evolves and changes, and that their acuteness and permanence is vigorously debated.

If one really insists on numbers, then the only sensible way out is to offer conditional estimates: *given* this combination of future choices, available resources, techniques, and social arrangements, *then* it is most likely that the Earth could support so many people. Naturally, terms of the debate change drastically if we were to be willing to live like average Bangladeshis or average Californians—and I do not advocate either choice. Nor do I call for a panicky retreat. But, given the undeniable evidence of human impact on the biosphere, I simply argue for as rapid a transition to stabilized counts as is humanely possible. To make a lasting difference, this shift would have to be combined with giving up the idea of steadily increasing material consumption.

Catastrophists also raise very valid points about long-term futures of modern agriculture. Farming increasingly dependent on unceasing inputs of fossil fuels, and hence inherently unsustainable on a civilizational timescale (10^3 years), is now a universal reality. Continuous intensive monocropping—repetitive planting of one crop species supported by high applications of synthetic chemicals—may be profitable in the short run and in narrow monetary terms, but it is surely not the best way to promote longevity and stability of agroecosystems. Signs of worrisome soil degradation—ranging from obviously severe erosion to subtle deterioration of soil structure—are common, as is the misuse of irrigation water. In other locales and regions environmental pollution reduces yields of some sensitive crops. How long can these impacts continue before they start undermining the global capacity to feed ourselves?

But I cannot subscribe to the final catastrophist diagnosis of our civilization being caught in an unstoppable slide toward a global demise. Neither runaway population growth nor irretrievable degradation of agroecosystems is a preordained matter: we know how to take care of most of the problems we face today, be it through often surprisingly simple technical fixes or through better socioeconomic arrangements. Our ingenuity is far from exhausted, and we can do so much more with so much less.

Cornucopians—with their robust faith in admirable adaptive capacities of our species and with their recitals of impressive examples of human inventiveness overcoming seemingly impossible challenges—see existing problems and looming threats merely as matters awaiting effective solutions. Again, there is much I find persuasive and appealing in their diagnosis—but I also have serious doubts about the inevitability of forthcoming solutions.

In spite of our rich repertoire of technical fixes and social adaptations, we have a tendency to wait too long before we commit these resources in a decisive, and effective, manner. Nor can we be certain that we will always judge correctly the acuteness of a problem demanding a timely solution. For example, what would have happened to our crops and to our health if the emissions of chlorofluorocarbons will have continued for another generation, thinning the protective layer of stratospheric ozone by 20 or 40 percent?

Acting as an advocate, I have no difficulty in arraying selected arguments in support of either of the two antipodal camps of catastrophist and cornucopian futures. Acting as a scientist for whom the ancient Roman *sine ira et studio* is not an empty phrase, I have to distance myself from both of these extremes. But a more realistic appraisal cannot be found by a mechanistic conversion to the middle ground. The only intellectually honest account must be pieced together by discarding a variety of myths found across the whole spectrum of our understanding of food, nutrition, agriculture, and the environment, by retaining and amplifying every solid, nondoctrinaire position of extreme explications, and by carefully acknowledging intricate interdependencies as well as clear primacies of various factors.

Consequently, it is not the middle ground I seek but rather the truth, be it uncomfortable or encouraging. When writing on these matters long before the two labels came into common usage, Alfred Sauvy noted crisply that "lack of precision in data and in method of analysis allows shortcuts toward reaching an objective predetermined by prejudice, shaped largely either by faith in progress or by conservative scepticism" (Sauvy 1949).

Shared Concerns

The first guiding principle of this book is to steer clear of either of these two rocks and to show that we have quite a few incremental, unglamorous, but ultimately highly effective means to deal with the challenge. In chapter I will present the catastrophists's argument, laying out basic concerns about our capacity to feed the world population. As I progress to discuss environmental constraints and opportunities, it will become clear that I share an acute concern about a number of unfolding changes that narrow or undermine our path of effective action. At the same time, I will show why I disagree with the catastrophists: the main reason is what I call the slack in the system, an enormous range of opportunities of doing things better.

Obviously, few of these improvements could be accomplished without widespread and effective economic, political, and social transformations. I would never underestimate the roles played by these diverse human factors in food production: it is self-evident that these influences—from government subsidies to the effects of rising incomes and health concerns on food choices—have had a profound impact on agricultural output. But this book will offer no normative solutions, no policy advice in these regards. I have several reasons for limiting my inquiry to the assessment of biophysical constraints and opportunities.

In the first place, we already have two recent assessments of global agricultural potential—FAO's very broadly conceived *World Agriculture: Towards 2010* (Alexandratos 1995), and the World Bank's more narrowly structured study *The World Food Outlook* (Mitchell et al. 1997)—which were edited by economists, and which deal primarily with economic factors and policy issues, as well as Dyson's analysis of global population and food trends which has strong social and economic components (Dyson 1996). In the second place, by focusing on critical biophysical determinants of crop yields and animal productivities, the book probes the lasting fundamentals that can be changed and modified only on time scales very different from those required to discard or to adopt yet another set of ephemeral policies.

For example, a large part of a policy-oriented book written just ten years ago would have to be devoted to the pervasive effect of truly monstrous governmental subsidies setting the rich world's agriculture onto often counterproductive and environmentally damaging courses. Some of the worst excesses of these subsidies are now gone, and many more may be removed in the near future—but concerns about soil erosion, water supply, and biodiversity cannot be disposed of by legislative fiat.

Although economic and social fixes cannot transcend the biophysical limits, they must provide the means for tapping many opportunities for increasing efficiencies and optimizing performances within those constraints. Some of these constraints—such as the total volume of water in ancient aquifers tapped for irrigation—are fixed. Others—such as the maximum water use efficiencies by plants—can change as a result of the changing environment or plant breeding. Too few of these opportunities will be realized unless modern societies pay much greater attention to their food production, which must begin with sustained commitment to both basic and applied research.

When judged by the allocation of labor force, ours are predominantly service economies. They depend, however, no less than millennia ago, on adequate food production. I find it astonishing that this truism is so widely, and so easily, discounted. Saying, as so many economists do, that agriculture does not matter as much as it used to because it now accounts for just a few percentage points of the GDP betrays a touchingly naive trust in arbitrary accounting procedures and the most profound ignorance of the real world. Our "postmodern" civilization would do quite well without Microsoft and Oracle, without ATMs and the WWW—but it would disintegrate in a matter of years without synthetic nitrogen fertilizers, and it would collapse in a matter of months without thriving bacteria. Our first duty is to take care of these true essentials.

Foundations of Agroecosystems

Growing crops is nothing else but a more or less astute management of peculiarly simplified ecosystems. Soils, waters, and biota, above all microorganisms, provide irreplaceable biophysical foundations of these agroecosystems. Chapter 2 is devoted to their critical assessments. The chapter evaluates the claims concerning the potential availabilities of cultivable land, water and nutrients, looks at the gap between typical field performances and maximum photosynthetic efficiencies, and assesses the risks of declining biodiversity.

I conclude, as did de Vries, that none of these basic biophysical factors—availability of farmland, supply of water and plant nutrients, limits on photosynthetic efficiency of crops, and biodiversity required for permanent food production—will pose early and unmanageable restrictions to further growth of food production (Penning de Vries et al., 1995, 1996). Naturally, this conclusion does not imply that there will be no regional shortfalls of some of these fundamental resources. Local and regional scarcities of land, water, and diverse biota are already common, and they may become increasingly severe without appropriate management.

And although the availabilities of cropland, water, and plant nutrients, as well as the provision of essential environmental services by bacteria, fungi, and invertebrates, may more than suffice in global terms, local scarcities will be made much worse due to unusually high rates of anthropogenic change affecting soils, waters, and biota in agroecosystems. Chapter 3 offers a critical look at these changes affecting the biophysical underpinnings of properly functioning food production. It will consider the complex, and far from satisfactory, evidence regarding the rates of soil erosion and its impact on crop yields: selective arguments support either very worrisome conclusions or a relatively relaxed outlook, but it is very difficult to offer any confident long-term prognoses. Similar lack of clarity is evident in appraising qualitative soil decline, above all losses of soil organic matter and salinization.

Whatever the verdicts concerning the current situation and likely prospects might be, I will show that we know how to prevent these kinds of ecosystemic degradation and how to maintain productive soils. We also know what to do in order to reduce one of the most common, and still rising, kinds of environmental pollution generated by agriculture, the introduction of excessive amounts of reactive nitrogen lost from inorganic fertilizers and animal manures. Banning chlorofluorocarbons removed a potentially major threat to agricultural productivity arising from the thinning of the Earth's protective stratospheric ozone layer, but controls of tropospheric ozone, an aggressive oxidant reducing crop yields in all areas affected by photochemical smog, are much more difficult. With rising urbanization and spreading car ownership these effects will be especially worrisome in Asia's populous nations.

Possibility of a relative rapid global warming induced by higher concentrations of anthropogenic greenhouse gases poses by far the greatest potential threat to future agricultural production. The closing part of chapter 3 reviews the best available evidence of possible environmental consequences and their likely effects on crop yields and livestock. These impacts may range from barely noticeable departures from long-term means to gradual changes to which we may rather easily adapt (by introducing new cultivars or by adjusting agronomic practices) to shifts occurring at rates unprecedented in human history and resulting in serious degradation of local or regional production capacity.

Unfortunately, our limited prognostic abilities do not allow us to pinpoint with confidence where such drastic changes are most likely to occur. We are on much safer ground concluding that unless global warming proceeds much faster than expected global food production should be able to adjust to most of its consequences, and our adaptations should be able to keep its negative impacts within tolerable limits.

Efficiency Gains in Food Production

The second guiding principle of my systematic look at our capacity to feed the still-growing human population is to focus attention on the enormous slack in global food production and to review, carefully and conservatively, numerous opportunities for reducing these inefficiencies. All but a small share of increased food production during the twenty-first century will have to come from intensified cropping. In chapter 4 I argue that this inevitable intensification should not rely primarily on new inputs, but rather on more efficient use of existing resources. Tightening the production slack should be particularly rewarding as far as the two key inputs to intensive cropping are concerned: both fertilizer and water use efficiencies are well below the realistically achievable rates. Chapter 4 reviews the best available evidence of this performance slack and details major opportunities for increasing efficiencies. Although no single measure can produce major savings, combinations of appropriate actions can yield impressive gains.

Prevailing fertilizer applications are accompanied by large nutrient losses; nitrogen leakage is particularly large due to leaching, erosion, volatilization, and denitrification. As a result, as little as 20–30 percent of nitrogen used in rice paddies, and often no more than 40 percent of the nutrient applied to rainfed fields, are taken up by crops. Direct measures reducing these losses include regular soil testing, choice of suitable fertilizing compounds, proper ratios of applied plant macronutrients (N, P, and K), adequate supply of micronutrients, correct timing of fertilizer applications, and proper placement of fertilizers. Better agronomic practices, ranging from the cultivation of green manures and organic recycling to reduced tillage and good weed control, would further improve the efficiency of fertilizer use. Combination of these measures has the potential to increase nitrogen fertilizer use efficiency in affluent nations by another 20–25 percent during the next twenty to thirty years, and gains of up to 50 percent are possible in those modernizing countries where the current use is most wasteful.

Similar efficiency improvements are possible for the use of water, both in irrigated and in rainfed fields. Capital-intensive substitution of traditional irrigation methods (flooding and furrow distribution) by pressurized systems (sprinklers, center pivots, lateral lines, and drip irrigation) can more than double average irrigation efficiencies, but simple management measures can be surprisingly effective. They include the planting of inherently more water-efficient C_4 crops, optimized timing of irrigation (taking into account a host of measured environmental conditions), and widespread use of agronomic practices promoting water conservation (ranging from reduced tillage to residue recycling).

Indeed, the delivery of production inputs on time and without excess and the optimization of essential agroecosystemic services that sustain high yields will be the key ingredients of a more precise farming of the coming generations. Although field farming can never become as precise as manufacturing, we must strive for an appreciable increase in the overall performance of cropping. I see the emerging GPS-guided (Global Positioning System) precision farming as only a very small part of this effort: more precise farming must rely on a much broader spectrum of information-rich management approaches ranging from recurrent soil testing to optimal timing of production inputs.

Omnivory is an important part of our evolutionary heritage, and the general affinity for consumption of animal foods has meant that with rising standards of living larger shares of crop harvests have been fed to large animals, poultry, and fish; for cereal grains the global share used for feed is now close to 50 percent and rising. In chapter 5 I present detailed calculations of feeding efficiencies for all important animal foods: milk, eggs, poultry, pork, and beef, as well as for representative herbivorous and carnivorous fishes (carp and salmon). I express the efficiencies per unit of edible product, a more revealing measure than the common comparisons per unit of live weight.

All comparisons—per unit of feed, or in terms of overall feed energy and feed protein conversion—favor milk followed by herbivorous fish, eggs, and chicken. Efficiency of pork production is much closer to that of chicken rather then cattle feeding, and the pig's omnivory is an added advantage in societies with a limited supply of feed grain. Chicken is the most efficient producer of protein per unit of space (including the area needed to grow the feed), and eggs require the least amount of water per unit of edible energy as well as per unit of dietary protein. All these considerations have played a limited role in affluent countries enjoying high intakes of animal foods—but they should become increasingly important as the populous countries of Asia and Africa strive to improve the quality of their diets.

A combination of dairy products, aquacultured herbivorous fish, and eggs, chicken, and pork offers the most efficient way to produce animal protein, but appreciable opportunities to increase reproductive and feeding efficiencies are present for all domesticated species. Universally applicable measures include better use of organic wastes, better harvesting, better processing and storage of roughage and concentrate feeds, and better use of feed additives to improve quality and palatability and to boost nitrogen utilization. More attention should be also given to the breeding and management of animals best suited for the tropics and to sustainable ways of both fresh- and salt-water aquaculture. As in the case of crop production, a large part of the increased animal food output during the next generation should come from a more efficient use of existing feeds.

Food Consumption and Requirements

The quest for tightening up the slack in the food system must go beyond fields and barns. Appraising this potential requires us to investigate the entire food chain. I have long been puzzled at why food—human needs for it, its availability at the retail and household level, its actual consumption, and its consequences—is largely absent from most publications and meetings on the achievements and prospects of the world's agriculture. To dwell solely on factors, conditions, and opportunities for increasing crop harvests and for boosting the productivities of domesticated animals is to offer a badly truncated view of the world's food prospects.

This omission implies that the understanding of food flows beyond the farm gate, appreciation of nutritional needs, and the rational management of the demand for food are matters of little consequence for the global food supply. The very opposite is true, but the task of integrating these considerations into appraisals of possible futures of food supply is complicated by our surprisingly poor understanding of many critical realities: we cannot be confidently prescriptive about how much people should eat. Nor do we know with a great accuracy how much they actually do eat.

In chapter 6 I will show that our statistics on outputs of agricultural commodities and on average food intakes are not as accurate as commonly believed: whereas the uncertainties are marginal in affluent countries, they are substantial throughout Asia, Africa, and Latin America. Major reasons for this include general underreporting of cultivated areas, rough estimates of staple crop harvests, and omission of food produced in kitchen gardens and gathered from the wild. Considerable uncertainties also surround the inexplicably neglected matter of harvest and postharvest food losses: in poor countries they amount commonly to 10–15 percent of the initial yield, and their reduction deserves at least as much attention as the quest for higher yields.

Commonly used food balance sheets are thus often based on imperfect information, and as they summarize food supply available at retail level they do not tell us what we actually eat. Although availabilities and intakes correlate, large variations preclude the use of any reliable scaling to derive the latter rates from the former values, and only repeated dietary surveys can find the typical levels of food consumption. However, the paucity of such studies and inherent weaknesses of such commonly used survey methods as single-day recalls of food intake make it difficult to come up even with good averages. Both underestimates and overestimates of actual intakes are common.

Averages of per capita food energy consumption range mostly between 2,000 and 2,300 kcal/day. This means that in affluent countries, whose daily per capita food supply averages more than 3,500/kcal, the gap between availability and consumption amounts to as much as 40–45 percent of all produced food; reducing this enormous retail, household, and institutional waste should be yet another essential component of the quest for more efficient food systems.

In chapter 7 I demonstrate how our understanding of human nutritional needs, a quest begun during the last decades of the nineteenth century, is still evolving as every decade brings new emphases, new recommendations, and new puzzles. Micronutrient shortages, above all those of iodine, vitamin A, and iron, can be relatively easily remedied by inexpensive supplements. Quantifying and eliminating shortages of food energy and dietary proteins is a much greater challenge.

Most of our knowledge of human energy requirements has been derived from studies in the affluent nations of the northern hemisphere. Standard equations of basal metabolic requirements allow for excellent predictions of energy needs in children, are of limited use for adults (particularly men), and overpredict the needs of populations living in the tropics. Recent studies also demonstrate considerable variability of energy needs not only among individuals but also among groups and populations. Average needs are surprisingly low in some agricultural and pastoral populations adapted to low food availability, as well as in affluent societies dominated by sedentary life styles (particularly for females).

Although the recommendations of desirable protein intakes have undergone many changes during the twentieth century, we still do not have definite standards. Earlier recommendations resulted in exaggerated totals of malnourished people, while recent studies are concerned proper ratios of individual essential amino acids and with dietary protein digestibility. Both energy and protein studies have a number of important implications for assessing the global nutritional adequacy.

Calculations based on the best available evidence show that global food needs average only about 2,000–2,100 kcal/day, substantially below the conservatively estimated supply of 2,800 kcal/day. Even a complete elimination of stunting and higher food needs for more active lives would raise the mean consumption totals by no more than 10 percent. And because of admirable human adaptability, including very low energy cost of pregnancy and adjustments to relatively low intakes, it is dubious to set universal nutritional standards. Existing malnutrition, whose extent may be considerably smaller than the commonly quoted FAO estimates, is overwhelmingly the consequence of unequal access to food.

Nutrition and Health

In chapter 8 I will first show how typical diets, relatively stable for millennia, have been transformed with industrialization and urbanization in all affluent societies. These dietary transitions are now particularly rapid in many populous modernizing low- and middle-income countries. They have included an overall increase in total energy supply attributable largely to higher consumption of animal foods (both meat and dairy products) and dietary fats. Animals foods now provide more than 30 percent of all food energy in affluent countries, and lipids contain more than 40 percent of all calories; in traditional cultures both shares were usually below 10 percent of the total energy intake. Intakes of simple sugars and salt have also increased, as has the consumption of fruits.

In contrast, modern dietary transitions have been marked by pronounced declines in the average consumption of carbohydrate staples (be they cereal grains or tubers), and by even greater decreases in eating of legumes. Much lower intakes of indigestible roughage and, in spite of general plenty, shortages of some minerals and vitamins have been also part of this dietary shift. There is no doubt that this profound dietary transformation and increasingly sedentary life styles have been the two key causes of the rising frequency of obesity. This condition, associated with a number of serious health problems and with premature mortality, is now widespread not only in North America and in parts of Europe but also in societies as different as Mexico and Egypt. Modern high-energy, high-fat, high-sugar, low-fiber diet is also implicated in the etiology of the coronary heart disease, the leading cause of Western mortality, and in higher occurrence of several kinds of cancer.

In spite of numerous uncertainties and controversies regarding the composition of diets best suited to promote healthy lives and to increase longevity, there is now a basic consensus about their makeup. These no-regret recommendations begin with breastfeeding as the best nutritional start for infants. Adequate intakes of good-quality protein are essential to ensure proper mental and physical growth in children. Traditional Mediterranean diets—plenty of complex carbohydrates, fruits, vegetables and olive oil, pulses and nuts, and dairy products, with moderate consumption of meat and fish accompanied by wine—combine best the foodstuffs associated with low rates of chronic diseases and with long life expectations.

But the diets of Mediterranean countries have been shifting toward the less desirable standard modern pattern, and a similar shift away from nutritionally beneficial traditional diets has been under way in more prosperous parts of Asia. Future dietary shifts involving billions of people will have profound effects on the composition of the global food demand. Because education and public awareness can greatly influence the eventual makeup of these diets, the quest for optimized nutrition should become an integral part of long-term goals and guidelines for agricultural production.

If China Can Do It . . .

Naturally, all of the key universal considerations that must be included in any revealing appraisal of the world's capacity to feed itself—basic biophysical realities, opportunities for higher cropping and feeding efficiencies, reduction of postharvest losses, and the quest for rational diets—will have to be addressed at local, regional, and national levels. As the world's most populous country, China offers perhaps the best illustration of possibilities arising from such a systemwide rationalization.

China's combination of relatively constrained natural endowment, already very intensive cropping, substantial postharvest losses, and a large (and still growing) population whose diet is undergoing a rapid transition adds up to an enormous challenge for the country's agriculture. Yet in chapter 9 I will demonstrate that there is little doubt that China has the potential to feed itself, a conclusion of immense importance for the world's agricultural prospects.

A Persistent Theme

Consideration of the entire food chain, from soil bacteria to optimum diets, makes this book different from virtually all "can we feed ourselves" writings. The efficiency argument is its most persistent theme. In order to see how much more could be done with what we already have, and how much less we may actually need, we have to start by appraising the existing productive levels and their plausible improvements. Indeed, this message is the book's mantra: our response to higher demand should not be primarily the quest for higher supply through increased inputs, but rather the pursuit of higher efficiency.

Analogy with our evolving approach to securing adequate energy supply is obvious. When pressed by rising demand we kept on looking, most unimaginatively, for ways to expand the extraction of fossil fuels and generation of electricity. Only eventually, as world energy prices rose by an order of magnitude from their historic lows (from around $2 a barrel in 1972 to nearly $40 a barrel in 1979) we shifted our emphasis to consider whole energy systems and to use the available energy more efficiently.

These approaches—drilling for oil in our buildings, car and jet engines, and industrial plants—brought some stunning savings in a very short period of time before the further aggressive pursuit of these rewarding strategies was undercut by the collapse of the world crude oil price in 1985. They will be coming back, not because of any imminent supply shortages but because of environmental constraints on our ever-growing use of fossil fuels. Higher conversion efficiencies are our best prevention and defense against the potentially worrisome perils of environmental change, be they aggressive oxidizing photochemical smog or potentially highly destabilizing global warming.

I foresee the same pattern in the case of higher food-chain efficiencies. Encroaching environmental constraints will be the main force behind the pursuit of higher efficiencies throughout the entire food system, including gradual transformations of prevailing diets. As per capita availability of farmland declines and as production further intensifies, we will have to take more steps to maintain the quality of soils and waters and to optimize the efficiencies with which we use the principal inputs.

We will have to make sure that soil bacteria have enough recycled biomass to feast on, that soil structure is conducive to soak up rainwater as a sponge; we will get more nitrogen and more irrigation water not just from new fertilizer plants and new wells but increasingly from reducing the losses in field applications of these essential inputs. Acting as smart risk minimizers, rather than just adapting to unwelcome realities, we should be actively taking many of these steps today, without waiting for crippling price signals or environmental breakdowns to begin their work.

Indeed, in some regions and countries we are already fairly advanced along this path of quality-and-efficiency farming: the combination of market pressures and sensible regulations based on long-term concerns about unique resources has taken care of that. But in most places, and not just in poorest countries of Africa, this work is yet to begin. When largely accomplished, this change, inevitable in my judgment, would amount to a new agricultural revolution as a new mode of precise, information-rich farming becomes the norm rather than the exception.

Conditional Outlook

Interpretation of complex realities is always open to major distortions as cultural and individual biases color perception and as dubious explanations become widely accepted truths. Appraisals of our capacity to feed the world's growing population are particularly prone to these misperceptions. Many people see the task as a looming crisis: that term carries an ominous feel as the word's meaning has long time ago darkened from its Greek meanings of decision, or a turning point, to apprehension barely concealing fear. The Chinese interpretation is much more realistic: the word crisis, *weiji*, is made up of two characters, the first one meaning danger, the other one opportunity.

Few great human challenges embody better this dynamic unity than the world's continuing quest for expanded food supply. Catastrophists dwell, not without a great deal of justification, on the first half of the whole, on dangers weakening the Earth's capacity to feed additional billions of people. But by largely neglecting the second half they distort a much more complex reality. They leave out the dimension that completes the image by going beyond apprehension and scares, and that reveals the opportunities renewing realistic hope.

There are no guarantees that we will succeed: irrational policies and misplaced priorities may lead us astray. Inaction, late action, or misplaced emphasis may bring us plenty of future troubles. But we do have the tools needed to steer a more encouraging course, and, as I will try to demonstrate, there appear to be no insurmountable biophysical reasons why we could not feed humanity in decades to come while at the same time easing the burden that modern agriculture puts on the biosphere.

In short, the prospects may not be as bright as we might wish, but the outlook is hardly disheartening. I hope that if you persevere and finish this book that you will come to agree with this appraisal. If so, you can count yourself among the true Malthusians, for the final message of that famous English cleric was not one of despair, but one of cautious hope. Of course, that designation has been firmly connected with depressing forebodings—but such an attribution is misplaced. When a man, shortly before his death, deliberately changes the conclusion and the emphasis of his previously much-publicized work, then it is this final version, rather than the initial product, that should be seen as the intellectual bequest to be associated with his name. Unfortunately this has not been the case with the work of Thomas Robert Malthus.

The English clergyman is known today for his observation, included in the first edition of *An Essay on the Principle of Population . . .* (1798), that "the power of population is indefinitely greater than the power in the earth to produce subsistence for man" and that "this natural inequality . . . appears insurmountable in the way to the perfectability of society." But Malthus closed the second, and so curiously rarely read, edition of his great essay with an appraisal whose eminent sensibility makes it enduring:

> On the whole, therefore, though our future prospects respecting the mitigation of the evils arising from the principle of population may not be so bright as we could wish, yet they are far from being entirely disheartening, and by no means preclude that gradual and progressive improvement in human society. . . . And

although we cannot expect that the virtue and happiness of mankind will keep pace with the brilliant career of physical discovery; yet, if we are not wanting to ourselves, we may confidently indulge the hope that, to no unimportant extent, they will be influenced by its progress and will partake in its success (Malthus 1803).

I do not feel that anything we have learned since Malthus penned these sentences could improve on his finely balanced and eminently sensible judgment. The arguments gathered in this book will make it clear why "we may confidently indulge the hope" for a decent future. I have thus no qualms feeling quite Malthusian—and hence reasonably confident about the prospects of feeding the world during the coming generations.

Feeding the World

1 Reasons for Concern

If we could confidently protract the trend lines of our accomplishments into the future, there would be hardly any cause for concern about humankind's ability to feed itself during the twenty-first century. Perhaps the most impressive way to underscore these achievements is to trace the growth of humanity, a process impossible without providing the requisite amount and quality of food. Prehistoric figures are highly uncertain, but around 10,000 B.C., when the glaciers of the last Ice Age were finally receding, the global population was not much larger than some three million people, and it took it about eight thousand years to reach thirty million (McEvedy and Jones 1979).

Two thousand years later, at the beginning of the common era, with the great empires of Han China and Rome civilizing the opposite margins of Eurasia, there were some 300 million people and that total had not doubled until nearly 1,600 years later (Durand 1974). Then the populations alive during the twilight of the Ming dynasty and the aggrandizement of the Turkish empire took only another three centuries to double, and the 1850 total of 1.25 billion doubled just a hundred years later, shortly after World War II (figure 1.1).

The next doubling, to five billion, was accomplished by 1987, in less than forty years. The six billion mark was surpassed during 1999. Annual growth rates of global population were below 0.1 percent until 1600, rising to 0.3 percent for the subsequent two centuries. The nineteenth-century rate doubled to nearly 0.6 percent, and the average annual growth rate during the twentieth century more than doubled again to just over 1.3 percent (United Nations 1998). That our

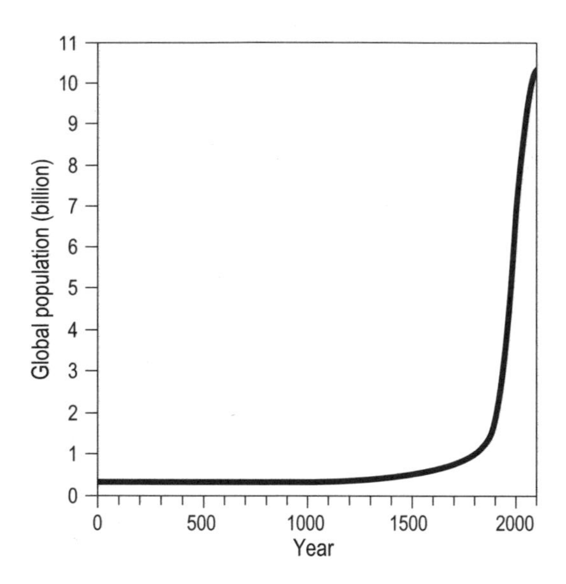

Figure 1.1
Two millennia of global population growth; plotted from data in McEvedy and Jones (1979) and United Nations (1998).

ingenuity has been able to feed this ever larger and ever faster-growing population has been humanity's most important existential achievement. Yet another way to highlight this impressive accomplishment is to look at the two critical trajectories of our evolution—at population densities brought by this growth, and at energy flows that have made this expansion possible.

Our species has evolved from small, scattered, vulnerable, and environmentally inconsequential groups of overwhelmingly vegetarian foragers to the most numerous population of large, substantially carnivorous mammals on the Earth. Typical human densities—measured per unit of land used for foraging or for crop cultivation—have increased from rates lower than those of similarly sized primates to rates vastly surpassing that of any large vertebrate. Chimpanzees—primates with whom we share 99 percent of our genome and whose adult body weight is just 10–20 percent lower than that of smaller-sized adults of our genus—inhabit their African forest environments with densities of about two individuals per km^2 (Clutton-Brock and Harvey 1979).

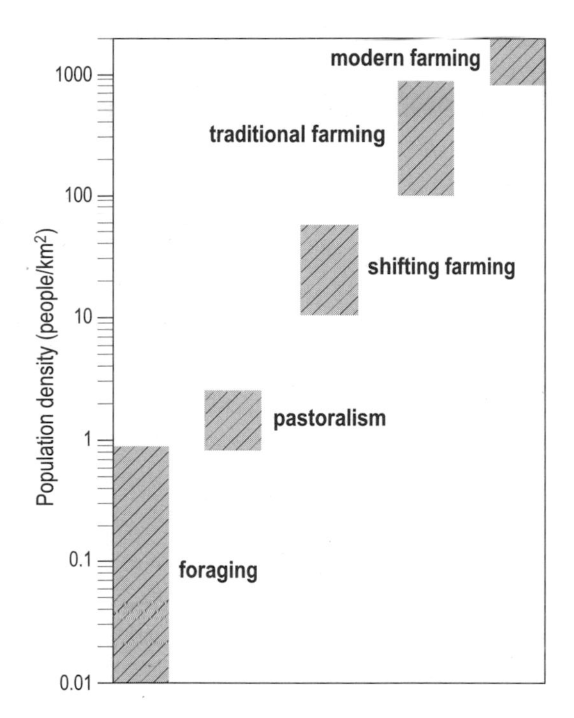

Figure 1.2
Comparison of carrying capacities of foraging, pastoralist, and agricultural societies (Smil 1998).

Our genus had spent most of its evolution in small groups of gatherers and hunters whose population densities, in environments climatically much less equable and much less productive in terms of edible biomass than equatorial forests, were usually well below one person/km^2 (Smil 1994b). But even the least productive shifting cultivation could support at least 10 people/km^2 of planted area, and traditional farming could sustain several hundred people/km^2. Modern agriculture can support well over 1,000 people/km^2 of arable land (figure 1.2).

This thousandfold expansion of carrying capacity could not have been accomplished without first investing ever more human and animal energy into crop cultivation: multicropping, recycling of organic wastes, and irrigation required more animate labor, which was aided, in some societies, only marginally by wind and water machines. During the

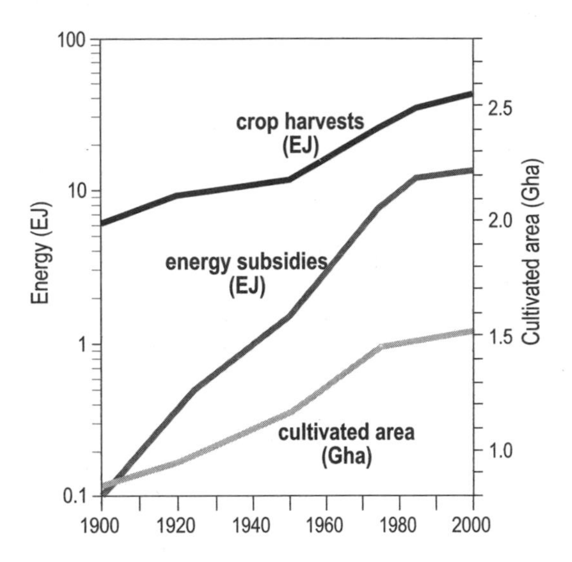

Figure 1.3
Global cultivated area, total crop harvest and energy subsidies, 1900–2000 (Smil 1994).

twentieth century these animate exertions became dwarfed by massive reliance on extrasomatic energies of fossil fuels and electricity. Their introduction began a profound transformation of traditional agricultures, a worldwide conversion from extensive to intensive cropping. Since 1900 the world's cultivated area increased by only about one third, but because of more than a fourfold increase of average yields the total crop harvest rose almost sixfold. This gain has been due largely to a more than eightyfold increase of external energy inputs, mostly fossil fuels, to crop cultivation (figure 1.3) (Smil 1991).

These energy subsidies have been used to build field machinery, power its operations, synthesize farm chemicals (above all energy-intensive nitrogen fertilizers and even more energy-costly pesticides), and support scientific research that has led to development of new crop varieties and more productive domestic animals. This shift from subsistence farming powered by animate energies to agriculture critically dependent on large inflows of fossil fuels has brought a surfeit of food too rich in lipids in affluent countries, and, on the average, adequate per capita supply of basic nutrients in all but a few war-ravaged poor nations. Remaining malnutrition—in both rich and poor countries—is overwhelmingly a matter of access to food rather than one of insufficient availability. Moreover, this substitution of human and animal energies by fossil fuels and electricity has made it possible to embark on a large-scale industrialization, and hence to create a predominantly urban civilization enjoying unprecedented levels of economic growth reflected in better health, greater social opportunities, female emancipation, higher disposable incomes, expanded mobility, and overwhelming flow of information.

Several obvious questions arise when looking ahead. Can we continue producing enough food for the population, which may undergo yet another doubling during the next two generations? Can we satisfy not only the basic metabolic requirements but also the demand generated by rising disposable incomes, which bring major dietary shifts from overwhelmingly vegetarian and monotonous diets to consuming a greater variety of foodstuffs in general and more meat in particular? And can we do so without greatly increasing the extent and the intensity of undesirable environmental changes and without compromising many irreplaceable biospheric services?

Some observers find it easy to give negative answers to all of these questions. Undeniably, the price paid for our agricultural achievements has included massive transformation of natural ecosystems and deepening dependence on fossil fuels, the processes which have resulted in an already significant alteration of biospheric cycles and in growing environmental pollution. Some indications that the rate with which we have been able to extract the food from the biosphere has begun to slow down are seen as a manifestation of biophysical limits already circumscribing the world's food production capacity. If true, this would mean that we have entered a new era of, initially, lowered returns and, eventually, of stagnation and decline, that our recent successes have been just ephemeral accomplishments, and that the Earth could not support, and sustain, yet another doubling of human population.

A brief look at every one of these trends and concerns—at demographic, dietary and productivity transitions—is in order before I will proceed with outlining first the basic requirements and environmental constraints of food production (chapters two and three), and then discussing the impressive opportunities for major efficiency gains all along the food chain (chapters 4–9), as well as the matters impeding and complicating their effective adoption.

Demographic Imperatives

As the most extraordinary century of population growth is coming to its end, there are signs that the global count is moving toward an asymptote faster than was anticipated even a decade ago (United Nations 1998). In 1900 the world's population totaled just over 1.6 billion after it took about 150 years to double. The next doubling, to 3.2 billion, happened in just over sixty years, and by that time it seemed almost certain that yet another doubling would take place before the end of the century. In retrospect, these were exaggerated expectations.

The first notable turning point was reached during the 1960s when the global rate of population's increase peaked at just over 2 percent a year. Absolute annual additions kept on increasing during the next two decades because of the growing global base, but gradual decline in fertility rate ended this trend sooner than anticipated (United Nations 1998). The world population grew by only 1.46 percent a year during the first half, and by 1.33 percent a year during the latter half of the 1990s. Consequently, its absolute increase peaked at 85 million people a year during the latter half of the 1980s, and it fell to 78 million by 1995. Further decline will bring it to 64 million after 2015 and to about 30 million during the 2040s.

These projected annual increases define the minimalist challenge fairly reliably at least for the next generation: mere preservation of today's average per capita food supplies (including the large inequities in food availability) would require the world's rising agricultural production to secure additional food for more than 70 million people a year during the first decade of the twenty-first century and more than 60 million a year during the second decade of a new millennium, substantially lower annual additions than had to be accommodated during the past generation.

Long-range population forecasts are notoriously unreliable, but the medium version of the 1998 UN projection calls for just 8.9 billion people by the year 2050, down from 9.4 billion forecast in 1966, and 9.8 billion in the 1994 revisions (United Nations 1998). Even the latest high variant is well below 12 billion people by the year 2050, and there are now increasing indications that yet another doubling of human population to 12 billion people is not at all inevitable (Lutz et al. 1997).

But even this combination of good news—accelerating decline of fertilities, peaking of absolute annual increases, and the possibility of avoiding yet another doubling of humanity with the world population total stabilizing at less than 10 billion people—will still leave us with much larger populations during the coming two generations: the total of 8.9 billion people in the year 2050 would be 48 percent larger than the global count in 1999. Moreover, the distribution of the global increase will be a major challenge to agricultural production as the future population growth will happen almost exclusively in Asia, Africa, and Latin America.

Populations of all but a few affluent countries will remain basically unchanged: compared to 1998, the UN's medium variant puts it almost 30 million lower in 2050, which means that a number of European nations and Japan would see appreciable population declines. Most notably, Russia is expected to have some 30 million fewer people in the year 2050 than in 1995. The U.S. population will, most likely, approach 340 million, but its net addition of almost 80 million people will be more than negated by declines elsewhere in the industrialized world. Virtually all of the world's population growth of some 3 billion people will be thus in low-income countries, and most of it will be concentrated in nations whose agricultural resources, although absolutely large, are already relatively limited. Brazil is the only modernizing populous country (with more than 100 million people) with abundant reserves of arable land and water.

By the year 2050 India, after adding nearly 600 million people, would have the population more than 50 percent larger than today and it would be, with just over 1.5 billion, the world's most populous country. China, after adding more than a quarter of a billion people to its 1995 total, would be a very close second. Five other countries would add more than

100 million people each: Nigeria, Pakistan, Indonesia, Congo, and Ethiopia. Only Congo and Nigeria have a relatively low population density per hectare of cultivated farmland and large untapped agricultural potential. China and Indonesia are already the paragons of highly intensive cropping, India and Pakistan are close behind, and natural aridity affecting large parts of its territory limits Ethiopia's food-production capacity. And yet, as a group, these nations would have to increase their food harvests by two-thirds merely to maintain their existing, and in many respects inadequate, diets.

Actual increments needed to reach desirable rates of food supply in these and scores of smaller, modernizing nations are not so easily determined. As I will explain in chapter 7, several factors—above all the increase of average body mass, relatively rapid aging of currently very young populations, and the need to eliminate chronic undernutrition and micronutrient deficiencies—will push up the demand beyond the rate dictated merely by extending the existing rates to larger populations. But even in aggregate these additional demands will be surpassed by demand arising from dietary transitions.

Dietary Transitions

Even when liberally adjusted for additional metabolic and nutrient needs, projections of future food requirements based just on expected population totals would be very misleading. Such static assumptions would be largely correct only in the world of frozen purchasing power, social inertia, and economic autarky. In the real world, rising disposable incomes, social transformations, and intensifying global trade are powerful prime movers of dietary transition. Moving up the food chain—that is, eating directly less staple grain and more meat, fish, eggs, and dairy products, and also more plant oils, fruits, and vegetables, while drinking more alcoholic beverages—has been a universal trend clearly discernible in all modernizing countries with rising personal incomes. This process encompasses several universal trends. (More information on dietary transition will be presented in chapter 8 before we look at links between diets and diseases.)

In general terms, the dietary transition ushered in by the industrialization during the nineteenth century has involved the extension of adequate supply of staples to all but some disadvantaged segments of the population, considerable enrichment of previously monotonous diets, and democratization of taste evident in such diverse ways as mass consumption of tea, coffee, and chocolate and a greater attention to the appearance of food. This transformation advances in several stages. Demand elasticity is particularly high at relatively low income levels. Rural populations emerging from subsistence farming and increasing their earnings by selling their farm products or by participating in a variety of nonfarm activities initially spend a large part of their rising disposable income on buying both more and higher quality staples and consuming more nonstaple foodstuffs. Higher family earnings have the same effect among poor urban populations.

As a result, in countries whose premodernization intakes of staples were barely adequate, the increased disposable incomes usually bring first an increase of average per capita cereal consumption ranging from a slight rise among better-off groups to appreciable amounts among the poorest families (only the highest income classes do not participate in this shift). India and Vietnam are at this stage of dietary transition: they still do not enjoy comfortable food supply as their average daily per capita food energy and protein availability are not at least 20–25 percent above the metabolic requirement compatible with healthy and vigorous living.

During the next stage, the saturation of staple intakes culminates in a brief plateau which is then followed by sometimes gradual, but often a surprisingly rapid, decline of staple food intakes. Further increases in disposable incomes are accompanied first by a leveling of staple intakes and by a continuous rise in the consumption of animal foods, vegetables, fruits, plant oils, and sugar. Eventually the elasticities for staple carbohydrates begin turning negative, and later on any income gains change only the relative shares of individual foodstuffs in stabilized, or even slightly declining, overall food intakes. Taiwan and South Korea, whose typical diets were barely above the subsistence level two generations ago, are now in this stage of dietary transition.

Economists' favorite consideration—opportunity cost of time—is yet another important factor driving dietary transition. For example, studies from both West (Burkina Faso) and East Africa (Kenya) demonstrate that the shift from traditional coarse grains to more convenient (but often imported) wheat and rice is related more to the frequency with which women are working outside the home than to household income (Kennedy and Reardon 1994). Countries accomplish their dietary transition at different paces. Perhaps the only useful generalizations, bearing great similarities to the pattern of demographic transition (from high to low birth and death rates) is that the progression becomes fairly fast once the process advances beyond a certain stage, and that late starters move commonly much faster than the pioneers (the European transition proceeded slowly in comparison to shifts in many Asian countries).

Although some segments of populations in a number of European countries had experienced slow improvements of their nutrition after the middle of the eighteenth century, these advances brought only uneven and modest gains in health and longevity of most of the people before 1890. After all, the average French food energy intake did not reach the rate prevailing in today's India until the second quarter of the nineteenth century, and it then went on to increase by half between 1830 and 1880 (Dupin et al. 1984). Great gains affecting food intake—and hence the health and longevity of people in lower socioeconomic classes—were concentrated in three generations between 1890 and the 1950s (Fogel 1991).

In contrast, the transition from barely adequate subsistence to comfortable food supply took only two generations, or even less, in a number of recent modernizers, particularly in Asian countries. In 1950 Japan's average per capita food supply was actually slightly below the Chinese mean—but by 1980 it approached 3,000 kcal/day, and it had one of the world's highest levels of animal protein consumption (Statistics Bureau 1950–1998). China's recent experience has encompassed this whole spectrum in just a single generation. The country's per capita rice consumption first rose, from less than 140 kg of unmilled grain in 1975 to a peak of 170 kg/capita in 1984, but since that time it fell to about 150 kg (State Statistical Bureau 1998). And a careful observer of today's Chinese eating habits knows that in every richer part of the country a great deal of rice actually ends as pig feed, so the real intake rate is almost certainly even lower. In contrast, demand for feed corn will keep rising for a long time to come.

Quantification of long-term impacts of dietary transition obviously depends on the rate of the process and on levels at which the demand will saturate. In aggregate, the transition from largely vegetarian to fairly meaty diets can have enormous effects on grain production. In traditional societies that enjoy basically adequate nutrition (average per capita supply of 2,400 kcal/day) but derive less than 10 percent of food energy from animal foods, such relatively small amounts of meat, eggs, and milk require little concentrated feed. In contrast, the same society with per capita food energy supply at 3,000 kcal/day deriving 25 percent of all food energy from animals will have to resort to large-scale feeding of grain. Even when this is done fairly efficiently (see chapter 5), supply of 750 kcal of animal food would require at least four or five times that amount of plant feed. This would call for additional output of 3,000–3,750 kcal of crops, or more than doubling of net crop harvests used directly for human consumption (2,250 kcal/day).

This is not an extreme case. There are countries where animal foods contribute well over 40 percent of all dietary intake. At the same time, this example does not portray the minimum saturation level as in many other countries dietary transition will never shift average diet to such a high share of animal foods. I will address the question of how far and how fast the process is likely to go in chapter 8. But even in an unlikely case of only a relatively modest impact of dietary transformations, there would be plenty of reasons for concern. Most of them arise from observing the processes that weaken the biophysical foundations of agricultural productivity—and catastrophists argue that this change has already begun to cause worrisome declines in crop productivity.

End of an Era?

As far as persistent catastrophists are concerned, a food crisis has been always imminent, and the end of an era of relative food abundance has been always near. I have arranged the following quotes from Lester

Brown's writings chronologically—but they could be clearly presented in any order because their message remains constantly alarming:

> Farmers . . . can no longer keep up with rising demand; thus the outlook is for chronic scarcities and rising prices (Brown 1974);
>
> Global food insecurity is increasing . . . the slim excess of growth in food production over population is narrowing (Brown 1981);
>
> Population growth is exceeding farmers' ability to keep up. . . . Our oldest enemy, hunger, is again at the door (Brown 1989);
>
> Humanity's greatest challenge may soon be just making it to the next harvest (Brown 1995b).

But recently the catastrophists' arguments gained a new edge as they have been tied more convincingly to some indisputable production slowdowns. Not surprisingly, Lester Brown and Paul Ehrlich have been the leading proponents of the end of an era of food production growth. Their reasoning is quite straightforward. Past difficulties and setbacks with food production were just temporary lags that were soon overcome by new techniques and better management. Now—"because our collective consumption has finally overtaken some of the planet's productive limits" (Ehrlich et al. 1993)—the world is into a new era when the proverbial pie stops growing and, inevitably, the question of how the constant, or even shrinking, pie is to be divided assumes great prominence.

Evidence cited for the limits closely approached, reached or even already surpassed—an epochal shift of the world's food economy from a prolonged state of overall abundance to one of a deepening scarcity—is made up of a litany of items that have been reappearing in the catastrophist literature for decades (Brown 1974, 1989, 1995b; Meadows et al. 1972, 1992).

The very biophysical foundations of agriculture, the argument goes, are at risk. Soil erosion is affecting a large share of cultivated land, removing billions of tonnes of topsoil and millions of tonnes of nutrients and organic matter. Other land degradation problems, including desertification, salinization, waterlogging, and pollution, are adding to this insult, and their combined effect results in lower yields. Farmland available for crop cultivation has been shrinking as farmers abandon marginal and depleted lands, and as all modernizing countries are losing prime arable land to growing cities, industries and transportation.

Water tables are falling in all principal grain-growing regions, be it North America's Great Plains, the North China Plain, or the Punjab, as depletion of many fossil aquifers continues and diversion of more surface water for irrigation becomes questionable as cities and industries compete for that diminishing resource. Intensification of cropping is reaching its limits in many regions where further additions of fertilizer would have marginal effect on yields. The food system is losing genetic heterogeneity, and in many places local food production is moving into a more nonlinear domain where it is disproportionately affected by environmental stresses brought by growing populations (Daily et al. 1998).

Catastrophists use the recent slowdown in the global output of grain as an indisputable proof that the crisis is already upon us. Statistics appear to confirm that message. After adding 450 million tonnes (Mt) of grain during the 1960s and 490 Mt during the 1970s, the world harvested only 260 Mt more in 1990 than in 1980; the 1995 harvest was actually about 2.5 percent below the 1990 total, and the 1998 crop was just over 5 percent larger than the 1990 harvest (figure 1.4) (FAO 1999). Reduced growth rate of the global grain harvest (figure 1.5) has been translated into stagnation of average per capita production: the global cereal harvest is now approaching 2.1 Gt, compared to 1.9 Gt in the

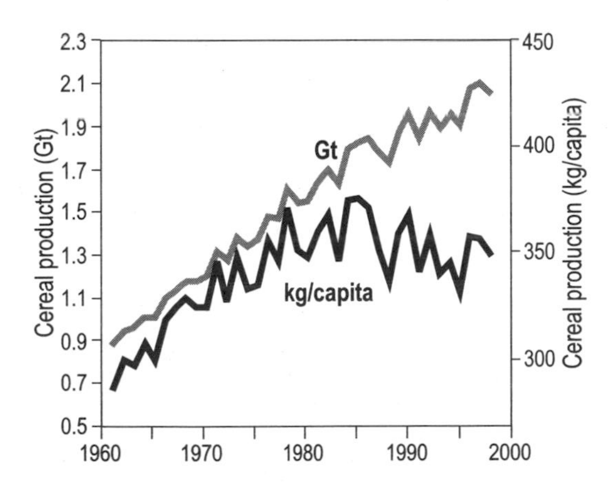

Figure 1.4
Global production of cereal grains, 1961–1998; plotted and calculated from FAOSTAT data (FAO 1999).

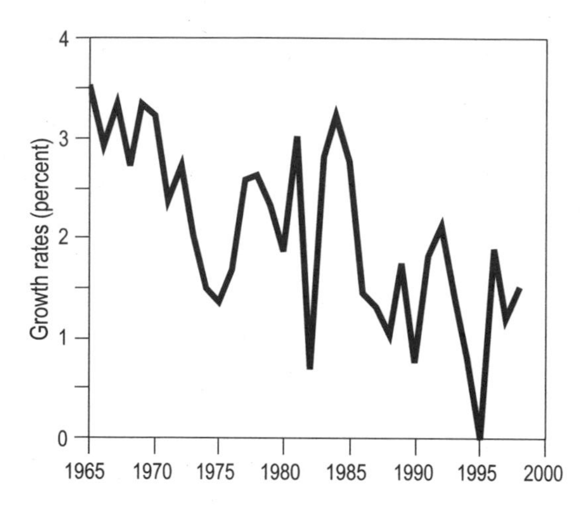

Figure 1.5
Annual growth rates of global cereal production; five-year moving averages calculated from FAOSTAT data (FAO 1999).

early 1990s, but per capita output has fluctuated around 350 kg/year (330–370) since the early 1980s.

Yet there are perfectly noncatastrophic explanations of a large part of this change: nearly all of the slowdown of the 1980s and the stagnation of the early 1990s can be explained by reduced cereal harvests in North America, Europe, and the former USSR (shifts resulting from a combination of domestic supply saturation, long-overdue reduction of extravagant farm production subsidies, and economic breakdown), and by Africa's continuing dismal performance (caused by wars and ubiquitous mismanagement). In contrast, Asian grain harvest was up by more than 13 percent between 1990 and 1998 (slightly ahead of the continent's population gain during those years), and Latin American crop was 28 percent higher, more than twice as high as the overall population gain.

In addition, catastrophists point to an apparently greater annual variability of global staple-grain yields, a development of great importance for the stability of world cereal markets and for the food security of importing nations. But a detailed analysis of growth rates of corn, wheat, and rice yields—done on a continental basis, and as aggregates for developed and developing economies for three times periods (1950–1964,

1965–1979, and 1980–1994)—concluded that any general notion of increased variability in global grain yields is wrong (Naylor et al. 1997). Increasing instability of U.S. corn yields, rather than a generalized fluctuation of all staple grain yields in every major producing region, has dominated the crop's global variation.

This conclusion is both reassuring and worrisome. The concern arises from the fact that the main reason for higher variability of North American corn harvests is the fact that a large share of the crop is now being produced at rates approaching the yield ceiling. At that level effects of adverse weather are not so easily counteracted by a steady rise of productivity, and cultivars selected for the best performance in optimum weather will be particularly affected.

Even more persuasively, catastrophists can also point to notable declines in growth rates of grain yields (figure 1.6). After going up by about 24 percent during the 1970s and 30 percent during the 1980s, the world's average wheat yield remained basically unchanged during the first half of the 1990s, and it was only about 2.5 percent higher in 1998 than in 1990 (figure 1.7). Again, part of this decline can be explained by reduced inputs (above all nitrogen fertilizer) on less subsidized crops.

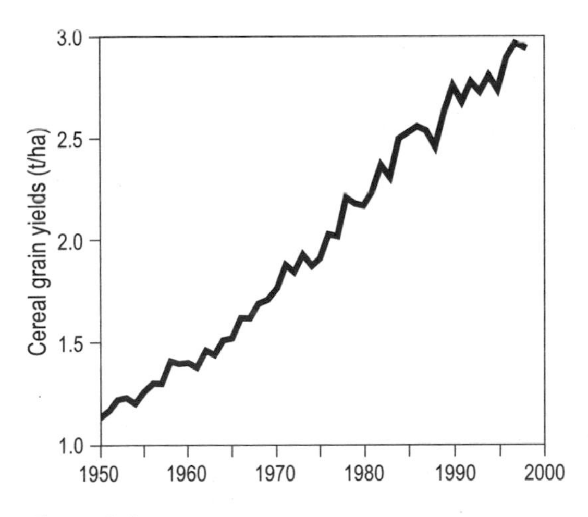

Figure 1.6
Average global yields of cereal grains, 1950–1998; plotted from FAOSTAT data (FAO 1999).

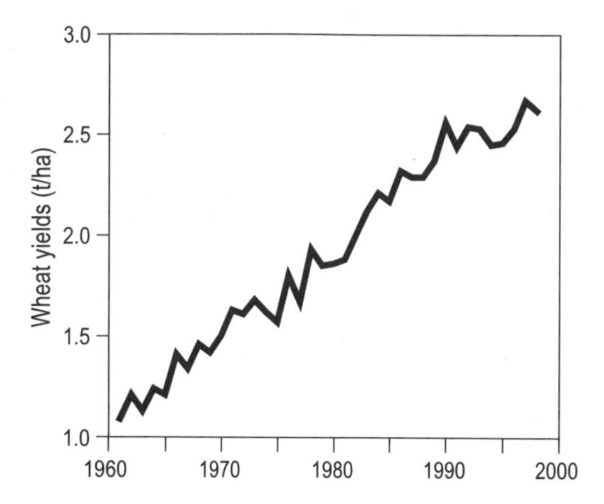

Figure 1.7
Average global yields of wheat, 1961–1998; plotted from FAOSTAT data (FAO 1999).

Corresponding growth rates for rice are about 20 percent, 27 percent, and less than 6 percent. Early adopters of high-yielding rice varieties (South Korea, the Philippines) have seen substantial declines, but relative latecomers such as Bangladesh and India are still on an ascending trend: the Bangladeshi growth rate was nearly three times as high, and India's twice as high, during the 1980s as it was for the previous two decades (Evenson and Rosegrant 1995).

But the broadly reduced growth rates of average cereal yields appear worrisome even to some mainstream agricultural researchers (Cassman and Harwood 1995; Cassman and Pingali 1995; Mann 1999; Cassman 1999). Agronomists and crop breeders have been particularly concerned about yield plateaus reached by staple grains in some of the most intensively farmed regions. Most notably, rice yield increases in Asia, whose annual mean of almost 3 percent during the 1970s spearheaded the continent's Green Revolution, dropped to less than 2 percent during the 1980s and to less than 1 percent during the 1990s (FAO 1999). Some national and regional trends display even more pronounced declines. The rate of increase for the Philippinese rice yields fell from close to 4 percent during the 1960s and 1970s to less than 2 percent during the 1980s and to a fraction of a percent during the

1990s; harvests of Punjabi rice have fluctuated around 5.5 t/ha since the early 1980s.

And long-term experimental plantings at the International Rice Research Institute (IRRI) in the Philippines, as well as elsewhere in tropical Asia, have shown that the yields of IR8, the pioneering high-yield cultivar introduced during the 1960s, have been declining when inputs and management practices have been held constant (figure 1.8; Cassman and Pingali 1995). In actual field plantings increasing fertilization and better management have been able to offset this underlying decline attributable to a combination of causes ranging from declining nitrogen content in paddy soils to increased incidence of crop diseases—but marginal returns of these higher inputs have been clearly diminishing (figure 1.9). This decline is also clear for nitrogen applications to grain crops at the global level. An additional tonne of fertilizer nitrogen helped to produce some 25 tonnes of cereals during the 1960s; the ratio remained just below 20 during the 1960s and the 1970s, but it fell to only about 13 during the 1980s.

Should these trends continue, and even deepen, then the period between the early 1950s and the late 1980s—when yields kept on

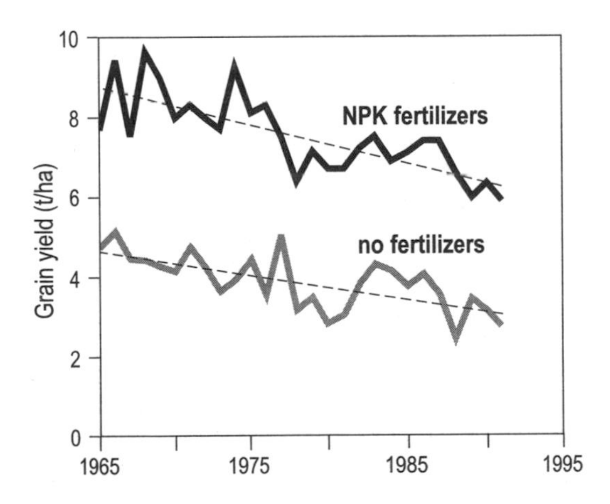

Figure 1.8
Trends in grain yield in a long-term experiment with continuous doublecropping of rice at the IRRI Research Farm in the Philippines; based on Cassman and Pingali (1995).

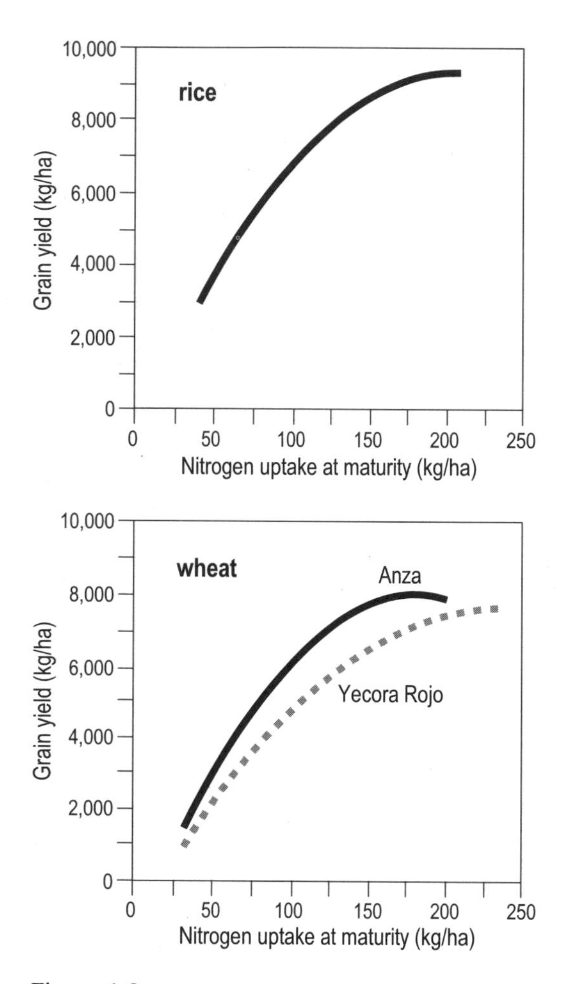

Figure 1.9
Diminishing response to nitrogen fertilizer: aboveground plant N uptake declines and levels off with higher fertilizer applications; based on Cassman et al. (1993).

increasing at impressive rates—would have to be seen as an aberration, as a brief era of unrepeatable growth. But if substantial, albeit attenuated, production increases could be sustained in decades ahead, then the process could be seen merely as an inevitable decline of growth rates common to all maturing systems, and one that could be possibly supplanted by a new growth cycle.

Moreover, some plant breeders point to recent advances in developing a new hybrid rice and a new breed of wheat whose yields (respectively 15–20 percent over existing hybrid rice, and 50 percent above today's best wheat varieties) as clear signs that feared yield plateaus are not imminent. And transgenic crops, already able to withstand pests and resist spoilage, may eventually incorporate traits that will directly increase their productivity—but their widespread diffusion will also carry environmental risks (Gasser and Fraley 1992; Dekker and Duke 1995; Paoletti and Pimentel 1996; Snow and Palma 1997).

Catastrophists also stress that stagnating or falling crop production could not be augmented any more by higher fish catches as most major fisheries have been overexploited and the global catch, after rising from 19 Mt in 1950 to 86.4 Mt by 1989, has been stagnating (FAO 1999) (figure 1.10). In spite of decades of research, scientific understanding of world fisheries is surprisingly poor, but there is a broad agreement that more than two-thirds of traditional commercial species are in trouble: they are being fished to capacity, have been overfished, or are slowly recovering from heavy overfishing.

The collapse of Canada's Atlantic cod fishery off the coast of Newfoundland in the early 1990s—resulting in a near-complete abandonment of a four-centuries old practice and a loss of some forty thousand jobs—was just the most spectacular demonstration of a virtually global trend. And another ancient fishery is also at risk: North Sea cod fishing proceeds at levels much higher than the pre-collapse rates in Canadian Atlantic waters, and if spawning were to fail for two consecutive years that fishery, too, could be rapidly reduced (Cook et al. 1997).

The best outlook for global catch is a stabilization around 90 Mt a year for the next twenty years, which means that the shortfall created by higher demand will reach no less than 20 Mt by the year 2010. Catastrophists will dispute FAO's belief that this gap can be filled by

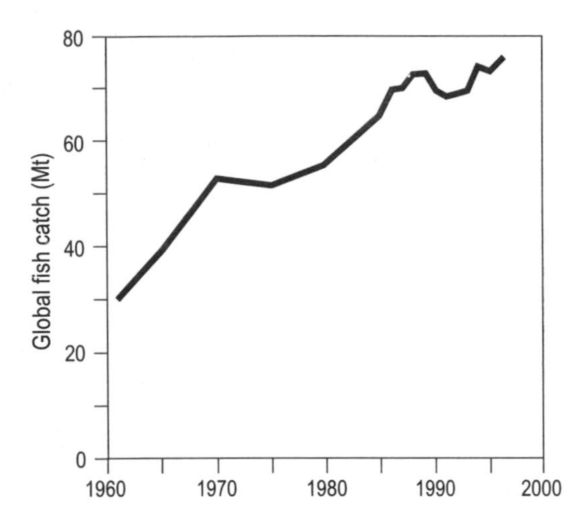

Figure 1.10
Global ocean fish catch, 1961–1998; plotted from FAOSTAT data (FAO 1999).

aquaculture. Nor is there much hope for more food from grazing, insofar as too many grasslands have been overexploited and degraded. At the same time, growing populations with rising expectations and higher incomes drive consumption up the food chain, demanding more grain to feed animals.

Ehrlich's diagnosis that "the extraordinary expansion of food production since Malthus's time has been achieved at a heavy cost—the depletion of a one-time inheritance of natural capital crucial to agriculture" is thus hardly surprising (Ehrlich et al. 1993). And his prognosis? Eventually, he believes, the Earth's sustainable population will number "far fewer than today's 5.5 billion" (written in 1993), and 10 billion people cannot be nourished even temporarily. Brown's conclusion: "With neither fishermen nor farmers able to keep up with population growth, most of the responsibility for achieving a humane balance between food and people rests with family planners" (Brown 1995b).

Needless to say, catastrophists see more problems in the future even if population growth were to slow down significantly: intensifying environmental stresses might make it very difficult to meet even relatively limited new demands, and a relatively rapid global warming could worsen the situation immeasurably: food scarcity, writes Brown, "may well become the defining issue as we exit this century and enter the next" (Brown 1995b).

What is needed at this point is an unprejudiced reality check. How correct are catastrophist perspectives, and to what degree are their conclusions influenced by neglecting those realities that do not fit the preconceived pattern? Answers to these questions should start with a closer look at the natural foundations of agriculture, at land, water, nutrients, photosynthesis, and biodiversity.

2

Appraising the Basics

Essential requirements of food production are easy to list. All of our food derives, directly or through animal metabolism, from photosynthetic conversion of solar radiation into phytomass. In the field this process, whose biochemistry is now minutely understood, is rarely limited by the flux of radiant energy: restrictions on its performance come from the inherent inefficiency of the conversion and from often inadequate supply of water and nutrients. So far we have been unable to raise the overall productivity of photosynthesis, but we have been very successful in channeling higher shares of newly formed phytomass into the tissues we want to harvest.

In spite of steady increases in average crop yields, more agricultural land will be needed to satisfy the food demand arising from growing populations and from higher intakes of animal foods. Although a broad consensus sees a large global potential for expanding cultivated land, there is also a recognition that these land reserves are quite unequally distributed among continents. How much of this land will be, or should be, eventually converted to agricultural use is highly uncertain.

An even more fundamental problem in appraising the adequacy of land for farming are our surprisingly poor inventories of the existing farmland: more land is actually cultivated than has been officially acknowledged, and there are also appreciable reserves of fallowed and underused land. At the same time, continuing farmland losses are also poorly known. But looking just at the area of farmland gives an unrealistic impression: quality of the resource is of immense importance. Our mismanagement can worsen it, while proper agronomic steps can improve it quite impressively. (These qualitative considerations will be taken up in chapter 3.)

In contrast to farmland, freshwater resources appear to be in much tighter supply. The growth of total annual water use has us on a course that would put some of the world's largest food producers close to the limit of their rational water extraction in perhaps less than two generations. In about a dozen less populous arid countries, withdrawals from aquifers are already surpassing the rate of natural replenishment; at least a score of countries, mostly in Africa, will join this unfortunate group in a decade or two. Better management of water use, and not only in irrigating fields, remains the best option for easing this food production constraint (see chapter 4 for details).

As far as the nutrients are concerned, we have solved the challenge of their natural shortages by massive applications of inorganic fertilizers. As a result, annual supply of the three macronutrients now greatly surpasses their natural provision rates through biofixation, mineralization, weathering, and atmospheric deposition. Proper agronomic management, aimed at higher efficiencies of fertilizer use and at optimizing the rates of nitrogen biofixation and recycling, will have to remain a key ingredient of modern cropping (for more on this, see chapter 4).

Although it is an ecological truism that maintaining adequate biodiversity is a key to productive agriculture, defining the adequate species diversity is far from simple. Addressing the changes that undermine it, and taking steps to improve it, involves decisions often resting on a very imperfect understanding of agroecosystem dynamics. Monocultures of annual crops are widely perceived as a highly undesirable way of field farming—but relationships between biodiversity and food production stability are not a simple case of the more the better.

Moreover, discussions of a genetically constrained basis of modern cropping almost invariably leave out the fundamental considerations of microbial diversity, the primary means of assuring smooth functioning of biospheric cycles. Unfortunately, our knowledge of microbial ecology is still rudimentary, as is our understanding of agricultural impacts on bacteria, fungi, and soil invertebrates.

Perhaps nothing illustrates this ignorance so well as the fact that only recently we came to realize the ubiquity of archaea in farm soils. We now know that these organisms—morphologically indistinguishable from bacteria but genetically closer to eukarya—form the third domain of life (figure 2.1). On the positive side, there are many well-proven ways to maintain the desired microbial diversity of farmed soils, and many plant species, ranging from neglected traditional cultivars to new candidates for commercial crops, are available to broaden the basis of our cropping.

Photosynthesis and Crop Productivity

Photosynthesis is by far the most important energy conversion on the Earth. Our high-energy civilization now extracts annually about six billion tonnes (Gt) of carbon in the form of fossil fuels and releases it, as CO_2, into the atmosphere during coal and hydrocarbon combustion. In contrast, global photosynthesis withdraws every year more than 100 Gt of carbon in CO_2 from the atmosphere and incorporates it into new plant tissues (phytomass). This flux is eventually balanced by a return flow of carbon originating in plant and animal respiration and in organic decay.

Cultivated plants account for only a small fraction of global photosynthesis. Almost 2.5 Gt (dry weight) of harvested crops (including grass from managed pastures) contain just over one Gt of carbon, or less than 1 percent of all annually produced phytomass. Of course, our direct

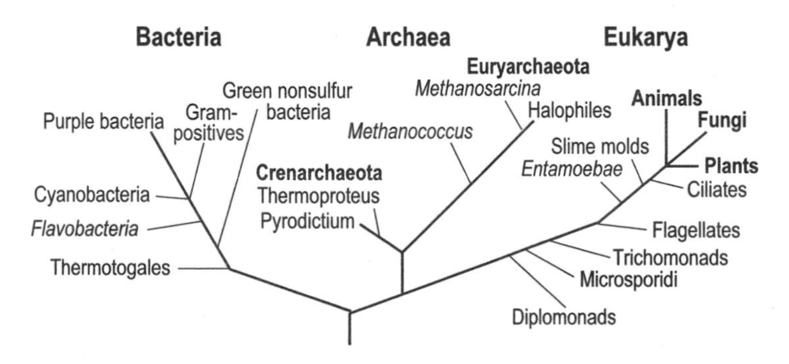

Figure 2.1
New taxonomy of life: Archaea, Bacteria, and Eukarya; based on Service (1997).

food-related claim on global phytomass is much higher, as it also includes grasses grazed by domesticated herbivores and phytoplankton eaten by freshwater and ocean fish, crustaceans and mollusca. But no matter what the final product of the process may be, photosynthesis is not a particularly efficient means of converting energy of electromagnetic radiation to chemical energy in phytomass.

Photosynthetic Efficiency

Photosynthesis is energized by blue and red light, as the light-sensitive pigments, dominated by chlorophylls, absorb solar radiation in two narrow bands, one between 420 and 450 nm, the other one between 630 and 690 nm. Photosynthetically active radiation (PAR) accounts for slightly less than one-half of total insolation. Theoretical efficiency of the conversion can be calculated by adding light energies needed to split molecules of water and CO_2 and by assuming that glucose is the principal initial product: light energy required to produce one gram of glucose is about 46 kJ, and the maximum theoretical efficiency of converting electromagnetic energy of sunlight to chemical energy of new phytomass is about 33 percent.

As the waveband of the PAR corresponds to monochromatic light at 575 nm, the conversion efficiency of PAR to glucose is at best 28 percent—and because PAR amounts to only about 45 percent of all incident radiation, no more than 12.6 percent of solar energy reaching the ground can be theoretically converted to new phytomass. No real plant can do so well. Reflection and transmission of light by leaves reduce the maximum by about one-tenth: about 4 percent of the incident radiation is transmitted and about 6 percent is reflected by a single leaf. Multiple leaf layers reduce these rates, and healthy dense plant canopies may absorb as much as 98 percent of the photon flux.

Part of light absorbed by chlorophylls (commonly 20–25 percent) will be reradiated as heat because the pigments cannot store the sunlight and enzymatic reactions cannot keep up with the incoming energy flux. And respiration consumes a portion of the fixed phytomass in order to energize conversion of simple photosynthates to more complex compounds—carbohydrate polymers, proteins, and fatty acids—needed for supporting structures, and for photosynthetic and reproductive tissues. Efficiency of these conversions is high, between 80–90 percent (Bugbee and Monje 1992).

Total respiration rates of individual species are determined primarily by their photosynthetic pathway. C_3 species follow a photosynthetic pathway beginning with the production of phosphoglyceric acid (PGA), a compound containing three carbons, and they include all but 5 percent of all terrestrial plants. Major C_3 crops include wheat, rice and barley, all legumes, tubers, and sugar beets. All C_3 plants respire during the night as well as during the day. In contrast, C_4 plants concentrate CO_2 in their cells prior to its photosynthetic reduction, produce first the four-carbon acids, malate and aspartate, and have little or no photorespiration. Consequently, their net photosynthetic rates are appreciably higher than those of C_3 species. C_4 plants also require higher temperature and full sunlight to optimize their photosynthetic rate, which can be some 40 percent higher than in C_3 species. Corn, sorghum, and sugar cane are the only major C_4 crops.

Reaction and respiration losses lower the maximum actually achievable photosynthetic efficiency to slightly more than 5 percent. The highest recorded short-term maxima of net photosynthesis in some highly productive plants under optimum growing conditions are between 4–5 percent for C_4 varieties and no more than 3 percent in C_3 species (Evans 1980). For most plants the achievable rates are further limited by shortages of water and macronutrients, especially of nitrogen and phosphorus, and for species growing in higher altitudes and latitudes also by low temperatures. Combination of these natural restrictions puts a clear limit on the rate at which the terrestrial biosphere can take up atmospheric carbon.

Phytomass formed per unit of sunlight is essentially constant, about 1.7 g/MJ of intercepted solar radiation for corn, 1.4 g/MJ for rice and wheat and 1.2 g/MJ for soybeans (Sinclair and Horie 1989). Only under controlled, optimized hydroponic conditions does growth become limited by radiation. Consequently, Salisbury (1988) was able to quadruple wheat's maximum daily field productivity by increasing inputs of photosynthetically active radiation to four times the normal summer mean. Best conversion efficiencies for well-fertilized and well-watered field crops are between 2 and 3 percent. Maximum daily growth rates reported for field crops (all in g/m^2) range from just over 50 for corn and sorghum to around 40 for sugar cane, above 35 for rice and potatoes, between 20 and 25 for most legumes, and just short of 20 for wheat.

Zelitch (1982) showed that when properly measured—that is, not over short periods of time, and not just on a single occasion under ideal conditions, but as seasonal averages in field—net photosynthesis has a close relationship with crop yield. Increasing its rate would boost our harvests, but, so far, we have been unable to do that. Intensifying research into the biochemical and biophysical foundations of photosynthesis has, so far, made little difference to production of food and feed crops under field conditions. We now understand the reaction's biochemistry in exhaustive detail and can follow its progress by increments of 10^{-15} second (Fleming and van Grandelle 1994). Yet the crops we cultivate today are not photosynthetically any more efficient than their traditional predecessors, or their wild relatives, and their growth rate has not increased as a result of domestication and breeding.

Harvest Index

Consequently, impressive yield improvements during the twentieth century have not resulted from higher rates of photosynthesis and from faster plant growth, but overwhelmingly from increases in the proportion of photosynthate accumulated in tissues we wish to harvest, be they seeds, leaves, stems, or tubers (Gifford and Evans 1981). This change is conveniently expressed in terms of the harvest index, the ratio of edible seeds, leaves, stalks, or roots and the crop's total phytomass. Donald and Hamblin (1976) gave a detailed background of the concept and described many historic changes of the index.

For example, traditional wheat varieties cultivated at the beginning of the twentieth century were around one m high and had harvest indexes between 20 and 30 percent. Mexican semidwarf cultivars of the 1960s measured no more than 75 cm, and their harvest index was around 35 percent. By the late 1970s many wheat cultivars had harvest indexes close to 50 percent, producing as much grain as straw. Such plants have relatively short, light stems and a few short leaves per stem but many ears with a high yield of grain.

Harvest indexes of field crops vary both among major cultivars and for the same cultivar grown in different locations. Typical averages are now between 40 and 47 percent for semidwarf wheats and 40 and 50 percent for high-yielding rice. The highest ratios recorded for wheat are around 50 and for rice up to 57 percent. Sweet potato is the only major crop with harvest index at 60 percent. There is a less welcome consequence of rising harvest indexes: as less of the phytomass becomes available for recycling, larger shares of nutrients are exported from the fields.

Due to considerable variations of the total plant yield, the correlation between grain yield and harvest index is far from perfect but still unmistakeable. For example, at the turn of the century Japanese rices produced less than 3.5 t/ha while the mean of the 1980s was about 6 t/ha. During the same period of time American soft winter wheat yields rose from just over 2 to 3.5 t/ha (Eastin and Munson 1971; FAO 1999).

The necessity of producing indispensable structural and photosynthetic tissues puts a clear limit on the rise of harvest index. The most likely maxima for cereals are around 65 percent, and values up to 80 percent may be achieved with some root crops. Values above those levels would compromise the integrity and function of plant's roots, stems, and leaves. For grain legumes a major constraint for achieving higher harvest indexes is their protein-rich seed. But neither the natural limit on photosynthetic efficiency nor the approaching maxima of harvest indexes prevent the narrowing of the gap between today's typical yields in high- and low-income countries.

And optimized agronomic practices (efficient provision of water and nutrients, elimination of pests and diseases) can also reduce often very large differences between the usual and the maximum yield attainable when water, nutrients, and pests do not impose any limitations (figure 2.2). The intent of the following comparisons is not to imply that the gaps between recorded maxima and national or global averages could be eventually closed, merely to indicate the plentiful room for improvement. Interestingly, most of the recorded maxima are not of very recent origin (Wittwer 1974; Gilland 1985; Fageria 1992).

The highest corn harvest, 22 t/ha, was gathered in Michigan in 1977 (with the addition of 426 kg N/ha). The highest wheat crop in North America, 14.1 t/ha, was recorded in 1965 in Washington (with

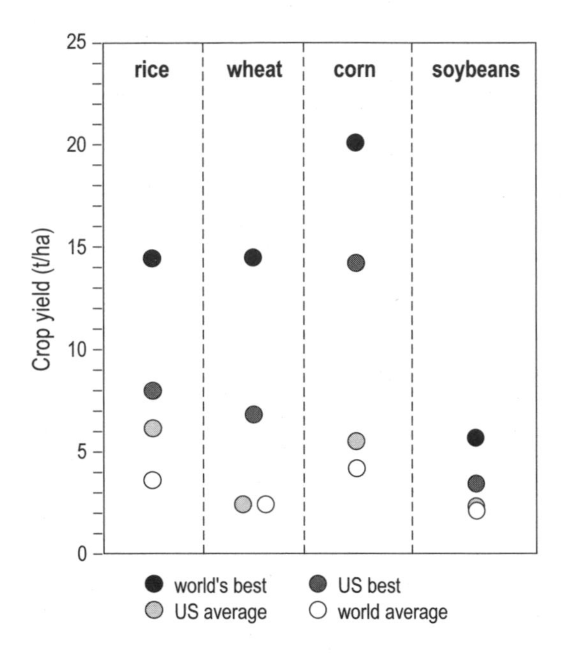

Figure 2.2
Gaps between the world's best, U.S. best, U.S. average, and world's average staple grain and soybean yields; plotted from data in Fageria (1992) and FAO (1999).

135 kg N/ha), and in Asia in China (in 1978) at 15.2 t/ha. And the highest rice yield came from Japan in 1981 at 13.2 t/ha (with 832 kg N/ha). Regional and national maxima of the 1990s were around 8 t/ha both for corn (United States) and for winter wheat (the Netherlands, Denmark) and between 6 and 7 t/ha for rice (Japan, Korea), while the global means are 3.5 t/ha for corn, just 2 t/ha for wheat, and 3 t/ha for rice (FAO 1999).

Land, Water, and Nutrients

Perhaps the most important weakness shared by the increasing number of fashionable end-of-the-century evaluations of long-term food prospects is the tendency to treat the triad of land, water, and nutrients in a rigidly static accounting manner. The basic formula rarely varies: this much land (or water or nitrogen) is currently available, that much can be accessed, extracted, or synthesized by a certain date, which, depending on the postulated demand for food, may or may not be enough to produce the required harvest.

A good illustration of the static approach is the common assumption that virtually all of the best cultivable land is already cropped and hence any additions will be necessarily of lower quality. In reality, currently cultivated good farmland is being degraded by excessive erosion, salinization, nutrient mining, and reduction of soil biota (see chapter 3), whereas the productivity of fairly poor but potentially cultivable land can be increased substantially by drainage, irrigation, terracing, land leveling, and vigorous organic recycling.

Simply put, the quality of these resources is a highly dynamic variable—and so is the efficiency of their use. Assiduous agronomic management can raise fertility of marginal soils and reduce the need for conversions of natural ecosystems to cropland; efficient water use can extend water resources that seem to be used to their full capacity; and in most intensively farmed regions optimized applications can stabilize the use of fertilizers without foregoing further growth of yields. This section is devoted almost solely to the basic accounting of the three indispensable resources; the next chapter will consider some undesirable qualitative changes involving land, water and nutrients, and the following chapter will take up in some detail numerous opportunities for their more efficient use.

Agricultural Land

Land's importance for food production is self-evident: only some 2 percent of the world's food energy and no more than 7 percent of all dietary protein come from waters. That we do not know with satisfactory accuracy something seemingly so easily measurable as the total area of cultivated land is not surprising. Official statistics, which most of the poor countries submit to the FAO for compilation of continental totals and of the global sum, commonly underestimate the actually cultivated area. The underestimates shown in the following two Asian examples may be larger than is usually the case, but they illustrate both the ubiquity and the possible extent of these errors.

Ongoing cadastral surveys of Nepali hills have revealed that the cultivated area of the region is almost four times as large as shown by the official decennial National Agricultural Census figures, with subregional multiples ranging from more than two to more than eight (Gill 1993). And while China's official statistics show 95 Mha of cultivated land, the actual nationwide total is at least 140 Mha, or roughly 50 percent higher (for more details on China's farmland, see chapter 9). In addition, in most low-income countries there is a far from insignificant amount of cultivated lands within villages, and also within cities.

Nutritional contributions of small, but intensively cultivated, kitchen gardens are important in countries ranging from small Caribbean islands to large economies of Asia (Hoogerbrugge and Fresco 1993). Yet, not surprisingly, estimates of the total area under home gardens are highly uncertain. Differences in definition are less of a problem than the sheer absence of representative regional data. In Java anywhere between 20 and 35 percent of village area may be cultivated, and in Sri Lanka two-fifths of households without any farmland cultivate their own home gardens. Areas of urban home gardens are even more underestimated.

Accounting is complicated by the fact that the extent of cultivated land keeps changing not only due to the losses of nonagricultural uses and gains from converting natural ecosystems, but also because of outright abandonment of both very good and very marginal fields. Although the extent of annually cultivated land in core farming areas may remain very steady from year to year, there may be considerable additions or subtractions on the margins (both in spatial and qualitative sense), some of them very short-lived, others long-lasting or even permanent.

Underusing or abandoning good land in the midst of intensive farming areas is also not uncommon. One would expect that a combination of the fixed amount of farmland and growing population in densely inhabited areas would lead to further intensification of crop cultivation. Yet in many places just the opposite has happened. Perhaps most notably—given the island's position as the paragon of human crowding—Preston (1989) found that underutilization of farmland exists even in south-central Java, one of the world's most densely populated and highly intensively cultivated areas.

Diversified household income strategy is the major explanation in most cases of abandoned or underutilized farmland. The combination of a higher standard of living (offered by long-distance migration to large cities or by more lucrative full- or part-time jobs in newly established manufacturing or in local or nearby urban services) and the desire to keep the land in family ownership is a common cause of this paradoxical change of land use.

Unplanned industrial development contributes to the underuse or abandonment of good farmland by creating isolated patches of land sandwiched between new factories and roads: some of them continue to be cultivated, but abandoned patches of this oddland can be seen all around the world, now perhaps most commonly in rapidly modernizing regions of Asia stretching from India to Korea and including perhaps most prominently China's coastal provinces, with their double-digit rates of annual economic growth.

In all industrialized regions there is also a great deal of land abandoned because of soil contamination. Although it would be very expensive to restore some of this land, other plots can be cleaned up quite economically. Plants themselves can often do the job: phytoremediation is emerging as an effective method for cleanup of many contaminated soils. Plants can do this by taking up and accumulating contaminants in their tissues, by transpiring volatile organic hydrocarbons through their leaves, by releasing root exudates stimulating microbial activity in soil, and by increasing the rate of mineralization within the rhizosphere (Burken and Schnoor 1997).

The reality of diverse farmland categories—currently cultivated and officially recorded land, cultivated but unacknowledged fields, formerly cultivated but now fallowed land, recently cultivated but now abandoned land, and farmland abandoned long time ago but potentially cultivable—presents a difficult classification and accounting challenge in preparing informative national and global accounts.

Only small European countries with set cultivation patterns, low population growth, and excellent statistical services know the extent of their actually cultivated and fallowed land with high accuracy. In the rest of the rich world errors on the order of 5 percent are the norm, while in most Asian, African, and Latin American countries the real totals may differ by anywhere between 10 and 20 percent from the official figures, and in quite a few counties that difference may surpass 30 percent. The accuracy of these figures cannot be easily improved by using readily available multispectral satellite imagery: in the absence of a great deal of reliable ground data, spectral confusion between various kinds of land use may result in misclassifying up to a third of the total area (Pratt et al. 1997). Whatever the actual extent of currently cultivated land, it is being reduced through conversions to built-up areas needed to house growing populations and to put in place new industrial and transportation infrastructures. This process will most likely only accelerate during the next two generations, as land claims for housing additional billions of people will be much increased by the expansion of economic capacities needed to meet rising developmental aspirations. With greater affluence the growing cities will also require more space for solid waste disposal and for water treatment, and they may also claim more suburban land for recreation.

These are, inevitably, worldwide phenomena, but their consequences are naturally most worrisome among low-income populous countries whose average per capita farmland availability is already well below 0.2 ha (figure 2.3). China offers perhaps the best example of this concern: during the 1990s the country added about 150 million people, but its arable land was reduced by at least another five Mha, or just over 3 percent of its total. In absolute terms this loss was equivalent to all farmland in Hungary or in Malaysia (but, as I will show in chapter 9, about 60 percent of this loss was due to conversion to forests, pastures, orchards, and ponds, rather than to urban, industrial, and transportation uses).

Besides China, the highest absolute farmland losses due to growing population and economic modernization will occur in India, Nigeria, Pakistan, Bangladesh, Brazil, and Indonesia, the nations where half of the projected increase in the world population is to occur during the next generation (UNO 1998). And farmland losses continue even in the most affluent societies with low population growth rates. By the early 1990s urbanized areas covered 15 percent of

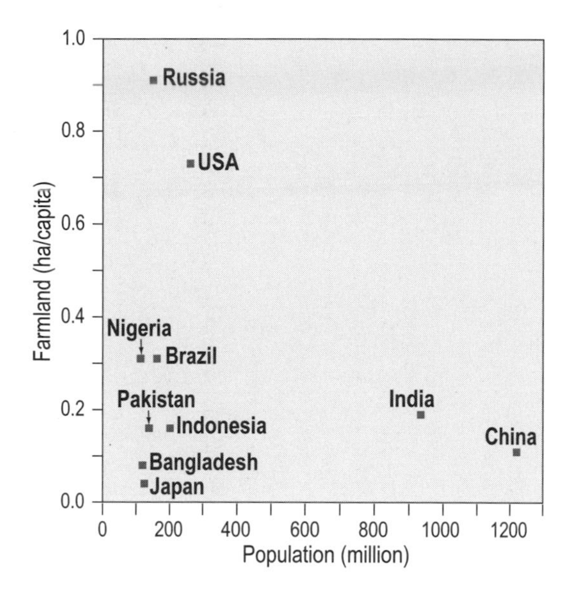

Figure 2.3
Per capita farmland availability in populous countries; calculated from official cropland statistics in FAO (1999).

England. If their annual advance of 10,000 ha a year remained the long-term mean, a quarter of the country would be taken up by cities by the year 2075, a share that is already surpassed in the country's north-western region.

Agriculture itself is a substantial consumer of arable land: irrigation needs about 0.1 ha of land for access, canals, pumping stations, and pipes for each hectare of effectively watered land. Consolidation of livestock production into larger enterprises will claim more farmland, as will better access roads to remote regions, new machine sheds, fertilizer and grain storage structures, and food and feed processing facilities. The aggregate values of these losses depend heavily on average assumptions, but relative rates equal to between 0.5 and 1 percent may be conservative estimates for most populous countries during the next two generations.

To what extent can these losses be compensated by bringing new farmland under cultivation? Given the land's pivotal role in food production, virtually all appraisals of long-term agricultural prospects have offered estimates of potentially cultivable farmland that could be added by converting forests, grasslands, and wetlands to cropping. Some of these studies have also tried to estimate the fraction of that overall potential that will be most likely converted to fields or pastures during a specified period of time. These are understandable and helpful efforts, but they require careful interpretation.

To begin with, estimates of potential agricultural land are based on only very rough mapping of the world's soils. *Soil Map of the World,* compiled at 1:5,000,000 scale (1 cm equals 50 km) by the FAO and UNESCO during the 1970s (FAO 1971–1981) remains the most detailed global systematic map of soils—and it has served as the base for a more recent map of world soil resources, prepared at 1:25,000,000 scale (FAO 1993). A more detailed effort, at 1:1,000,000 scale, has been under way, and by the mid-1990s two-thirds of the world's countries completed their national maps. Those maps, however, cover less than a third of the continental surface, and regions containing less than two-fifths of the world's population. Inevitably, any totals derived from these maps will have considerable, but hard to evaluate, margins of error.

More than a generation ago the panel on the world food supply of the President's Science Advisory Committee (PSAC) came up with an often cited estimate that 3.19 Gha, or about 24 percent of the Earth's ice-free surface, can be cultivated (PSAC 1967). At that time there were at least 1.45 Gha of actually cultivated land. Revelle (1976) multiplied the PSAC total by the number of crops that could be cultivated during a growing season in various climatic zones to come up with the total potential gross cropped area equal to 4.23 Gha. A decade later the Dutch Model of International Relations in Agriculture built up the estimates of potentially available farmland from 222 different soil regions based on the FAO/UNESCO 1:5,000,000 soil map (Buringh 1977). This mapping ended with a total almost identical to the PSAC's value: 3.419 Gha, or 25 percent of the land area. Consequently, Buringh (1977) concluded that feeding twelve billion people is no problem from an agricultural point of view.

Twenty years later a IIASA project used a similarly detailed bottom-up approach to estimate the extent of land with crop production potential in the world's less developed regions by inventorying agroecological variables determining the productivity of rainfed agriculture (Fischer and Heilig 1997). Digitized soil and landform characteristics and climatic data were combined to create an inventory of several thousand land units, each of which was then tested for its suitability to grow any of the twenty-one principal crops ranging from rice to olives. The land found suitable for rainfed cropping was then divided into very suitable, suitable, and marginally suitable categories. Finally, areas protected in national parks and other reserves, and the land currently under cultivation and in urban, transportation, and industrial use was subtracted to obtain the potentially available totals.

The study concluded that, in addition to roughly 1.5 Gha of existing farmland, there are slightly over 1.6 Gha of additional land with rainfed cultivation potential, with about 300 Mha located in areas with sufficient rainfall. Because about three-fifths of this potentially best land is covered by forest and wetlands, converting it into farming would inevitably entail major deforestation and wetland losses. Post-1950 intensification of cropping has already helped to save large areas of natural ecosystems from conversion to farmland, and only further continuation of this trend will be able to preserve more of the nature's legacy (Avery 1997; Waggoner et al. 1996).

Undoubtedly, the total area of potential farmland is quite large, but its spatial distribution is highly uneven and its initial quality will be generally inferior to the existing cropland. National rates, and hence the eventual global extent, of converting natural ecosystems to future farmland depend on a multitude of constantly changing factors ranging from domestic policies to terms of trade, and from newly tapped national resources to the speed of demographic transition. Most of the potential cropland in low-income countries is roughly split between South America and Africa, with Brazilian *cerrado* and grasslands of the sub-Saharan Africa representing the greatest arable land reserves. Opening up this land would mean, above all in the latter case, further extensive tropical deforestation, and, because of limited transportation links, production from much of this new land would be, certainly at least in the initial stages, overwhelmingly only for local consumption.

Total land requirements in a world of nine to ten billion people will depend heavily on the composition of their average diets and on the intensity of cultivation. An overwhelmingly vegetarian diet produced by high-intensity cropping would require no more than 700–800 m^2/capita. A fairly balanced Chinese diet of the late 1990s—supplying about 2,800 kcal/day with about 15 percent of this total coming from animal foodstuffs—is produced from an average of 1,100 m^2/capita by methods ranging from highly intensive cultivation to extensive single cropping (the nationwide multicropping index is about 1.5). In contrast, the Western diet, with its high share of meat and dairy products, claims up to 4,000 m^2/capita; but using that total is inappropriate because very large shares of the available food supply (averaging 3,400–3,700 kcal/capita) are wasted (for details see chapter 6).

Average daily supply of 2,500 kcal/capita would be quite adequate for affluent nations, and if the 30 percent share of animal products were composed largely of dairy products, poultry, and pork, then even a moderately intensive single cropping would not need more than about 1,500 m^2/capita to provide it. In contrast, high shares of beef could push this rate up to 3,000 m^2/capita. Consequently, the extreme land requirements needed to feed ten billion people would range from just 800 Mha to 3 Gha, and the most reasonable range would be between 1.1 and 1.5 Gha. This would call for no more farmland than we are already cultivating, even without further intensification of cropping.

A clear conclusion is that in global terms the availability of farmland is not a limiting factor in the quest for decent nutrition during the next two generations. If there were no other natural constraints on food production, and if all land-scarce nations had adequate purchasing power, then land-rich regions should be able to produce enough food even without new major land reclamation efforts. Combination of a moderate increase in cultivated area and higher intensity of cropping would assure an appreciable safety margin for the global food supply.

This means that the affluent countries should not experience any weakening of their food production capacity because of the declining availability of farmland. For example, when assuming no major shifts in demand, the USDA projected that the country's crop area needed to meet both domestic and export food supply will decline (USDA 1989). Only high export demand would keep the total area steady, and push the irrigated area by about a third.

Regional and national assessments for low-income societies tell a different story. Combination of continuing population growth and uneven spatial distribution of potentially available farmland will only increase already substantial differences in per capita availability of arable land in those countries. FAO's (1981) detailed appraisal of long-term agricultural prospects in ninety poor countries (excluding China) found that fifty-one of them had abundant or moderately abundant reserves of arable land—but their population was just one-third of the assessed total. In contrast, eighteen countries with extreme land scarcity, already cultivating an average of 96 percent of the potentially arable land, supported half of all the population. In regional terms, per capita land availability remains high in Latin America, and more than adequate in sub-Saharan Africa. The greatest concerns exist, and will intensify, in the Middle East and in South and East Asia.

Combination of limited reclamation opportunities and growing space requirements for larger populations means that countries in these regions are facing continuous declines in per capita availability of arable land. Because many of these countries already experience, or will soon experience, serious scarcities of water supply, the extent to which the falling land availability could be compensated by increasing intensity of cultivation will depend greatly on more efficient use of water.

Water Use In Farming

Water's importance in cropping is best summed up by an accumulation of superlatives: in most agroecosystems water is the most limiting growth factor during most years. This is not only because of seasonally, or chronically, low rainfall, particularly in subtropical regions, but also because of inherently high water demand of crops: besides being a rather mediocre converter of solar radiation, photosynthesis is also a very inefficient user of water. C_4 species lose at least 450–600 moles of H_2O per mole of absorbed CO_2, and the ratio is commonly two, and even three, times higher for C_3 plants.

Water-use efficiency of an individual plant is almost directly proportional to the level of atmospheric CO_2: assuming identical regimes of temperature and humidity, plants had to transpire twice as much water during the last glacial maximum twenty thousand years ago, when CO_2 levels of 180 ppm were half of today's concentration (Farquhar 1997). Conversely, crops would have to transpire only half as much in an atmosphere with 700 ppm of CO_2. (For more on the effects of climatic change on agriculture, see chapter 3.)

These transpiration rates capture only a part of water demand. Evapotranspiration rates, including evaporation from the surface of the soil and from crop canopy, are more appropriate for expressing water needs of field crops. They are usually related to the fresh weight of crop yield, and are commonly expressed in the reciprocal form, that is in harvests (in kg) per hectare-millimeter of water (Stanhill 1986).

Plants with high resistance to transpiration—pineapples whose stomata are closed during the day, citrus trees with their waxy leaves—have by far the lowest evapotranspiration rates. For most of the common field crops the differences under identical climatic conditions are due mostly to crop planting date and rate of growth, plant height, their roughness and reflectivity, and ground characteristics. Strong and hot winds and low humidity obviously increase water loss. In absolute terms, ranges of seasonal crop evapotranspiration go from just 250–500 mm/ha for low-yielding small-grain cereals and pulses, to 800–1,000 mm/ha for corn, and to more than one meter per hectare for alfalfa, sugar cane, and high-yielding rice.

Where climate and nutrients do not limit productivity, annual, or seasonal, evapotranspiration of leafy crops is strongly correlated with the total yield of dry phytomass. Where the harvested plant part consists of grains or fruits, or where the yield is determined by output of oils and sugars, varietal differences are much more important. For example, with identical evapotranspiration modern rice cultivars can yield four times as much as grain as their traditional counterparts.

Extension of cultivation to drier environments and adoption of high-yielding cultivars have resulted in steadily rising shares of food production coming from irrigated land. The world's total irrigated area rose from only about 8 Mha in 1800 to 48 Mha in 1900 and it just surpassed 90 Mha in 1950. By 1980 it more than doubled to

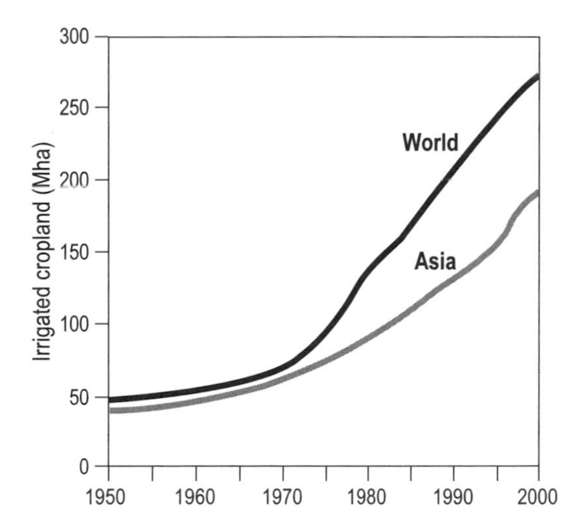

Figure 2.4
Global and Asian totals of irrigated cropland; plotted from data in FAO (1999) and Rozanov et al. (1990).

just over 200 Mha and in 1998 it stood at 250 Mha (figure 2.4) (Rozanov et al. 1990; FAO 1999). This amounted to about 17 percent of all cultivated land, but these irrigated fields produced roughly 40 percent of the world's food output, compared to less than a third of the total in 1950. In countries deriving large shares of crop production from irrigated farming more than half, and up to 90 percent, of all water is used in agriculture. The worldwide share of water withdrawals for irrigation was between 65 and 70 percent in the mid-1990s. Contradictory trends will act both to increase and to lower this share.

Not unlike in the case of farmland figures, most of the commonly quoted totals and shares concerning agricultural use of water in general and irrigation in particular are not uniformely derived. This means that available figures are difficult to interpret and to compare, and calculations of future water needs rest on arguable cumulative assumptions.

A closer look at irrigation statistics for China, the world's largest user of fresh water in agriculture, illustrates the need for caution in interpreting the meaning of grand totals. Nickum (1995, 1998) argues persuasively that the aggregate figures obscure more than they inform about China's complex and diverse state of irrigation. The basis of China's accounting, the effectively irrigated area, is defined as a level land with water resources and irrigation facilities capable to provide adequate volume of water for crops under normal conditions. During a rainy year, such a plot of land may need no irrigation, while during a prolonged drought it may receive far from adequate moisture. Nor does the aggregate figure inform about the number of yearly irrigations and their effectiveness.

These are, of course, universal problems. In hyperarid regions irrigation must replenish all of the soil moisture taken up by crops, while supplementary irrigation in temperate climates may be used to supply just 20 or 30 percent of evapotranspired water. Unfortunately, there are no reliable global or regional figures apportioning these shares. Not surprisingly, different authors estimate that the average amount of water applied per hectare of irrigated land is as low as 10,000 and as high 12,000 m^3. Uncertainties concerning existing water supply make global appraisals of present agricultural water claims even more imprecise.

Annual runoff estimates published during the last two decades have ranged from 33,500 km^3 to 47,000 km^3, a difference of 40 percent! In two recent supply estimates, Postel et al. (1996) put the total of globally available fresh water at 40,700 km^3, while Shiklomanov (1993) used 42,650 km^3. On the demand side the two sources estimate the total agricultural, industrial, and urban use (including reservoir losses) at, respectively, 4,430 km^3 and 3,578 km^3 in the year 1990, while Raskin (1997) calculated the total of 3,698 km^3 for the year 1995.

These figures imply withdrawals between just a bit over 8 percent and about 11 percent of the available fresh water runoff. Agriculture is by far the largest user of this water, claiming about two-thirds of the total use, or between 5 and 7 percent of the global fresh water runoff. Neither share would seem to signal a high degree of concern about the future adequacy of water supply—but a closer look at the global water cycle leads to a much less sanguine conclusion.

Of the roughly 110,000 km^3 of precipitation that falls on the continents nearly two-thirds are evapotranspired and some 40,000 km^3 end up in streams (Gleick 1993). About a fifth of this flow is in locations too remote for economic capture, and only a portion of either temporarily or continually available streamflows (usually no more than 60 percent) could be tapped in order to maintain other expected river flow functions, including silt-clearing, waste removal, navigation, and electricity generation. This leaves no more than about 12,000 km^3 of water for withdrawal.

Worldwide use of 3,600–4,400 km^3 of fresh water by agriculture, industries, and households already amounts to about a third of all available flows. Even if the growth rate of future extraction were merely proportional to the overall population increase, the withdrawal share would surpass one half of total runoff during the third, or the fourth, decade of the twenty-first century. Published forecasts of total water withdrawals for the year 2025 range between 5,000 and 6,400 km^3, with agriculture taking anywhere between 3,200 and 4,200 km^3. Continuation of this trend would see the overall demand at least topping two-thirds of accessible fresh water runoff after the middle of the twenty-first century.

Potential withdrawals could be appreciably higher, but estimates of future agricultural water needs have a considerable range of uncertainty due to the assumptions concerning the composition of prevailing diets. Small grains and pulses grown with 1,200 mm/ha will (after adjusting for milling and storage losses) need only about 0.25 m^3 of water per MJ of food energy. Although the direct water consumption by livestock accounts for only a small fraction of all water used in farming, indirect requirements for roughage and concentrate feeds make animal foods much more water-intensive. For example, producing mixed corn-soybean feed for broilers and pigs will require anywhere between 2.5 and 4 m^3/MJ of meat, and the rate for beef may be, depending on the method of its production, two to three times as high (see chapter 5).

Overwhelmingly vegetarian diets—dominated either by small grains and legumes or by rice and averaging 2,500 kcal a day per capita—can be thus produced with as little as 900–1,200 m^3 of water per person per year. In contrast, well over 2,000 m^3 of water is needed annually in order to produce the diet now prevalent in rich countries, and no less than 2,000 m^3 would be required for less meaty, well-balanced nutrition.

If the current share of about 40 percent of all food production originating from irrigated areas did not change, feeding a largely vegetarian world would require no more than about 4,800 km^3 of irrigation water. When assuming that 2,000 m^3 of water would be required every year in order to assure decent nutrition, a global population of ten billion people would need 20,000 km^3 of fresh water to produce its food. With the unchanged share of the total output coming from irrigated fields, this would call for some 8,000 km^3, or two-thirds of all accessible fresh water runoff, and higher irrigation needs could lift the total close to 10,000 km^3. Addition of higher industrial and urban requirements would then put the overall total very close to the supply maximum.

These simple calculations and comparisons show persuasively that the human use of fresh water resources is approaching an era of increasing supply stress even on a global scale. This is much unlike the case of farmland where business-as-usual scenarios carry the risk of serious land shortages in an increasing number of populous, land-scarce countries but indicate no crippling effect on the overall global capacity to produce food. Combination of a continuing neglect of opportunities for improved efficiencies of water use and of dietary changes resulting in higher water needs per capita would accelerate the onset of this supply stress and expand its spatial extent, while the combination of remedial actions could make it manageable (see chapters 4, 5, and 8).

At the same time, the uneven distribution of fresh water flows makes it inevitable that substantial shortfalls must already exist in a number of arid countries—and that the list of similarly affected countries, and regions, will only grow. In addition, water pollution and advancing depletion of some major aquifers will further limit the volume of water readily available for irrigation. Commonly used criteria recognize water stress when a country's average supply falls below 1,700 m^3 per capita, and water scarcity when the rate goes below 1,000 m^3. In the early 1950s only a handful of small island and desert countries (such as Bahrain and Malta) belonged to the latter category. By 1990 the list of water-scarce countries grew to twenty, the most populous ones being Algeria, Kenya, and Saudi Arabia (figure 2.5). Concurrent trends of urbanization, electrification, industrialization, and expanded cropping will inevitably intensify competition over available water resources.

Urbanization is major reason for increasing per capita use of water. So is the universal trend of consuming a rising share of fossil fuels indi-

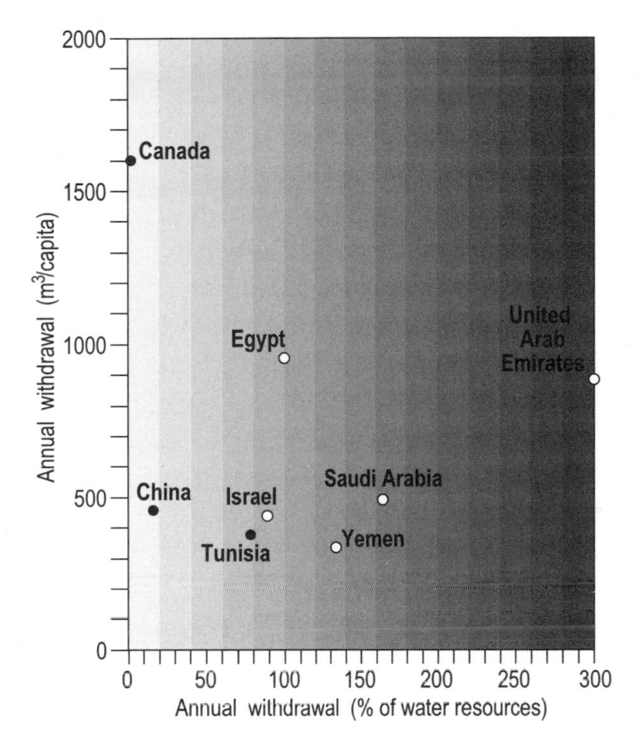

Figure 2.5
Countries with serious water scarcity; plotted from data in the World Resources Institute (1996).

rectly in the form of electricity generated in large thermal power plants, which requires large volumes of water for cooling. Many industrial activities—including metallurgy and petrochemicals—are highly water-intensive. Increasing costs of exploiting new water resources is a good indicator of this growing competition and increasing scarcity (Rosegrant 1997). By the year 2025 some thirty countries, with combined population of about 900 million people, are predicted to be in the water-scarce category (Engelman and LeRoy 1993). The outlook is particularly worrisome for large parts of Africa: two-thirds of the continent's countries are now developing serious water scarcities (Falkenmark 1989).

The most likely short-term consequences of this trend in most of the affected countries will be spreading deterioration of average food intakes and higher dependence on imports; long-term scenarios addressing this challenge can range from catastrophic outcomes (sharp increases in mortality and the loss of any remaining social cohesion) to welcome adjustments (as stark choices finally catalyze many long-overdue reforms of African governance). In any case, the challenge of water supply must rank at or very near the top of any long-term assessment of food production potential.

The third key natural input encompasses a whole class of elements. As even good-quality land has only a limited yield potential without adequate water supply, so the available water will not be fully used by field crops without the presence of adequate plant nutrients. Although there are no serious natural constraints on producing sufficient quantities of these elements in the form of synthetic fertilizers or extracted minerals, actual application rates of the two leading macronutrients will be increasingly limited by environmental consequences of nitrogenous and phosphorous compounds lost from the fields.

Crop Nutrients

Fluxes of CO_2 and H_2O dominate the input into photosynthesis, but complex organic compounds could not be formed without much smaller, but in their aggregates still substantial, amounts of the three plant macronutrients and without more than a dozen of micronutrients ranging from relatively common calcium to rather scarce cobalt. Micronutrient deficiencies can have profound effects on plant yields, but their rectification requires applications rarely exceeding a few hundred grams per hectare, which may not have be repeated every year.

In contrast, redressing the common shortages of the three macronutrients may require annual applications amounting to tens (for P and K) to hundreds of kilograms (for N) per hectare. Natural mobilization of these nutrients, augmented by recycling of a large part of organic wastes, was sufficient to support traditional harvests of relatively low-yielding crops—but even a complete recycling of all organic wastes from the currently harvested land and from all confined domestic animals would not be able to supply all macronutrients removed from soils by modern high-yield cropping.

The only way to support ten billion people by traditional cropping dependent solely on recycling of organic matter and rotations with legumes would be to double, or even to triple, the extent of currently cultivated land. This would require complete elimination of all tropical rainforests, conversion of a large part of tropical and subtropical grasslands to cropland, and the return of substantial share of labor force to field farming—making this clearly only a theoretical notion.

This nutrient disparity between nonsynthetic sources and the actual demand is most obvious in the case of nitrogen, the nutrient most commonly limiting crop yields. Nitrogen's importance in photosynthesis is perhaps best illustrated by the fact that more than three-quarters of the element present in leaves is associated with the photosynthetic apparatus. Nitrogen content of phytomass ranges widely, with C:N ratios varying from more than 300 for woody tissues to 20 for legumes. Common food and feed grains contain between 1.1 percent (some rices) and 2.4 percent (best durum wheats) of the element (that is, between 7 and 15 percent of protein); nitrogen makes up nearly 6.5 percent of the soybean mass, but only 0.25 percent of sweet potatoes.

The element is in no short supply in the biosphere: there are some 75,000 tonnes of atmospheric nitrogen above every hectare of arable land—but the two atoms of the dinitrogen molecule are joined by one of the most stable triple bonds which in nature can be broken only by lightning and by nitrogenase, an enzyme possessed by only a limited number of microbial species.

Lightning supplies an order of magnitude less of reactive (or fixed) nitrogen than the reduction of atmospheric N_2 to NH_3. This vital process is performed enzymatically by at least sixty genera of cyanobacteria, fifteen genera of symbiotic actinomycetes, and more than twenty genera of free-living bacteria. But most of the naturally fixed nitrogen comes from two genera of symbiotic bacteria, *Rhizobium* and *Bradyrhizobium*, associated with leguminous plants.

In spite of the enormous research effort devoted to the study of symbiotic fixation, there is not even a single crop for which we have reliable estimates of annual fixation rates. Almost all published values have at least threefold ranges, and much larger differences are common. Measured outputs for cyanobacteria span more than two orders of magnitude. Good conservative approximations of contributions provided by symbiotic fixation would assume annual means of at least

150–200 kg N/ha for alfalfa, clover, and other leguminous cover crops, between 70 to 100 kg N/ha for all other legumes except for beans, and 50 kg for beans. Free-living cyanobacteria in rice fields can fix up to 30 kg N/ha a year, but those in dry fields in arid regions may contribute no more than a few kg N/ha.

Total supply of nonfertilizer nitrogen—from net mineralization of soil organic matter and crop residues during the growing period, from biofixation in floodwater and rhizosphere, from compounds dissolved in irrigation water, and from atmospheric deposition—can be considerable. Its annual rates can be as high as several tens of kg per hectare. But field studies by Cassman et al. (1996a) in the Philippines demonstrated that the supply and uptake of indigenous nitrogen from soil and floodwater in paddies is highly variable both among fields with similar soil types and in the same field over time.

Key processes transforming nitrogenous compounds within soils include assimilation (incorporating of the nutrient into crops), and microbially mediated decomposition, mineralization, immobilization, enzymatic oxidation, nitrate reduction, nitrification, and denitrification (figure 2.6). Fixed nitrogen is stored in soils either as ammonium or, after oxidation by nitrifying bacteria, as a much more water-soluble nitrate. But these inorganic compounds add up to only a small fraction of nitrogen stored in soils. The bulk of the nutrient is present in soils bound in organic matter: common rates in temperate ecosystems range between 2.5 and 6 t N/ha, but there is great deal of both spatial and temporal variability.

Aggregate human requirements for the element, supplied in plant and animal proteins, are relatively modest (for details see chapter 7): when assuming an average need of 50 g of protein per capita a day, they were equivalent to some 19 Mt N per year for six billion people alive in 2000. But in order to secure this rather small flux (our needs for fossil fuels are about five hundred times more massive), we are now intervening in the nitrogen cycle on a scale comparable to the largest natural flows.

Preindustrial civilizations—relying for their nitrogen supply solely on recycling of organic wastes and planting of leguminous crops—caused only marginal delays and modest local changes in nitrogen's flows. Post-1830 exports of Chilean nitrates brought into the cycle the long-

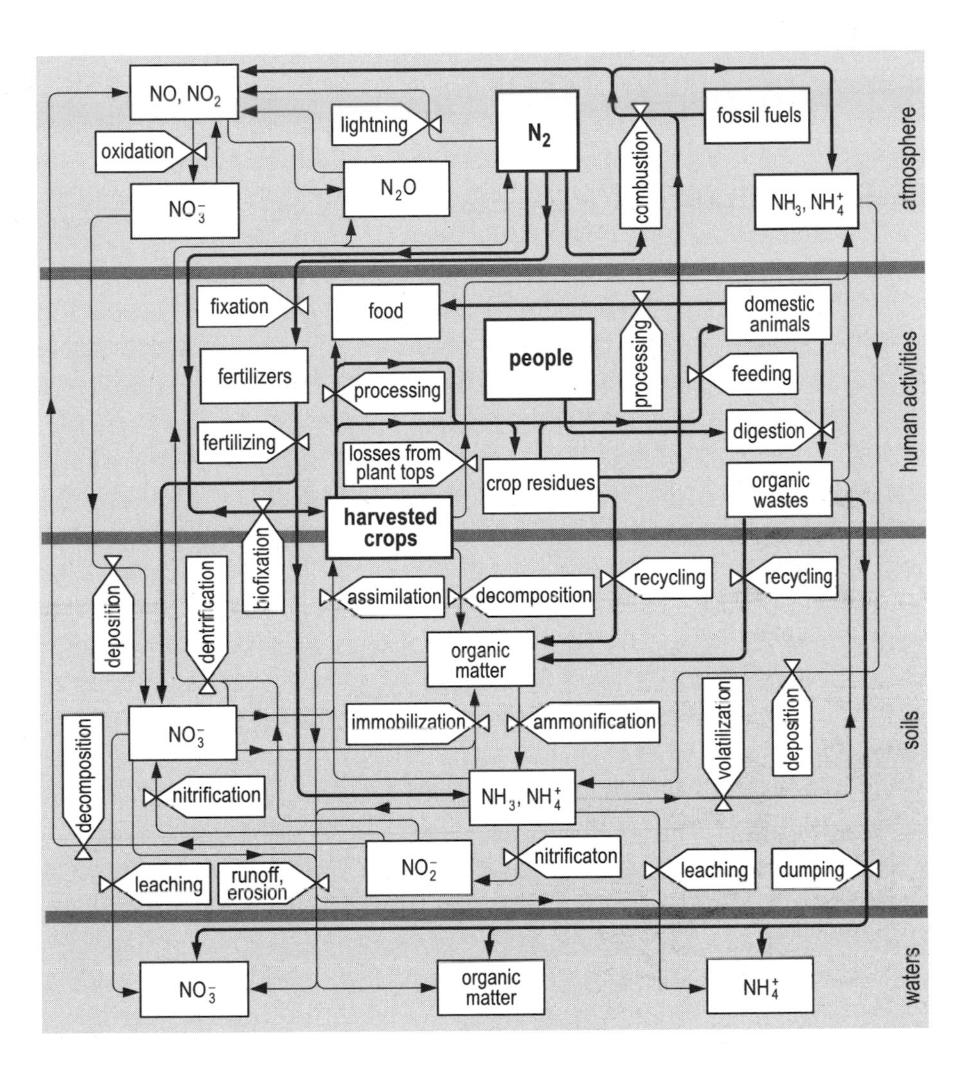

Figure 2.6
Biospheric nitrogen cycle centered on crops (Smil 1997).

sequestered lithospheric nitrogen, soon supplemented by the mining of organic guano deposits. Even the invention of ammonia synthesis by Fritz Haber and Carl Bosch had little immediate effect on agriculture. The first plant synthesizing ammonia was completed in 1913, but the worldwide use of nitrogenous fertilizers took off only during the 1950s.

Current production of about 80 Mt N a year is larger than the total amount of nitrogen received by farmland from atmospheric deposition, biofixation and recycling of organic wastes (Smil 1997a, 2000). Depending on the quality of diet we would be willing accept, our numbers would have to shrink by two to three billion in a world devoid of synthetic nitrogenous fertilizers. This dependence will only increase in the future: at least four out of every five children born during the next half a century in Asia, Latin America, and Middle East will synthesize their body proteins from nitrogen fixed by the Haber-Bosch synthesis of ammonia.

Nitrogen needed for the synthesis is taken from the atmosphere, and hydrogen now comes overwhelmingly from natural gas but can be secured in other ways (traditionally it was produced by hydrogenation of coal). Although much lower than two generations ago, the energy cost of ammonia synthesis in large modern plants is still substantial, averaging over 40 GJ/t N, of which 60 percent is feedstock and 40 percent is process energy (Smil 2000). Because ammonia requires special storage and application, most of it is converted to liquid or solid fertilizer for easy shipment and spreading. Urea, whose formulation needs additional 25 GJ/t N, is now the leading solid nitrogen fertilizer.

Using natural gas solely as the feedstock would mean that the world's current synthesis of nitrogenous fertilizers (80 Mt N/year) would require annually about 100 Gm^3. If the gas provided, as has been increasingly the case, both feedstock and fuel, its total consumption would rise to some 150 Gm^3, still less than 7 percent of the world's natural gas extraction in 1997 (British Petroleum 1999). While our dependence on synthesis of nitrogenous fertilizers is now truly irreplaceable, it is not constrained by the availability of natural gas. The only major constraints on higher applications of nitrogen fertilizers are their environmental impacts (for details see chapter 4).

Similar conclusions can be made concerning our critical dependence on liquid fuels for farm machinery. For example, merely to match the 1995 mechanical power of American tractors with horses would require us to build up the stock of these animals to at least 250 million, ten times their record count in the second decade of the twentieth century (Smil 1994b). About 300 Mha, that is *twice* the *total* of U.S. arable land, would be needed to feed these draft animals! And yet all of the U.S. agricultural field machinery consumes annually no more than 1 percent of the country's liquid fuels.

Phosphorus, unlike nitrogen, is not a doubly mobile element: because it is not volatile and does not form any stable gaseous compounds it cannot be cycled through the atmosphere. Its biospheric flows are thus dominated by a one-way flux of inorganic compounds present in the crust to the ocean. Inorganic phosphorus present in the top 50 cm of farm soil adds up typically to some 3 t P/ha; organic phosphorus adds roughly 500 kg P/ha. Cereals and legumes are among the plants containing relatively high shares of phosphorus (in excess of 0.3 percent).

Much like with nitrogen, even complete recycling of available organic matter would not be able to provide all phosphorus needed by modern high-yielding crops. But whereas ammonia synthesis is a relatively energy-intensive process, production of phosphorous fertilizers requires merely digging up suitable ores and treating them with sulfuric or phosphoric acid in order to increase the element's availability. By far the largest accessible stores of phosphorus are in sedimentary marine ores, and the rest is mostly in igneous apatites. Estimates of currently recoverable reserves vary, but even the most conservative total is no less than 35 Gt, and known resources add up to further 100 Gt (Fantel et al. 1989). Annual production of phosphate rock during the 1990s has been about 150 Mt, with 80 percent of the output going into fertilizers (FAO 1999). Consequently, there is no danger of any looming supply shortages.

Although availability of the nutrient may not be a concern, effects of human intervention in phosphorus flows definitely are. Moving the nutrient from phosphate rocks to farm soils has been the single largest anthropogenic interference in the element's biospheric flows ever since the widespread adoption of phosphatic fertilizers during the last decades of the nineteenth century. The traditional view was that the major part of applied nutrient is rather quickly tied up by the soil, piling up in unavailable forms (Karlovsky 1981). This misconception led to irrational use of fertilizers: global N/P ratio in applied fertilizers passed 2.0 only in the mid-1950s, by 1970 it was about 3.5 and now it is well over 5.0.

Potassium is needed by crops in amounts substantially smaller than those of the other two macronutrients. Augmenting its natural supply from weathering and organic recycling is fairly easy: all that is required is to mine and crush potash (K_2O), a mineral available in a number of rich deposits around the world. Annual potash production averaged less than 25 Mt during the 1990s. At this rate the extraction of known deposits could easily continue during the 21st century.

Agroecosystems and Biodiversity

The most obvious fact concerning biodiversity and modern agriculture is an extremely narrow base of our cropping. Unlike in the case of insects—whose roughly 750,000 species make up about two-thirds of all described creatures and whose eventual total can be multiplied severalfold by further discoveries (Erwin 1988)—we expect to find only a relatively small number of new plant varieties. Of the roughly 250,000 described higher plants, about 30,000 species are known to be edible, some 7,000 of them have been actually used as food, only 150 have ever become important crops, and only fifteen species produce more than 90 percent of all food.

The three leading cereal crops—wheat, rice, and corn—supply almost two-thirds of all food energy and slightly more than half of all protein derived from plants (Wilson 1986; Paoletti et al. 1992). Rice is a more dominant staple in monsoonal Asia than wheat is throughout the Western world: in some of the region's countries it supplies up to 65–70 percent of all food energy. But nowhere is the impression of low cropping diversity stronger than on the American plains and Canadian prairies: for an hour or two, that is for stretches of up to 1,800 km, a jet plane flies over fields planted largely with no more than two or three crop species.

In contrast, many traditional agroecosystems were highly diversified. By far the highest diversity of cultivated plants has been found in tropical home gardens. In Java, where they average about 1,000 m^2 (and could be only a fifth of that), they harbor commonly between 100 and 150 species per hectare, with recorded maxima between 250 and 350 (Hoogerbrugge and Fresco 1993). Tree densities of Javanese home gardens may approach 2,000 stems per hectare, and their total biomass surpasses 120 t/ha (figure 2.7) (Jensen 1993).

In addition, Clawson (1984) emphasized a less understood diversification strategy of traditional small-scale tropical farming, namely the practice of intraspecific polyculture in addition to interspecific multiple cropping. Growing different cultivars of the same species increases the chances of reaping a reasonable harvest of a particular crop. Planting of multicolored corn cultivars in Central America and cultivation of various potato kinds in the Andes are outstanding examples of this strategy. In contrast, not only is the specific diversity of modern cropping low, but there is also very little intraspecific variation as huge areas are planted with identical cultivars, now often bred to be resistant to a particular kind of herbicide.

There is no single satisfactory way to measure biodiversity. A reductionist approach favored by most biologists sees genetic diversity as the basic currency, with all other levels of hierarchy (from species to ecosystems) as surrogate expressions (Humphries et al. 1995). With this perspective it is inevitable to conclude that the drastic narrowing of cultivated plant diversity has put our food production in greater peril. Until quite recently most ecologists would have given an unqualified endorsement to this conclusion, and many writings on world food supply still contain blanket claims that greater biodiversity increases productivity, improves food security, and boosts economic returns (Thrupp 1997). Recent research offers a more nuanced view of this relationship.

Biodiversity and Productivity

Theoretical expectations of benefits arising from greater species diversity have been confirmed by experiments demonstrating higher net primary productivity and nutrient retention in ecosystems with a greater number of plant species. Complementarity of resource use is the best explanation of this effect as species differ in the ways they secure light, nutri-

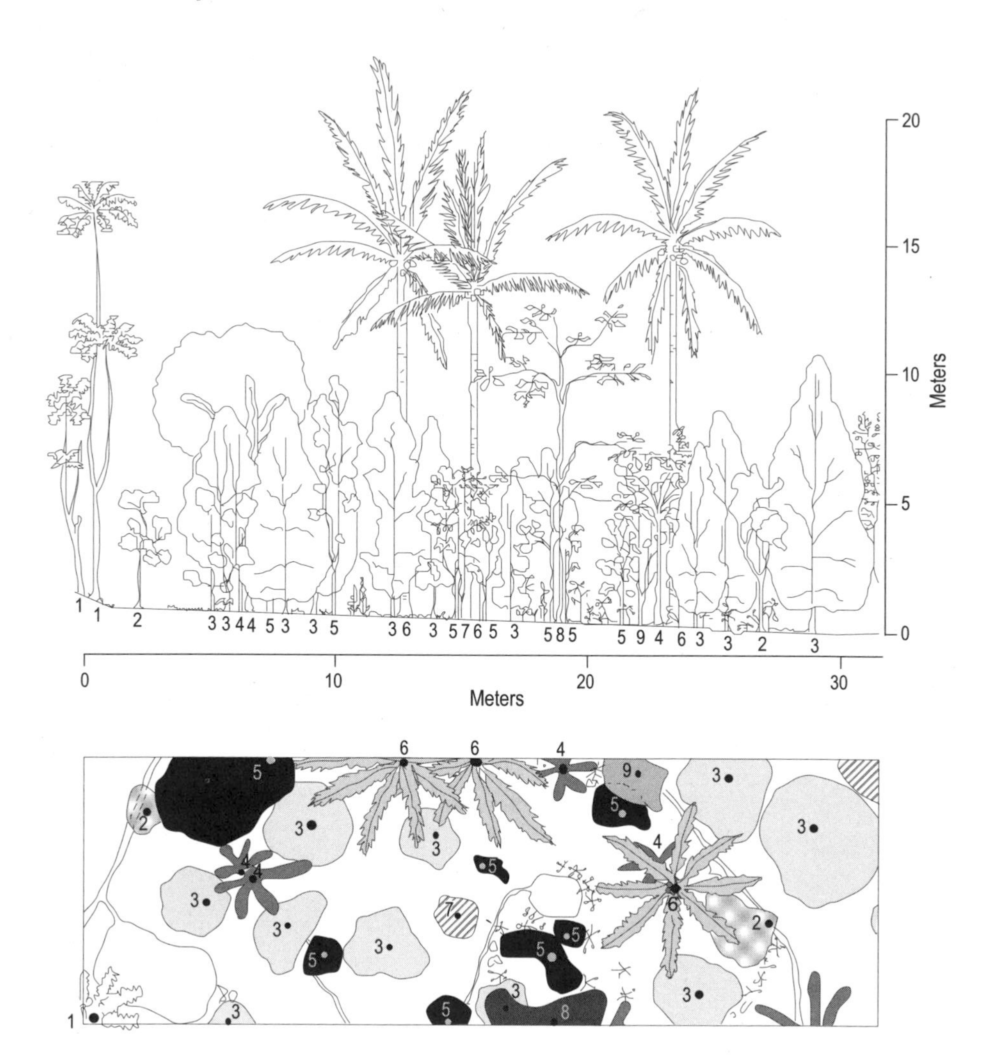

Figure 2.7
Plant density and diversity in a Javanese home garden; based on Hoogerbrugge and Fresco (1993).

ents, and water; hence, their greater diversity should allow access to a higher share of available resources and to improved productivity. Experimental evidence also shows that biodiversity may regulate ecosystem predictability and may represent a form of insurance against the loss or poor performance of selected species (McGrady-Steed et al. 1997; Naeem and Li 1997).

But today we also know that there is no simple link between species diversity and ecosystemic stability. Most obviously, some of the world's most extensive and ancient ecosystems, including boreal forests and bogs, are fairly species-poor. Recent experiments with grassland plots conclude that functionally different roles represented by plants are at least as important as the total number of species in determining processes and services in an ecosystem (Tilman et al. 1997; Hooper and Vitousek 1997). In many cases there is little doubt that functional characteristics of component organisms are more important than the total number of species. Clearly, all species are not equal, and hence the loss, or addition, of certain plants with essential functional traits will have a major impact, while the disappearance or introduction of other species will cause little or no overall change.

Even more revealingly, Wardle et al. (1997) found in an investigation of ecosystem properties of fifty islands in the northern Swedish boreal forest zone that such desirable traits as higher microbial biomass, higher litter quality, and more rapid rates of nitrogen mineralization coincide with lower plant diversity and an earlier successional state. This would mean that some fundamental ecosystem processes are not crucially dependent on higher levels of biodiversity. A long-standing assumption that more diverse landscapes should be more resistant to exotic plant invaders has been also shown to be invalid by recent research (Kaiser and Gallagher 1997). Comparison of several U.S. landscapes as well as a larger global sample set demonstrated that species-rich areas, be they small patches or whole biomes, had more exotic invaders than the poorer ones. Globally, temperate zones appear to be most invaded, and savannas and deserts least disturbed.

Appraisals of agricultural biodiversity must also take into account Michael Huston's argument that the assumption of competitive equilibrium is not very useful in explaining biodiversity of ecosystems, mainly because repeated impacts of extreme natural events do not allow sufficiently long periods of time needed to reach such states (Huston 1994). Natural disturbances make formerly inferior competitors viable, or even dominant, and species diversity is thus a complex outcome of orderly competitive displacement and disruptive disturbances.

The role of soils in this process is critical: sites with low soil fertility are slow to regain competitive equilibrium after a disturbance, allowing many species to coexist in a nonequilibrium state. High plant diversity is thus associated with slow growth and low phytomass productivity. Huston (1993) extended this conclusion to agricultural soils, arguing that fertile and highly productive soils are best suited for intensive food production while poor lands with high plant diversity is not suitable for farming.

At the same time, species diversity and functional diversity are correlated, and either one may be a good indicator of ecosystem functioning (Tilman et al. 1997). Consequently, these recent findings are clearly no argument in favor of more extensive monocropping inasmuch as functional diversity could hardly be provided by a single species. Monoculture may have temporary economic advantages for farmers—evaluations based solely on overall yield may show its superiority—but, genetic concerns aside, we have known for a long time that it is not an agronomic optimum because it tends to promote soil erosion and exacerbate weed and pest problems (Power and Follett 1987).

As far as the intraspecific diversity of today's major cultivars is concerned, Smale (1997) shows that the claims about genetic erosion and about the perils of narrow and alien genetic bases, undesirable shifts ascribed to the Green Revolution, are unfounded. Although there is no clear consensus about the breadth of the genetic base of bread wheat and about the rate of its change, there are clear trends showing that the shares of dominant wheat cultivars have been declining in a number of European countries and in the United States since the 1920s—while the average number of different landraces per pedigree of newly released wheat cultivars has been increasing. Recent advances in genomics make it even more likely that we should not see any perilous erosion of intraspecific diversity of major cultivars.

Practical steps to reduce the extent of crop monocultures and to enrich the diversity of soil organisms are easily taken. There is a long list of

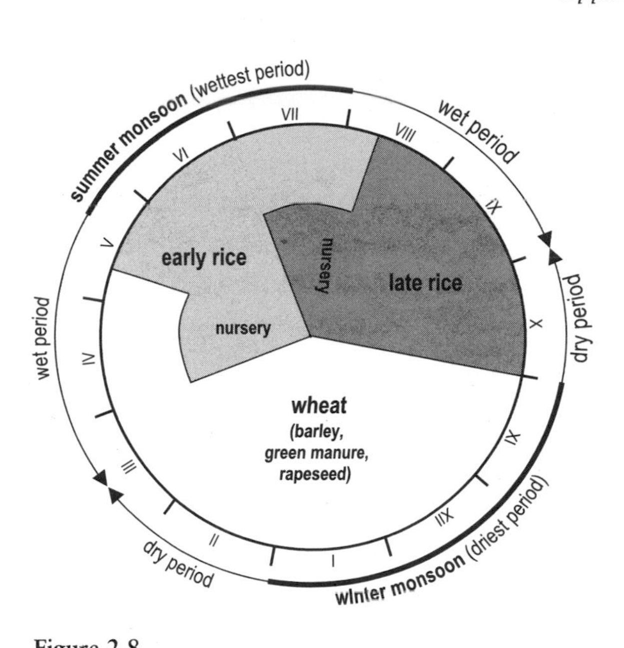

Figure 2.8
An example of traditional crop rotations in South China (Smil 1998).

management practices promoting these goals, including crop rotations, agroforestry, reduced tillage, mulching, and intensive recycling of organic wastes. The simplest means of assuring greater specific diversity in field production is the well proven practice of crop rotations. In the long run, global crop diversity could be enriched by diffusion of some traditional species, whose cultivation is now restricted to a dwindling number of locales, and by introduction of new crops.

Crop Rotations and New Crops

Crop rotations—sequential cultivation of different plants on the same land—have been an essential part of every traditional agriculture (figure 2.8). They have ranged from the simplest rotations of two crops to complex sequences of more than ten different crops extending over a period of several years. The practice got its first convincing scientific support with the discovery of nitrogen fixation by leguminous crops during the 1880s (Smil 1997a). Although the diffusion of intensive monocultures after World War II has reduced their frequency, rotations remain important even in the regions where monocropping is common. Most U.S. corn is grown in two-year rotation with soybeans or in two- to three-year rotations with alfalfa, cotton, and other crops.

But the shift from multiyear rotations to simple two-crop sequences (particularly with row crops, such as corn and soybeans) has generally contributed to soil degradation. Comparisons based on a fifty-year sequence in Alberta showed that soil organic matter content was about 20 percent higher with five-year rotations compared to a two-year wheat-and-fallow sequence (Juma et al. 1993). In contrast, rotations including leguminous cover crops help to develop favorable soil structure and result in the highest soil aggregate stability. Such rotations will also reduce soil erosion, while rotating corn with soybeans may result in much high erosion than continuous corn cropping (Bullock 1992).

Many beneficial effects of crop rotations have been well established (Higgs et al. 1990): the most important ones are reduced soil erosion; improved soil tilth; and interruption of weed, insect, and crop disease cycles (and hence reduced needs for pesticides and herbicides). As far as the crops are concerned, increased yield is the most obvious advantage attributed to rotations, even to the simplest sequences (Karlen et al. 1994). Long-term experiments on the Morrow Plots in Illinois are a perfect illustration of this benefit: the forty-nine-year average of corn yields was nearly 50 percent higher when the crop was rotated with oats and hay compared to its continuous monoculture (Runge et al. 1990). Rotation effect, resulting in higher yields, is the result of complex interactions of changed soil fertility and organic matter, weed and pest intensity, and water availability.

The earliest American rotation experiments, begun in the nineteenth century, demonstrated considerable yield increases (30–60%) of corn and wheat grown in rotations with clover over the monocropped harvests. Recent studies show gains of 5–20 percent for corn rotated with soybeans, and similar ranges have been reported for sorghum and wheat. The actual mechanisms and pathways responsible for this difference are poorly understood.

The other most important benefit is to reduce the degree of pest infestations. Not surprisingly, the benefits are highly variable: rotations are fairly effective against pests originating within the affected fields (such as nematodes and soilborne pathogens) but do not control highly mobile invaders. Similarly, rotations are effective against some weeds and fungal diseases. Rotations of crops with different rooting depths will be also able to use nutrients and water more efficiently. Again, leguminous cover crops are outstanding in this respect, with alfalfa removing nitrates from a depth of nearly two meters during the first year, and eventually from depths greater than five meters.

An alternative, or an adjunct, to rotations is intercropping. This ancient practice, now used widely only in small-scale cropping in low-income countries (Horwith 1985), depends on the balance between intra- and interspecific competition. The latter is commonly less than the former, but not necessarily for all resources. The net outcome is that one species does better when intercropped, the other one a bit, or even much, worse. The most obvious advantages is a reduced risk of total crop failure (Mead et al. 1986).

Introduction of new crops is a more radical way of enriching agricultural biodiversity. Because of a very large number of suitable candidates, the potential contribution of new crops appears to be very promising. But inertia of traditional cropping systems and, even more importantly, a deep entrenchment of food preferences act as powerful obstacles to introducing new species. A more likely change would seem to be a wider adoption of crops that have enjoyed long regional prominence.

Soybeans are the most spectacular example in this category. Until the middle of the twentieth century they were hardly grown outside of the East Asian region. The United States was the first country to embrace them: total American plantings of soybeans amounted to a few thousand hectares in the early 1930s, but they have been sown annually on more than 20 Mha since the early 1970s, producing more than 50 Mt of seeds a year. The increase in Brazilian soybean production has been even faster, from a negligible harvest in the early 1960s to more than 20 Mt in the early 1990s. These two countries now produce two-thirds of the global soybean harvest, virtually all of it for cattle and pig feed, an easier way to introduce a new crop than promoting a new food variety.

And yet the benefits of this crop diversification are not as large as widely assumed. Soybean cultivation enriched temperate and tropical farming with a leguminous crop suitable for rotations with cereal grains or oilseeds—but, as will be shown in Chapter 4, this nitrogen-fixing plant may be often a net user rather than a net supplier of the nutrient, and its cultivation on sloping land may result in relatively high rates of soil erosion.

In contrast, cultivation of leguminous forages provides both substantial surpluses of fixed nitrogen available to the subsequent crop and excellent anti-erosion protection, but it has been declining or stagnating. With some 30 Mha cultivated worldwide, alfalfa (*Medicago sativa*), traditionally grown throughout Europe and in the Americas, is still the largest forage crop, but its plantings have been recently shrinking by about one million hectares a year. Global plantings of clover and vetches have also been decreasing. This retreat has paralleled the steady increase in applications of inexpensive synthetic nitrogenous fertilizers and the expanding production of high-protein soybean-based feeds.

There is no shortage of traditional cultivars now grown in only a few restricted areas, that would deserve wider adoption, and there are many as yet uncultivated plants which should make valuable crops. An excellent example of a crop deserving regional resurgence is quinoa (*Chenopodium quinoa*), an annual grain crop producing masses of white and pink seeds in huge, sorghumlike clusters. The seeds are exceptionally high in protein (up to 19 percent) whose amino acid balance is superior to any other grain crop. Improved varieties of this traditional Inca staple can yield around 4 t/ha and can make notable contribution to better nutrition throughout the high-lying Andean region (Inness 1989).

In a review of potential new crops, Hinman (1986) concluded that buffalo gourd, jojoba, and kenaf were approaching commercial production, and he identified bladderpot, gumweed, and guayule as other potentially most promising sources of food or industrial materials. But, judging by the experience of the latter half of the twentieth century, probabilities of either a large-scale commercial diffusion of neglected regional crops or adoption of new plants for commercial production are not very high.

Widespread adoption of perennial mixes of grasses to replace the temperate zone's cereal monocultures would be the greatest step toward high-biodiversity farming. But the probability that such mixtures—favored by the Land Institute of Salina, Kansas, as the most appropriate solution to a variety of current agronomic problems (Pimm 1997)—

would even just begin displacing existing agroecosystems during the coming generation is very low.

In contrast to decidedly modest prospects of new crop introductions in temperate climates, the potential for new tropical crops should be much larger. After all, such common staples as corn, rice, and potatoes, and such important vegetable and fruit species as peppers, tomatoes, and oranges—all of them now grown far beyond the tropics—originated in tropical climates. Cultivation of several species of palms would appear to be especially appealing (Plotkin 1986).

Fruit of pupunha or peach palm (*Bactris gasipaes*) has nearly perfect proportion of carbohydrates, protein and micronutrients; the buriti palm (*Mauritia flexuosa*) produces fruit high in vitamin A; oil of pataua palm (*Jessenia bataua*) has fatty acid composition almost identical to that of olive oil and its protein is superior to that of soybean; babassu palm (*Orbignya martiana*) has more oil than coconut, and its seedcake contains 27 percent of protein.

Microorganisms and Agroecosystems

We must keep in mind that although most of our food output now comes from a handful of species belonging to no more than a few scores of major cultivars, the diversity of soil microorganisms sustaining crop harvests remains undiminished, although largely unknown. Recent molecular-phylogenetic studies relying on amplifying and cloning ribosomal RNA sequences, an approach pioneered by Carl Woese, give us accurate information about the relatedness and diversity of organisms. Use of these gene-typing techniques led to the dethroning of higher organisms as the main repositories of biodiversity; we now know that most of life's diversity is locked in single-celled microbia (Williams and Embley 1996; Pace 1997).

Two of the three life's domains—Archaea and Bacteria—contain only microorganisms, and microbes are also the most diverse part of the third domain, Eukarya. Archaea and single-celled eukarya are as genetically distinct from each other as animals are from plants, enough to qualify them as separate domains of life. Moreover, these analyses show conclusively that most of the life's diversity is not within the Eukarya domain (fungi, plants, and animals), but among single-celled organisms belonging to Archaea and Bacteria. Sequencing of the first archeon,

Methanococcus jannaschii, in 1996 showed it to be neither typically bacterial nor eukaryotic, with the bulk of its genes new and more closely allied, although not identical with, the eucaryotic composition.

Because less than 1 percent of known bacteria can be grown in laboratory cultures, we have had a badly skewed perception of soil microbial diversity. These organisms are vastly more abundant in soils than fungi or protozoa—surface soils of temperate regions harbor anywhere between 10^8 and 10^9 bacteria per gram, and their diversity is immense. A gram of soil may contain four thousand independent bacterial genomes, or up to forty thousand different species—yet so far we have described only some four thousand bacterial species, most of them not soil inhabitants (Brusaard et al. 1997). Similarly, we have described just 5–10 percent of soil fungi, protozoa, nematodes, and mites.

There is no niche in soil-and-root environments not colonized by some archaea, bacteria, or fungi, and these organisms provide irreplaceable environmental services through their critical roles in cycling of carbon, nitrogen, phosphorus, and sulfur (Smil 1997a). Although the soil microbial biomass makes up only 1–4 percent of total soil organic matter, it is the largest labile reservoir of plant nutrients. In arable soils it commonly contains 100 kg N/ha, and its relatively rapid recycling of nutrients is a key to maintaining agricultural productivity.

Symbioses of plants with microorganisms are another essential foundations of agricultural productivity, above all due to root associations with symbiotic mycorrhizal fungi and with nitrogen-fixing bacteria attached to leguminous species. Mycorrhizae are critical for channeling soil-derived minerals to plants, while nitrogen fixers provide the only reliable supply of the macronutrient in natural ecosystems. The restricted diversity of such microorganisms is in great contrast to the huge taxonomic diversity of hosts. A mere ten species of *Rhizobium* and *Bradyrhizobium* form symbioses with nearly 18,000 species of legumes, and only some 120 species of arbuscular mycorrhizal fungi can be found in association with at least 200,000 species of plants (Douglas 1995).

And because so many of them are also more robust than plants, microorganisms are able to maintain a large share of the original biodiversity under a variety of agricultural practices, including greater cropping intensification. For example, it appears that microbial decomposition, a key environmental service provided by bacteria, can proceed even if the species composition of a particular subsurface community is significantly altered (Levin 1995). Still, given the relatively short experience with such practices, a key unanswered question concerns possible critical thresholds in the maintenance of soil's biodiversity (van Noordwijk 1996). Because of our rudimentary knowledge of bacterial ecology, relations of soil microbial diversity to soil quality, cropping system, and yield are poorly understood.

In one of the few comparisons of conventional corn-soybean rotation with organic farming (animal manure or legumes as nitrogen sources) Buyer and Kaufman (1996) found that total counts of bacteria and fungi, their diversity and spatial distribution were not significantly different. But we cannot assume that this finding is applicable to all agroecosystems because we know that field operations ranging from surface flood irrigation to tillage can change the makeup and the abundance of subsurface microbial communities in both positive and negative ways (Madsen 1995).

Appraising such changes is difficult: complexity of microbial communities means that we rarely know the normal composition of affected biota and, even less likely, understand their interactive functions. Yet without such baselines it is impossible to evaluate the degree of degradation. In addition, most of our knowledge about environmental impacts on bacteria comes from laboratory cultivation of pure cultures, rather than from studies of complex assemblages grown under often radically different natural conditions. Perhaps the most encouraging conclusion that can be made on the basis of our imperfect understanding is that agricultural chemicals applied at recommended levels do not seem to have any long-term harmful effect on soil microbial activity (Hicks et al. 1990).

Notions of ubiquity, persistence, stability, and resilience of soil microbia have been strengthened by the recent discovery of archaea in ordinary farm soils (Bintrim et al. 1997). Previously, these organisms were thought to be present only in a limited number of usually extreme environments. Almost certainly, more surprises lie ahead in our studies of soil microorganisms.

3

Environmental Change and Agroecosystems

The historic brevity of modern intensive agriculture should make us cautious when assessing its long-term capacities. Such regions as China's Huang He valley or Iraq's Tigris-Euphrates alluvium have been farmed continuously by traditional methods for more than seven thousand years, and similar cropping and gardening were carried on throughout much of Asia, Europe and Americas for two to three thousand years. In contrast, large-scale intensive farming dependent on high inputs of nonrenewable energy supporting harvests of carefully bred cultivars is the creation of the second half of the twentieth century.

Some environmental problems caused by these practices are merely continuations and intensifications of undesirable changes that had nearly always accompanied traditional cropping: soil erosion and changes in soil quality are the two most obvious examples. But widespread substitution of mixed farming by large-scale specialized operations (either by commercial monocultures or by concentrated feeding of animals) has both aggravated many old problems and created new environmental realities.

None of these concerns is more important than maintaining the productivity of agricultural soils. Mere enumeration of changes brought by intensive cropping—shortening or elimination of traditional fallow periods; diffusion of double- and triple-cropping; introduction of faster maturing, high-yielding varieties; higher frequency of field operations leading to soil compaction caused by heavy field machinery; greater use of irrigation; large-scale salinization from mineral-laden groundwater pumped from deep wells; applications of agrochemicals resulting in greatly changed soil nutrient cycles and in the presence of persistent residues—makes clear that many soils had to endure much greater impacts during the past one or two generations than they had experienced during centuries of traditional farming.

Given the inevitability of further cropping intensification, such undesirable impacts are bound to increase. Fortunately, there is no shortage of preventive measures for dealing with nearly all of these concerns. Managing varied consequences of environmental pollution is generally more difficult. Modern farming itself is a source of many types of environmental pollution whose impacts reach far beyond the fields: pesticide residues can be found in the bark of distant trees and in animal and human tissues; leached nitrates from fertilizers and manures may transform aquatic ecosystems far downstream from farmed areas; and nitrous oxide (N_2O) finds its way all the way to the stratosphere.

Nonagricultural environmental pollution has a far greater impact on crop yields. Although water pollution, preventing or limiting the use of irrigation water, is a very serious concern in many localities, air pollution injurious to plants now blankets large, and increasing, areas of Americas and Eurasia, where it is already responsible for substantial yield losses. Almost inevitable expansion of areas experiencing recurrently, or chronically, high levels of ozone may become the most common cause of yield reduction due to environmental pollution.

Dealing with possibly rapid global climate change could be an even greater challenge. We have a reasonably good understanding of the direction of most of the anticipated changes, but our quantitative appreciation of long-term outcomes and negative as well as positive effects remains inadequate. In spite of these uncertainties, we can discern fairly well both the broad strategy, as well as many particular details, needed to minimize negative effects of changing climate. Neither the degradation of soils and pollution of agroecosystems, nor the consequences of a historically unprecedented climate change, spell inevitable weakening of the world's food production capacity. Only our continuing mismanagement of these challenges, rather than the problems themselves, could bring such an unwelcome outcome.

Changing Soils

Transformation of natural ecosystems has assumed dimensions apparent even on regional and continental scales. Crop fields start some 30 km east of the place I live, where the rocks and forests of the great Canadian Shield meet the prairies in eastern Manitoba—and their rectangular pattern ends only some 1,200 km to the west, in the foothills of Alberta Rockies. To add just one Asian example, the whole North China Plain—a region of some 400,000 km^2, an area larger than Japan—is just a mosaic of fields, settlements, roads, streams, and canals.

Between the middle of the nineteenth century and the early 1990s we converted close to one billion ha of forests, grasslands and wetlands to farmlands (Richards 1990). Only in those instances where the cut forest or cleared grasses were replaced by carefully constructed terraces did soil erosion remain negligible: in all other cases its rates had instantly multiplied. Qualitative soil changes have been also very fast and widespread. Physical and chemical alterations of soils have changed the composition of soil life, and continuing intensification of agriculture has also greatly affected most organisms living above ground.

During the latter half of the twentieth century increasing size of holdings and fields brought more visible changes to many American, European, and Asian agroecosystems: disappearance of many groves, hedgerows, trees and shrubs bordering the fields; drainage and filling of small water bodies; regulation of many water courses. As a result, formerly fairly diverse agroecosystems acquired a more uniform pattern over large areas (Goedmakers 1989). These changes almost invariably reduced or disrupted breeding and foraging opportunities for many invertebrate and vertebrate species, a trend reflected in falling numbers of animals ranging from songbirds to badgers (Terborgh 1992; Askins 1995). And too often they have led to further increases of soil erosion and to continuing decline of soil quality.

Soil Erosion and Its Effect on Yields

Soil erosion has been almost certainly the most frequently discussed form of agroecosystemic degradation during the second half of the twentieth century. This interest has been particularly evident since the early 1980s (Kelley 1983; Brown and Wolf 1984; Morgan 1986; Lal and Stewart 1990; Pimentel 1993; Agassi 1995). But these widespread concerns have not been matched by reliable information about the actual extent and intensity of the process, and understanding of its long-term effects on crop productivity is poorer still. Strictly speaking, the concern should not be about normal soil erosion, an unavoidable geomorphic change whose rates range from imperceptible to spectacular. The concern should instead be about rates chronically in excess of natural soil formation.

Naturally high rates of water erosion prevail on all exposed sloping land of humid mountain regions, be they in the tropics or in higher latitudes, whereas barren soils in arid subtropical and temperate regions (even where fields are flat) are particularly prone to wind erosion. Permanent vegetation cover, particularly dense forests, reduces both water and wind soil erosion to negligible amounts. In contrast, cropping leaves soils exposed for weeks (even for months) before closed plant canopies begin protecting most of the ground. Even then, roots and leaves of annual crops can never provide as much antierosion protection as dense root mats and overlapping stems in grasslands or as multistoried forest vegetation.

Erosion of fields under annual cover crops may be, even on a gently sloping land, easily ten times, and in row crops on steeper slopes several hundred times larger than in soils under closed-canopy forests. Short-term means of cropland erosion rates have been measured and estimated with varying degrees of accuracy for numerous locales around the world. Their extremes range from barely measurable losses (much below one t/ha) in bunded lowland paddies to several hundred t/ha on steep tropical slopes and in the world's most erodible soils of China's Loess Plateau in Shanxi, Shaanxi, and Gansu provinces (Smil 1993).

Because these rates refer to specific localities and because they were derived from very short time spans, they should be seen merely as approximate indicators of an undesirable process, not, as is sometimes the case, as bases for extrapolating in time or space. Average soil erosion rates for larger areas or for national territories are invariably much lower, but (their accuracy aside) such figures are available only for a very few nations. Maximum average annual losses compatible with sustainable cropping are also highly site-specific, ranging from a few t/ha to about 20 t/ha. The highest tolerance rate in the United States is about 11 t/ha.

Recent concerns about widespread impacts of soil erosion and soil quality decline have been greatly influenced by the Global Assessment of Soil Degradation (GLASOD) which produced the *World Map of the Status of Human-induced Soil Degradation* (Oldeman et al. 1990). GLASOD evaluated four major classes of soil degradation: water erosion, wind erosion, chemical deterioration (loss of nutrients, salinization, acidification, pollution), and physical deterioration (compaction, waterlogging, and subsidence). According to the study about 750 Mha of continental surfaces are moderately to excessively affected by water erosion and 280 Mha by wind erosion, with deforestation and overgrazing being major causes.

GLASOD estimated that mismanagement of arable land was responsible for excessive erosion on some 180 Mha of crop fields, or about a fifth of all land affected by erosion. Extensive areas of highly erodible croplands are found not only in arid parts of Africa and in the interiors of Asia and North America but also in European Russia and Ukraine, and in many parts of the humid tropics, from Madagascar to the Philippines.

But because of GLASOD's highly generalized mapping (with its scale of 1 : 10,000,000 at the equator, one cm^2 of the map covers an area of 10,000 km^2, roughly an equivalent of Cyprus or Lebanon) and because the lack of solid historic baselines needed for trend evaluations, the map is, much like the often-quoted high local erosion rates, just another indicator of a problem, not a reliable basis for appraising future erosion rates and for estimating crop productivity declines.

Obviously, the only reliable way to discern trends in soil erosion is to have representative baseline measurements followed by systematic monitoring. Few nations have large-scale, comprehensive, periodical inventories of these losses. A notable recent exception are nationwide surveys conducted in the United States in 1982, 1987, and 1992 (Lee 1990; Bloodworth and Berc 1998). In 1992 water (sheet and rill) and wind erosion did not surpass tolerable levels on 68 percent of the country's farmland, a 21 percent improvement compared to 1982. And,

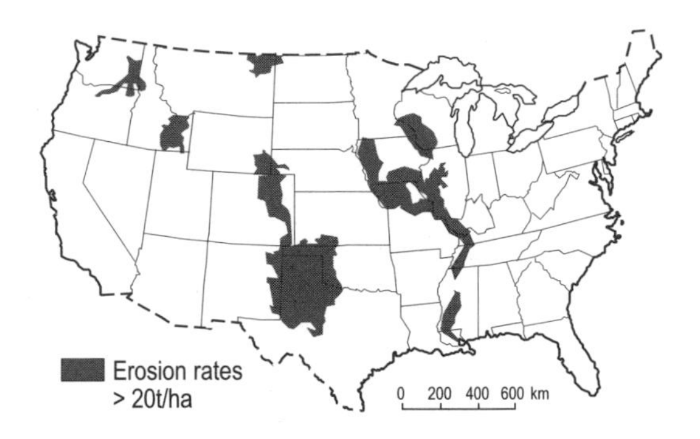

Figure 3.1
Areas of the highest soil erosion in the United States; based on Lee (1990).

encouragingly, even in the areas of excessive soil erosion (southern Iowa, parts of Texas, the Piedmont region) average rates were lower in 1992 than in 1982. These excessive losses range commonly between 10 and 25 t/ha (figure 3.1).

Even with this kind of information it is not easy to come up with realistic estimates of time lags and effect multipliers needed to quantify the eventual reduction of crop yields, particularly on a national scale. We know that even with assiduous recycling of crop residues and with regular manuring, two activities helping the process of soil formation, erosion rates in excess of 10 t/ha will, in most cases, result in chronic thinning of topsoils.

Where surface horizons are shallow, changes in soil's physical properties may become pronounced: rooting depth is reduced, surface crusting may lower water infiltration and decrease soil thickness, loss of fine particles and degraded soil porosity diminish water absorption capacity and soil aeration. Nonuniform soil loss, common on sloping fields, causes excessive field patchiness, resulting in waste of uniformly applied agricultural chemicals.

In deeper soils reduction of the rooting depth and the diminished volume of the medium for soil fauna may be less important than the loss of accumulated nutrients and reduction of soil organic matter. For example, common soil loss of 15 t/ha implies an annual removal of about 50–60 kg N/ha, and 5–8 kg P/ha. Major effects of erosion on soil's chemical properties include decrease of organic carbon and extractable phosphorus, and changes of acidity, nitrate soil profile, and cation exchange capacity (Cihacek and Swan 1994).

By far the most undesirable off-site impact is excessive siltation of canals, streambeds, and reservoirs. This process, often amounting to as much as a few percent of the total storage volume per year, causes problems with navigation and reduces availability of water for irrigation and electricity generation. Costs of these changes to municipalities, industries, and utilities may greatly exceed any costs to farmers (Higgitt 1991).

All changes caused by excessive erosion will tend to lower soil productivity and will result—sometimes almost immediately, often only after considerable delays—in reduced crop yields. Sensible indication of productivity decline cannot be obtained by generalized linear extrapolations based on absolute loss of soil thickness or by simply multiplying the eroded mass by an average nutrient content. There are three principal ways to assess the impacts of excessive soil erosion on crop yields: paired comparisons of selected eroded phases within a field (or between two adjacent fields eroded to different extent), experimental topsoil removal, and simulation modeling (Olson et al. 1994).

One of the first studies modeling erosion-induced yield declines concluded that in the Corn Belt, America's most productive crop region, continuation of soil losses prevailing during the late 1970s would reduce average yields by only about 4 percent in a hundred years, with the range from less than 2 percent on very gentle slopes to as much as 20–45 percent (mean of 18 percent) on steeper fields (Pierce et al. 1984). Putman et al. (1988) predicted losses of just 1.8 percent from water, and 0.5 percent from wind, erosion as the U.S. means after one hundred years. Shaffer's et al. (1994) projection of corn yield losses for major soil series across the north central region of the United States resulted in a mean decline of about 10 percent over four decades, with extremes ranging from 21 percent in twenty-two years to zero in eighty-nine years. Losses on severely eroded sites can cause such yield declines in much shorter periods.

When using the comparative approach, Schertz et al. (1989) found yield declines up to 17 percent for corn and as much as 30 percent for soybeans grown on severely eroded land in Indiana. Similarly, a ten-year retrospective study of corn and soybean yields in the same state found reductions ranging from 14 to 20 percent on severely eroded sites in comparison with slightly eroded fields (Weesies et al. 1994). And during a recent set of experiments in Tanzania, corn yield reductions per one cm of decreased topsoil thickness ranged from 38.5 to 87.7 kg (Kaihura et al. 1996).

Jarnagin and Smith (1993) reviewed more than a score of U.S. studies published before the 1970s whose results were expressed as yield reductions per inch (2.5 cm) of topsoil lost. For typical annual erosion rates (10–20 t/ha) on common agricultural soils such a depth would be lost in fifteen to thirty years, and the yield reduction ranges were between 4.7 and 8.7 percent for corn, and between 2.2 and 9.5 percent for wheat. In contrast, eight years of experiments in Montana spring wheat fields indicated that every inch of soil loss decreased average yield by less than 1 percent. Interestingly, this study concluded that more moisture, rather than higher fertilizer applications, is the best way to offset the decline (Arce-Diaz et al. 1993).

Experiments with mechanical removal of topsoil have measured wheat yield declines of as much as 61 percent with 30 cm removed—and as little as 17 percent for corn with 45 cm stripped away (Massee 1985; Gollany et al. 1992). Much higher yield reductions were found in tests with corn in Nigeria, where removals of 20 cm of soil reduced yields by 80–100 percent (Mbagwu et al. 1984). Interestingly, an experiment with adding 20 cm of soil brought a mere 1 percent increase of corn yield—and a 17 percent decrease of oats harvest (Mielke and Schepers 1986).

In the United States these local and regional studies have been used to prepare nationwide estimates of long-term impacts on crop productivity. Crosson and Anderson (1992) used three published models of erosion impacts to argue that the productivity decline due to erosion will not be a serious threat to U.S. farming inasmuch as it might amount to just between 3 and 10 percent in a hundred years. More recently, Crosson (1997) used the GLASOD global soil degradation study (Oldeman et al. 1990) and assessments of productivity losses in dry regions prepared by Dregne and Chou (1992) to calculate the average annual loss of global crop productivity at 0.18 percent during the forty-five years preceding

1990. He also concluded that there is no convincing evidence to prove that degradation-induced productivity losses are rising, and that the combination of population growth, technical advances, and increasing commercialization of farm products will increase the value of agricultural land, providing a powerful incentive for investment in soil conservation and proper agronomic management.

Not surprisingly, higher losses could be expected in more vulnerable agroecosystems. Aridity in large parts of Africa makes the continent particularly prone to erosion. According to Lal (1995) yield reductions due to past erosion average about 8 percent for the continent, with minima of less than two and maxima of 40 percent in the Maghreb, East African Highlands, eastern Madagascar, and parts of southern Africa. In absolute terms this translates to annual losses of some 4 Mt grain and 7 Mt tubers. Lal also predicted that if recent trends continue unabated Africa's crop yields may be reduced by about 15 percent by the year 2020.

Besides relying on a limited number of relatively short-term investigations or on inadequately verified models, these large-scale, long-term appraisals of erosion effects also ignore the overall complexity of the process. Clearly, much of the eroded soil is not lost forever to food production; it is merely stored somewhere else before its eventual deposition in the ocean. The world's extensive and fertile alluvial plains are the most obvious result of this process.

Transfer of soil within fields, deposition on neighboring fields or on downstream alluvial lands, and redeposition of windborne matter downwind are universal, continuous processes, but we have only a few localized quantification of their effect. For example, Larson et al. (1983) found that very little or no eroded sediment may be leaving the cultivated land in an area with gentle relief without any major surface outlet, the landscape type common in the north-central United States dominated by glacial-derived soils. Quantifying the net impact of such transfers on a large scale would require a much deeper understanding of the overall change.

Qualitative Soil Degradation

Although erosion is the most debated form of soil degradation, there are other, often very subtle, long-term soil changes. Anecdotal reports of drastic and rapid soil quality deterioration in many countries have created an impression of crisis. Undoubtedly, combined effects of subtle degradation are much greater in many soils than changes caused by erosion. But large-scale assessments of long-term changes in soil quality are even more difficult, and hence even rarer, than evaluations of erosion trends. This is hardly surprising given the nature of soils.

Soils are extremely complex assemblages of mineral particles and living organisms. Soil's texture arises from the physical nature of the material, but its structure results from interactions of these materials with soil biota. Soils provide both physical support for roots and the necessary porosity for their growth; they accumulate organic matter and mineralize its constituents in order to recycle plant macronutrients; they retain and supply water and mineral micronutrients; and their microbial fauna has an irreplaceable role in biogeochemical cycling, particularly in that of the three doubly mobile elements, carbon, nitrogen, and sulfur. Continuous cropping affects all these factors, but assessments of long-term changes are difficult.

Multiple measurements are needed because of high spatial and temporal variability of most soil characteristics and processes, and some methodological disagreements may result in diverging interpretations. For example, Richter and Markewitz (1995) argue that the traditional view of soil encompassing the relatively thin uppermost horizons—usually limited to 0.5–2 m, marked as O, A, E, and B and positioned above the C horizon parent material—is incompatible with the degree of microbial activity and root penetration which strongly imprint many soils to depths of five or eight meters (figure 3.2). The depth distribution of soil CO_2 and acidity would make the C horizon a fundamental component of soil, resulting in a huge increase of the world's soil volume.

But there is no dispute about what is perhaps the key indicator of soil quality: adequate levels of soil organic matter are essential both for the maintenance of soil fertility and structure. Its decline following the conversion of forest or grassland soils to arable land—and hence the reduction of natural stores of carbon and nitrogen, which fall by 30–50 percent just one to two decades after grasslands or forests are converted

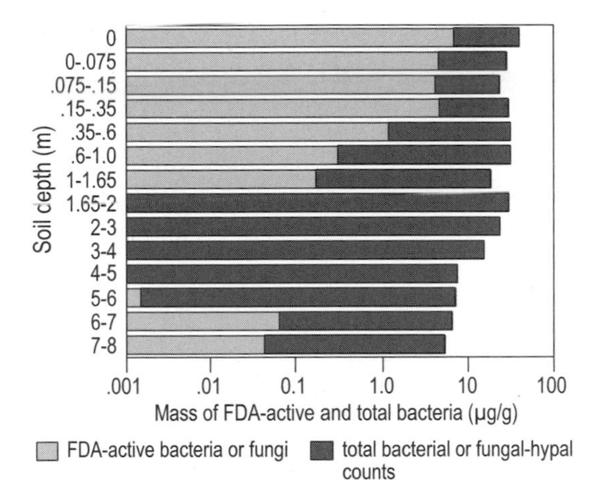

Figure 3.2
Presence of bacteria throughout the 8-m profile of the Ultisol soil in South Carolina (Richter and Mackenzie 1995).

to crop fields (Rozanov et al. 1990)—is perhaps the best documented long-term qualitative change in many soils. The reason is obvious: cultivation aerates the soil, promoting oxidative decomposition of organic matter.

Soil carbon level is usually a good indicator of soil fertility. Michigan studies indicate a more than 20 percent increase of potential yield with every one percent gain in soil carbon content (Lucas et al. 1977). Average organic carbon content of undisturbed soils is commonly around 2 percent. It falls below 1.5 percent in early stages of cultivation, and it is less than 0.5 percent in badly damaged soils. Dead biomass makes up more than nine-tenths of all soil carbon; the rest is made up largely of living roots, bacteria, fungi, and soil invertebrates. In chemical terms, soil organic matter is a complex mixture of acids, carbohydrates, and aromatic carbon compounds; some of them are readily decomposable, whereas others persist for centuries. Even relatively carbon-poor arable soils hold more of the element in the topmost one meter than the plants they support, and also more than is present in the atmosphere above them. Globally, these figures scale to at least 1,500 Gt C in soils and about 770 Gt C in the atmosphere.

Very few studies have tried to assess long-term trends in soil quality on a large scale. A rare recent study of this kind has been done by Lindert (1996) who attempted to identify long-term soil quality trends in China and Indonesia by using the best available survey data gathered between the 1930s and the 1980s. His surprising conclusions: although organic matter and soil nitrogen have declined in both countries over the course of recent decades, the reduction in many regions was only slight, and often statistically insignificant.

During the same time, total phosphorus and potassium stores have generally risen, and alkalinity and acidity have fluctuated with no overall worsening. However, Doberman (1996) criticized Lindert's conclusions because of inherent limitations of the data used (above all lack of spatial congruence) and methodological failures (use of abrupt soil horizon boundaries, and limited usefulness of total elemental content for assessing soil quality).

Proving the link between soil structure and crop yield has been very difficult (Karlen et al. 1994). Because of the lack of historic baselines it is also difficult to assess the long-term consequences of soil degradation caused by compaction resulting from the use of heavier machinery and from more passes across fields (Soane and van Ouwerkerk 1994). Compaction accelerates erosion and induces anaerobiosis and denitrification (these, in turn, promote CH_4 and N_2O emissions). Extreme compaction can reduce yields by as much as 70 percent.

On a less worrisome note, there seems to be also little doubt that virtually all early estimates have greatly overestimated the impact of desertification, mainly because they mistook the cyclical nature of these changes for steady degradation (Thomas 1993; Thomas and Middleton 1994). Desert margins contribute relatively little to global food supply: even in sub-Saharan Africa they account for no more than an eighth of the region's total grain production. The local impact of such a cyclical shift can be, of course, severe.

As with other soil degradations, we have no reliable inventories of land affected by salinization: global estimates range from 600 to 2,000 Mha. Elevated water tables in heavily irrigated areas of arid regions have been the leading cause of this condition, which reduces yields, inhibits groundwater recharge, increases risk of flooding, and brings toxic elements to the surface. Depending on its severity, salinization can be managed by crop choice, proper seed placement, careful water applications, and leaching (Chhabra 1996). Reclamation through planting trees and grasses is often very effective.

Asia has the world's largest area of cropland affected by salinization, but the process does not appear to be a primary threat to the continent's food security during the next generation (Cassman and Harwood 1995). Long-term impacts of nearly constant irrigation in multicropped rice fields are of much greater concern. Olk et al. (1996) found that the increased frequency of irrigated cropping tends to bring large increases in the phenolic content of soil organic matter, most likely a result of slower lignin decomposition brought by oxygen deficiency in submerged soils. If this change—not detectable by analyzing such gross properties of humic acids as carbon or nitrogen content—were to persist it could have important effects of nitrogen cycling and on the capacity of nitrogen supply to double- or triple-cropped rice.

Yet another long-term qualitative concern arises from accumulation of heavy metals in soils resulting from applications of phosphatic fertilizers and from an otherwise highly desirable practice of organic waste recycling. Small amounts of heavy metals are found in phosphate rock and hence are commonly present in commercial phosphate fertilizers; animal manures and sewage sludge are the main organic source of trace quantities of heavy metals. Repeated fertilizer, manure, and sludge applications may lead to accumulation of cadmium, arsenic, chromium, lead, mercury, nickel, and vanadium in soils (Mortvedt 1996). Because long-term health effects of this accumulation are unknown, some countries, acting as risk minimizers, have put limits on heavy metal additions.

Maintaining Productive Soils

Productivity declines caused by both overt and subtle forms of soil degradation are being constantly masked by planting of higher-yielding cultivars, by increasing rates of fertilization and by irrigation. In many instances this may be a proper choice. For example, Diemont et al. (1991) found that low inputs, rather than excessive soil erosion, are the main cause of low productivity of Java's upland crops grown in highly erosible soils. But in many cases further intensification may only add to soil deterioration. Fortunately, there is nothing inevitable about excessive erosion and the accompanying decline of soil organic matter and plant nutrients: many well-proven agronomic practices can not only prevent excessive erosion and maintain soil quality, but can also build up and restore degraded soils.

In some environments leaving soils alone may be an effective remedy. At Rothamsted plots returned to natural state began accumulating nitrogen from atmospheric deposition and biofixation at a linear rate of 49 kg/ha, so that after eighty years those previously low-N plots had levels equal to those of wheat plots receiving annually 35 t/ha of manure (Legg and Meisinger 1984). In the short term (less than ten years) planting of grasses and legumes (or their mixtures) increases pools of mineralizable carbon and nitrogen but may have negligible effects on total soil organic matter (Robles and Burke 1997). Fallowing may be impossible in densely populated Asian regions, but it may be the best choice in areas with plenty of marginal land where concentrating on yield increases on better soils may be a much better long-term strategy.

Leaving soils covered with crop residues is perhaps the single most effective postharvest practice reducing soil erosion inasmuch as it cushions the impact of raindrops. Kinetic energy of the largest raindrops will be roughly forty times their mass, making their impact two orders of magnitude more powerful than the resulting surface runoff, and the rate of detachment of eroding particles will be highly correlated with rainfall intensity (Smil 1991). Crop residues should be thus treated as a valuable renewable resource to be carefully managed in order to maintain soil quality and promote crop productivity (Smil 1999a). Besides reducing erosion and runoff rates, residual cover also enhances water infiltration. Even better, cover crops planted for feed or as green manures can intercept two to five times more precipitation than cereal grains.

There is also no shortage of traditional, indigenous soil—and, by extension, water—conservation measures practiced to this day by farmers in both humid and arid environments. One of the best compilations and evaluations of such simple but effective techniques contains examples from every region of Africa (Reij et al. 1996). A shorter compilation of a similar nature is available for India (Kerr and Sanghi 1992).

Contour planting and terracing are two ancient and highly effective means of reducing slope erosion. In Java on fields with 10° slope erosion rates on plots planted with potatoes, up-and-down slope is nearly 140 t/ha (and runoff is 17 percent)—but contour planting cuts it to about 40 t/ha, and terracing to less than 5 t/ha (Fagi 1996). Other common methods of indigenous soil protection include preparation of shallow planting pits, half-moon basins, and tied ridges, as well as mulching. Some of these measures are also effective for rehabilitation of badly eroded fields.

Universally applicable, and generally fairly inexpensive, ways of controlling wind erosion were formulated in the wake of the prolonged North American drought of the 1930s (Chepil and Woodruff 1963; U.S. Department of Agriculture 1972). They include formation of soil clods large enough to resist the wind force; roughening of the surface to reduce wind velocity and capture drifting soil particles; shortening of field widths across the prevailing wind direction; and maintaining as much vegetation cover or plant residues on the ground as practicable. Common practices designed to achieve these goals include stripcropping, minimum tillage, stubble mulching, planting of cover crops and wind barriers, and crop rotations.

Conservation tillage originated from stubble mulching introduced to combat North America's severe soil erosion of the 1930s, and it has become, particularly since the 1960s, an important component of the rich world's agronomic operations (Unger and McCalla 1980; Gebhardt et al. 1985; Little 1987). The practice includes a variety of techniques that reduce soil or water loss when compared to traditional moldboard plowing: it ranges from contour and ridge planting to no-till cropping where only narrow soil strips are loosened to cover the seed. Although the extreme form, no-till farming, is not suited for every crop and every environment, a great diversity of conservation tillage practices makes it possible to find appropriate ways of less disruptive soil preparation for every major commercial crop in every climatic zone.

Eventual effect of these measures could be felt even in extreme situations. GLASOD estimated that erosion on some 9 Mha (with 5 Mha in Africa) has gone beyond the point of economically rewarding rehabilitation, but farmers in some areas have demonstrated that this unrestorable land can be improved at modest cost with appropriate incentives and techniques (Norse 1994). The Machakos district of Kenya has been perhaps the most publicized case of such a restoration (Tiffen et al. 1994).

Replenishment of soil organic matter is possible with regular recycling of crop residues, manures, and other organic wastes. Keeping high levels of soil organic matter is particularly difficult in regions with long, warm, and wet growing seasons where decomposition proceeds rapidly. Continuing incorporation of straw in temperate regions can bring a 40–50 percent increase in biomass carbon and nitrogen in less than two decades. Levels of soil organic carbon appear to be a linear function of the amount of recycled crop residues and enrichment can continue for decades. Rothamsted experiments show soil carbon still increasing even after 130 years of regular manuring (Johnston 1997).

Environmental Pollution

Many kinds of environmental pollution have been reducing crop yields and livestock and aquacultural productivity on local or small regional scales for many generations. Dumping of untreated industrial wastes into streams and water bodies has led to the accumulation of unacceptable residues in crops and fish, and in many instances it has destroyed local aquaculture. Unusually high levels of both very common (nitrogen and sulfur oxides) and less often encountered air pollutants (arsenic, peroxyacetylnitrate) have reduced yields of crops cultivated downwind from large emission sources.

And agriculture itself is a major environmental polluter (Conway and Pretty 1991). Excessive and inefficient fertilization lets nitrogen and phosphorus escape into the environment, where they cause many undesirable changes. Recycling of manures can result in a worrisome buildup of heavy metals in some agricultural soils. And very soon after pesticide applications took off during the 1950s came concerns about their environmental consequences (Carson 1962). More than a generation later many of these concerns are still with us (Guerrero 1992).

Today's pesticide applications vary widely. In the early 1980s French farmers applied more than twice as many active ingredients per ha than the U.S. average, and the Canadian mean was about one third of German rate (OECD 1993). Both excessive and inappropriate applications are common problems in many poor countries. Pesticides, and their metabolites, are found not only in treated fields, but also in nearby air and waters. A growing number of research projects has been measuring these effects on scales ranging from local to national. Monitoring of ambient concentrations of pesticides in a California county with high application rates found detectable levels at several populated sites and even at an urban location (Baker et al. 1996).

During late spring and summer months herbicides can be detected in rainfall throughout the midwestern and northeastern United States, with the highest levels in the Corn Belt (Goolsby et al. 1997). The total mass of deposited atrazine and alachlor, two common herbicides, corresponded, respectively, to about 0.6 and 0.4 percent of all applications to crops in the two U.S. regions. Kolpin et al. (1991) called attention to the problem of pesticide metabolites in shallow aquifers. While the U.S. Environmental Protection Agency mandates maximum contaminant levels for many commonly used pesticides, this investigation of more than eight hundred wells across the Midwest discovered that metabolites are much more common in drinking water than their parent compounds. Moreover, the sampling also found DDE, a metabolite more toxic than the parent DDT whose use was discontinued in 1973!

Effects on human health are of the greatest immediate concern because hundreds of millions of farm workers and consumers are exposed to excessive levels of pesticides. Acute poisoning occurs with intolerably high frequency in many poor countries, but in terms of widespread and lasting impact the most worrisome effect of pesticides is as triggers of significant changes in the structure and function of the human immune system (Repetto and Baliga 1996). These changes can result in higher risks of infectious disease and cancers associated with immunosuppression even among otherwise healthy populations.

But quantifying these impacts with reliability high enough to allow for unequivocal deregistration of dangerous compounds or effective regulation of more tolerable ones has been exceedingly difficult (Wargo 1996), and overall increases in life expectancy demonstrate that the effects cannot be very severe. These concerns and controversies are here to stay—but tighter regulations, introduction of more readily biodegradable pesticides, and better ways of overall agronomic management should prevent any significant rise of existing impacts, which in any case have not had any demonstrably negative effect on crop productivity.

We also do not have any systematic comparative data about the effects of air and water pollution on food production. But, so far, the overall economic impact has been only marginal even in such a polluted country as China. The best available estimates of food production losses attributable to air and water pollution put their value at between 2 and 3 percent of China's total annual farm output of the early 1990s, with air pollution accounting for more than two-thirds of the total loss (Smil 1996a; Smil and Mao 1998).

In some instances greater damage has been prevented by fortuitous biochemical realities. Perhaps the most notable case of this kind is the bioaccumulation of hydrophobic pollutants—polychlorinated biphenyls, dioxins, dibenzofurans, and hexachlorobenzene. Their presence has been well documented in natural aquatic environments, but only recently McLachlan (1996) provided insights into the process in agricultural food chains. Concentrations of the chemicals in plants do not achieve equilibrium with air before harvests, and highly hydrophobic compounds are not well absorbed in the digestive tracts of domestic animals. As a result, significant biomagnification occurs only in humans, with differences between human and cow milk being twenty fold to fifty fold for most of the compounds, not enough to cause any major health effects.

At the same time, there is no shortage of looming concerns. Two of them—rising fertilization of the biosphere with nitrogen escaped from fields (and also from industries and transportation) and increasing levels of tropospheric ozone—are of particular import because of their potentially large impact, as far as both their spatial extent and their economic consequences are concerned.

Losing Nitrogen

Compared to the flows of the preindustrial era, we have already at least doubled the inputs of all reactive nitrogen reaching soils and the atmosphere. In global terms, these inputs now rival the total fixed by all bacteria in natural terrestrial ecosystems. In regions having both intensive agriculture and large urban-industrial areas, anthropogenic nitrogen inputs surpass severalfold the total natural flows of reactive nitrogen. And in many watersheds and fields our inputs dominate natural flows by more than an order of magnitude.

Applications of synthetic fertilizers are responsible for the largest share of this reactive nitrogen introduced by humans into the biosphere. During the 1990s synthetic fertilizers contributed annually about 80 Mt N, compared to over 20 Mt N from fossil fuel combustion. Given the multitude of pathways in nitrogen's complex cycling, rising applications of nitrogenous fertilizers have resulted in growing losses of the nutrient, both in soluble forms as water-borne nitrate (NO_3) and in volatile compounds, mainly as ammonia (NH_3) and nitrogen oxides (NO and NO_2).

In those parts of western Europe where heavy nitrogen applications predated World War II, nitrate concentrations began to rise quickly in both groundwaters and surface waters after 1950 (figure 3.3). By the early 1980s they were either near or above the European Union's limit of 50 mg NO_3/L in a number of regions, especially in England and the

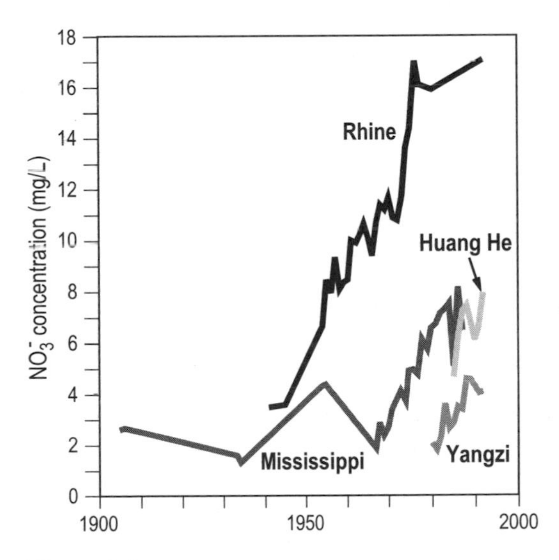

Figure 3.3
Nitrate concentrations in four major rivers (Smil 1997).

Netherlands. Van der Voet et al. (1996) calculated that 87 percent of nitrate accumulation in ground waters of European Union countries originates from leaching from agricultural lands, and noted that a substantial share of leaching from nonagricultural soils is also caused indirectly by fertilizers through deposition of volatilized ammonia.

Excessive levels of nitrate—originating largely from leaching of fertilizers and manures—have been present in water wells throughout the American Midwest for several decades. Concentrations above the maximum contaminant limit are particularly common (in anywhere between 10 and 28 percent of cases) in the Corn Belt states of Iowa, Illinois, Nebraska, and Wisconsin, as well as in Minnesota and Kansas (Nolan et al. 1997; GAO 1997).

High nitrate levels in drinking water potentially pose several health risks. Methemoglobinemia (blue baby disease) is the most dangerous condition: if not treated (with ascorbic acid or methylene blue) it can result in asphyxiation, but its incidence has been increasingly rare, with only a few thousand cases documented since 1950. Suspected chronic consequences of drinking water with high nitrate concentrations include higher incidence of some cancers (esophageal, stomach) in adults, but establishing causal links remains controversial.

As a result, concerns about nitrogen in water have shifted from early preoccupation with health effects to impacts on aquatic ecosystems: these are much easier to demonstrate and are certainly much more costly. Combined loadings of nitrogen from fertilizers and also from the dumping of urban sewage and from atmospheric deposition now amount to about 30 kg N/ha a year in watersheds of the eastern United States, and to more than 60 kg N/ha in parts of northwestern Europe and Japan. China's coastal provinces will have similar loadings soon.

Average NO_3 concentrations in the most affected rivers (the Thames, Rhine, Meuse, and Elbe) are now two orders of magnitude above the mean of unpolluted streams. In the early 1990s more than a tenth of western Europe's rivers had NO_3 levels above the maximum contaminant limit and nitrate levels are increasing in both of China's two largest rivers, the Huang He and Yangzi. High levels of dissolved nitrates cause eutrophication—the enrichment of waters with what is normally a growth-limiting nutrient—which promotes the growth of algae and cyanobacteria. Subsequent decomposition of these blooms creates hypoxic or anoxic environments that may even kill fish and shellfish species. Eutrophication has been particularly widespread in estuaries (such as Long Island Sound, San Francisco Bay, and Chesapeake Bay) which may receive commonly ten times as much nitrogen as heavily fertilized fields (Vollenweider et al. 1992; Valiela et al. 1997).

Anthropogenic nitrogen does not enrich only aquatic ecosystems. Continuous nitrogen inputs are now taking place on an unprecedented scale through atmospheric deposition of ammonia and nitrates. Recent rates of nitrogen inputs to natural ecosystems have become significant even by agricultural standards: in large parts of eastern North America, Europe, and East Asia they are around 50, and for some sites even over 80, kg N/ha a year, an equivalent to average fertilizer applications on American cropland! For most forests such rates are an order of magnitude higher than the means of the preindustrial world, and they equal or surpass the quantity of the element made available through decomposition.

Given the widespread nitrogen shortages in natural ecosystems, such an enrichment should initially stimulate photosynthesis and appreciably increase the carbon and nitrogen stored in plants, litter, and soil organic matter. But our understanding of how natural ecosystems will respond to nitrogen enrichment in the long run, particularly when accompanied by concurrent CO_2 increase (see the next section), is poor. Inevitably, the response will have to be self-limiting: total nutrient inputs will eventually surpass the combined plant and microbial demand, and saturation of ecosystems will limit the carbon sink stimulated by the nutrient and will accelerate the element's transfer to waters as well as its emissions as N_2O to the atmosphere (figure 3.4).

Before this happens, continued nitrogen enrichment may bring gradual and unwelcome changes. Mycorrhizal infection of roots may weaken, fine-root phytomass may be reduced (thus lowering intakes of water and phosphorus), nitrogen-fixing species will become less competitive, and lichens associated with nitrogen-fixing cyanobacteria may retreat. As trees develop their maximum canopies they may become more sensitive to water stress and more vulnerable to micronutrient deficiencies. Sensitive soils may acidify faster as nitrification increases.

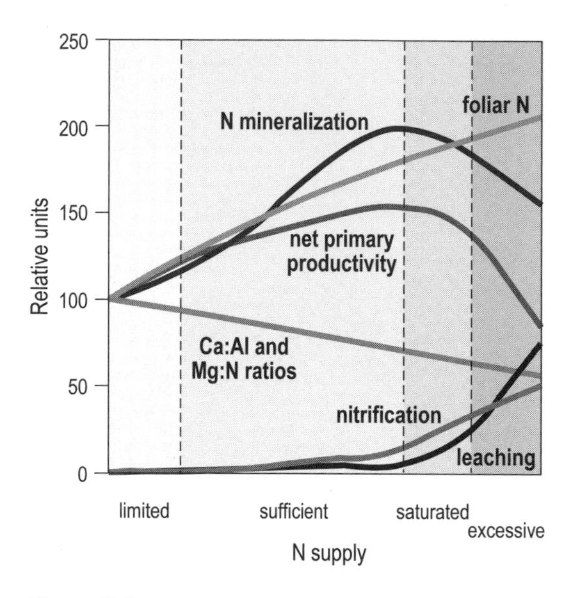

Figure 3.4
Stages of an ecosystem response to nitrogen inputs; based on Aber et al. (1998).

In ecosystems adapted to a low nitrogen supply, individual species or whole plant communities may become extinct because a few nitrophilic varieties, able to grow faster under high inputs of the nutrient, will dominate. Although the total phytomass of affected ecosystems may increase, their biodiversity could be much reduced. Ground cover may decline in some forests as denser canopies would reduce the incoming sunlight, or the ground cover may shift to more shade-tolerant species. The denser foliage of a nitrogen-rich world could support higher numbers of animals seeking cover but reduce the counts of those mammals and birds that prefer open vegetation.

Yet another consequence of chronically high nitrogen deposition, one with direct implications for cropping, may be rising N/S ratios. Sulfur deficiency is already a problem in some soils, particularly in Asia's rice fields, where sulfur-free fertilizers displaced previously common ammonium sulfate. Sulfur deficiencies can lower the cereal yield by reducing the number of shoots and panicles and the size of grains, and they can lead nitrogen to accumulate in nonprotein forms, mostly as nitrate.

Net effects of combined CO_2 and nitrogen enrichment are particularly uncertain when the two processes stimulate contradictory outcomes. For example, CO_2 enrichment may be slowing down decomposition by producing more abundant litter with higher C/N ratio—but nitrogen enrichment, producing a more nutritious litter, could have the opposite effect. But what the C/N ratios in enriched ecosystems will be like is also uncertain: will they decline with higher nitrogen inputs, or will the leaves contain less nitrogen so that folivorous insects would have to eat more to maintain their intake of protein? Lower quality and palatability of forages would be particularly unwelcome in tropical grazing where grasses are already low in protein.

The only fact that is reasonably certain is that the human enrichment of the biosphere with nitrogen will intensify throughout Asia and in Africa, the two continents where most of the future population growth will take place (for more, see chapter 4). The reverse is already happening in Europe: in order to limit further nitrogen losses, the European Union is now committed to undertaking substantial reductions of nitrogen applications, which will cut inflow of the element to the Baltic Sea by 45 percent and reduce atmospheric deposition of nitrogen compounds by about 20 percent (van der Voet et al. 1996). Will there be eventual limits on fertilizer applications in Asia, where food output has become critically dependent on the nutrient?

Changing Ozone Levels

Global warming has received much more attention than any other long-term environmental change. But loss of stratospheric ozone and higher levels of ultraviolet radiation reaching the biosphere would also have global consequences for food production, while increases of tropospheric concentrations of the gas are already causing damage to crops in some regions. Most of the human impact on stratospheric ozone concentrations has been due to decades of chlorofluorocarbon (CFC) emissions, but farming has been a unique source of relatively minor but highly destructive emissions of methyl bromide (CH_3Br), an important soil fumigant used to control nematodes, weeds, and fungi in tomatoes, peppers, ornamentals, and strawberries. Volatilization of the compound may be minimized by applying it when temperatures are cool, and injecting it deep into moist soil covered by tarps (Yates et al. 1997).

Once the gas reaches the stratosphere then every atom of bromine is about fifty times more efficient in breaking down stratospheric ozone than an atom of chlorine in CFCs. Fortunately, only some 4 percent of CH_3Br molecules survive that long passage. Even so, the gas is now responsible for about 10 percent of the ongoing destruction of stratospheric ozone layer. In 1997, a decade after the first agreement on phasing out CFCs, came the agreement about phasing out the gas. Rich countries will end its use by the year 2005; low-income nations will have another decade to do so (Spurgeon 1997).

These are welcome expectations as significantly higher levels of UV-B radiation could have appreciable effects on crop yields and on animal health (Worrest and Caldwell 1986; Biggs and Joyner 1994). Experiments performed under field conditions (using UV-B supplement in proportion to ambient UV-B) confirm the findings of laboratory studies that yield reduction may be caused by direct effects of radiation reducing number of stems, stem length, and total phytomass, rather than by lowered photosynthetic productivity (Mepsted et al. 1996). Fears about losing stratospheric ozone's protective function were contained by international treaties, but rising levels of tropospheric ozone are perhaps the most worrisome air pollution problem of the next generation as far as farming in populous modernizing countries is concerned. The growing presence of this highly oxidative gas is a clear sign of high seasonal, or increasingly semipermanent, levels of photochemical smog.

Essential precursors of this complex mixture of gases are volatile organic compounds (VOC) CO and nitrogen oxides (NO_x) released mostly from combustion of transportation fuels. Nitrogen oxides are released not only by internal combustion engines and by fossil fuel-fired electricity-generating plants but also by microbial nitrification and denitrification in heavily fertilized soils; consequently, intensive agriculture also slightly aggravates this spreading environmental problem.

Generation of photochemical smog begins with a rapid oxidation of NO to NO_2 involving a variety of reactive molecules (OH·, CO, hydrocarbons, aldehydes). Subsequent dissociation of NO_2 and oxidization of hydrocarbons produces rising levels of ozone, one of the most aggressive oxidants demonstrably implicated in the deterioration of human health, damage to various materials—and in decreases of agricultural productivity in a variety of crops (Lefohn 1991; McKee 1993). High ground levels of ozone cause pigmentation and bleaching of leaves, foliar chlorosis, and necrosis. Crops exposed to excessive ozone levels—particularly wheat, corn, barley, soybeans, and tomatoes—have yields at least 10 percent lower than those grown in unpolluted air.

Initially it appeared that photochemical smog would remain restricted to very sunny subtropical climates, but rising emissions of NO_x and VOC emissions brought seasonally elevated concentrations of ozone to all large mid-latitudes urban areas of the Northern hemisphere (as far north as 50°N), and made them a semipermanent feature of all large cities in tropical and subtropical urban regions and their surroundings. The combination of growing intercity traffic and more extensive conurbations has extended the areas of high ozone concentrations far into the surrounding countryside, particularly during summer, when high-pressure cells limit both vertical and horizontal atmospheric mixing.

Crop productivity in three semicontinental regions combining high population density with concentrated industrial production and intensive agriculture—or "metro-agro-plexes" (Chameides 1994)—is already affected by excessive levels of ozone. Western Europe, Eastern North America, and East Asia now consume about three-fourths of the world's fossil fuels and nitrogen fertilizers and produce about a third of the world's grain harvest and even higher shares of vegetables and fruits—and they are already affected by O_3 levels above 50–70 ppb, the threshold for damage caused by cumulative exposure during the growing season.

Both the extent and the intensity of these impacts will increase. Car ownership is very close to saturation in the eastern North America (1.8 people per car in 1995), but the average distance driven per year is still increasing (MVMA 1996) and continuing exurbanization of the region will further lengthen many commutes and extend the area of higher photochemical smog concentrations even further into the previously less affected countryside (Chameides et al. 1997; Finlayson-Pitts and Pitts 1997). Air traffic in the region also keeps increasing.

In East Asia, Japanese and Korean emissions may increase only marginally, but in eastern China all factors promoting higher NO_x emissions are already undergoing a truly explosive growth. The core of China's emerging metro-agro-plex is southern Jiangsu (Sunan) between Shanghai and Nanjing, which is being transformed into one of the world's most concentrated belts of export-oriented manufacturing. Although the relative population growth of the region is even lower than the country's impressively low average (only about 0.8 percent a year compared to the nationwide rate of about 1.1 percent in 1995), the region will continue to be a magnet for immigration from the interior, especially as the recent trend of increasing income disparities shows now signs of abatement.

Further intensification of cropping in Eastern China will release additional NO_x emissions. Average annual rates of nitrogen fertilization are already among the world's highest, and are still increasing. During the next generation the Chinese metro-agro-plex will undoubtedly be much richer because of its thriving manufacturing and aggressive urbanization. But it will be also intolerably polluted by photochemical smog extending over areas at least as large as those currently affected in North America. Solutions of this problem are not easy. In spite of considerable expense devoted to smog controls, scores of cities in North America are exceeding the ambient air quality standard for ozone, and many places have seen no appreciable improvement during the past decade.

What Could Climate Change Do?

Although the realization that rising atmospheric concentrations of CO_2 could eventually lead to global warming is more than a century old (Smil 1997a), sustained scientific interest in the matter dates only since the late 1950s, when Revelle and Suess (1957) published their now classic paper on anthropogenic generation of CO_2 as an open-ended, global-scale geophysical experiment. Widespread public and policymaking interest has been evident only since the 1980s. Assessment of possible impacts rests on global climate models (GCMs) whose increasing sophistication still falls far short of reproducing the intricacies of several critical feedbacks, especially those involving clouds and ocean-atmosphere interaction (GAO 1995; IPCC 1996). Consequently, the best available GCMs disagree substantially even about such key variables affecting agricultural productivity as the future amount of precipitation or soil moisture in particular regions.

Most long-term simulations consider the effects arising from the doubling of preindustrial atmospheric CO_2 concentrations, which are to reach levels around 600 ppm. However, by 1995 gases other than CO_2 accounted for more than a third of the global warming effect. Hence, more accurate question is what the consequences of tropospheric warming equivalent to the one brought on by the doubling of preindustrial CO_2 will be (figure 3.5). The best, and the most detailed, consensus summary of these effects has been contained in reports of the Intergovernmental Panel on Climatic Change (IPCC 1996).

Key changes affecting agriculture would include an increase of annual surface temperatures by an average of anywhere between 1°C, and 3.5°C, with 2°C now considered most likely (IPCC 1996). The fact that both the mean and the maximum in the first IPCC report were about 30 percent higher underscores the existing uncertainties. The temperature rise would be more pronounced on land; it would increase

greenhouse gases	CO_2	CH_4	N_2O
molecule			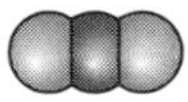
atmospheric concentration in mid-1990s	360 ppm	1720 ppb	313 ppb
annual atmospheric concentration increase (%)	0.4	0.6	0.25
global warming potential during the next 20 years	1	56	280

Figure 3.5
Greenhouse gases.

in the poleward direction and be greater in winter than in summer. Warming would intensify global water cycle, and, again, this effect would be more pronounced in higher latitudes and during wintertime. At the same time, some arid regions may get even drier as warmer ascending air reduces the formation of low-level clouds. So far, precipitation trends since 1900 show a general tendency toward more moisture in higher latitudes and less rain in subtropical regions (Karl et al. 1997).

Increased surface temperature and evaporation would also bring more precipitation, but also more interannual variability, to the Asian monsoon, a change of immense importance for nearly half of the world's population, depending on its regular returns. Thermal expansion of seawater and gradual melting of mountain glaciers could result in an appreciable rise of sea level: the mean rise could be as low as ten centimeters and as high as one meter. In addition, on a warmer planet with intensified water cycle more CO_2 would be released from arable land, because the emissions of the gas from decomposition of soil organic matter are not primarily determined by soils' carbon stores but rather by soil temperature and precipitation, the two key variables controlling organic decay.

Diminishing stores of natural phytomass also mean a reduced of biospheric sink for anthropogenic carbon, and for corresponding amounts of N, P, and S. For every one hundred carbon atoms phytomass typically sequesters fifteen atoms of nitrogen and one atom each of phosphorus and sulfur. With phytomass density diminished, more N and P will be available to cause eutrophication of waters, and more sulfur may add to local or regional acidification.

Agriculture as a Source of Greenhouse Gases

Before outlining the variety of effects these changes might have on agriculture, I must point out farming's major contributions to global climate change. Farming generates all three leading greenhouse gases, and three distinct sources account for these emissions: conversions of natural ecosystems to croplands and grazing land, releasing large amounts of CO_2 from removed phytomass and from affected soils; higher rates of N_2O emissions from the denitrification of nitrogenous fertilizers; and rising emissions of methane (CH_4) from rice fields and from enteric fermentation of growing cattle herds.

Conversion of natural ecosystems—mainly forests and grasslands—to croplands and grazing land (and also to urban, industrial and transportation uses) is one of the two leading causes of anthropogenic increases of atmospheric CO_2. While the mature forest may store more than 150 Mt C/ha, cattle-grazed grassland growing in its place will rarely store more than one-tenth of that mass, and most arable lands will tie up less than one-twentieth of the forest's reservoir of the element in their crops (sugar cane, with maxima above 15 t C/ha, is a notable exception).

Although the best historic estimates of this conversion process cannot be as accurate as our calculations of CO_2 emitted from fossil fuels, there is little doubt that the total amount of CO_2 emitted annually from fossil fuels during the early 1950s was still below the quantity released from the conversion of forests and grasslands and from the burning of biomass. Annual CO_2 emissions from coals and hydrocarbons surpassed those from phytomass only during the 1960s—but more forests and grasslands were converted to arable land and grazing during the second half of the twentieth century than during the first. In addition to releasing CO_2, these conversions also emit CO, CH_4, particulate organic carbon, and elemental carbon in soot. They also produce NO_x in fires, followed by further gradual C and N losses from decomposition of the partially burned phytomass, and from higher decay of organic matter, and erosion and leaching of exposed soils.

Changes in the soil carbon reservoir have a major impact on atmospheric concentrations of CO_2. Most of that carbon is tied up in organic matter, about nine-tenths of it in dead biomass derived from plant litter and from roots through decomposition, and the rest in living roots, bacteria, fungi and soil invertebrates. Global size of this reservoir has been estimated to be between 1,200 and 1,800 Gt C (Smil 1997a), or at least twice as much as all terrestrial phytomass. Releasing mere 0.1 percent of the world's soil carbon pool raises the atmospheric concentration of CO_2 by one ppm. Future extensions of arable land will thus continue to be major sources of the gas both from the removed vegetation and from exposed soils.

Bulk of the rising emissions of agricultural methane can be ascribed to growing populations in rice-eating parts of Asia and to a widespread human desire to consume more beef and dairy products. Even though most of Asia's doubled rice output since 1950 has come from higher yields, the area occupied by paddy fields has grown by more than 40 percent. Soils in these flooded fields are anoxic, a perfect environment for methanogenic euryarchaeota (formerly classed as bacteria) producing the gas by the reduction of CO or CO_2. The best Chinese estimate of CH_4 emissions from what are the world's most extensive rice fields is about 15 Mt/year. In spite of an additional 200 million Chinese to be added during the coming two generations, this rate is expected to change little in the near future as most of higher rice production will come from intensification of existing cultivation (Yao et al. 1996).

Higher demand for beef and dairy products has boosted the global cattle count by about 40 percent since 1950. And because modern animals are heavier than their predecessors, they also produce more CH_4 per head. Substantial variation of CH_4 emissions from both paddy fields and cattle preclude any highly reliable estimates, but the two processes appear to contribute roughly similar amount of the gas (around 100 Mt a year). Besides CO_2 and CH_4, farming is also a growing source of N_2O released from microbial denitrification of nitrates. The process is the opposite of nitrogen fixation, and it reduces nitrates to nitrogen oxides and ultimately to N_2, but when the reduction does not go all the way it produces highly variable shares of N_2O. Emissions of N_2O from agricultural soils are particularly high with applications of anhydrous ammonia, organic fertilizers, or combinations of organic and synthetic materials.

Because of the enormous temporal and spatial variability of N_2O emissions global estimates of the flux contributed by the denitrification of inorganic fertilizers and manures applied to farmland are highly uncertain. Totals attributed to agriculture during the 1990s have ranged over an order of magnitude, from mere 0.03 Mt to 5.5 Mt N/year (Watson et al. 1990; Nevison et al. 1996; Mosier et al. 1998). Since the regular measurements of the gas began in 1977 its tropospheric concentration has been rising by about 0.25 percent a year (figure 3.6). At just above

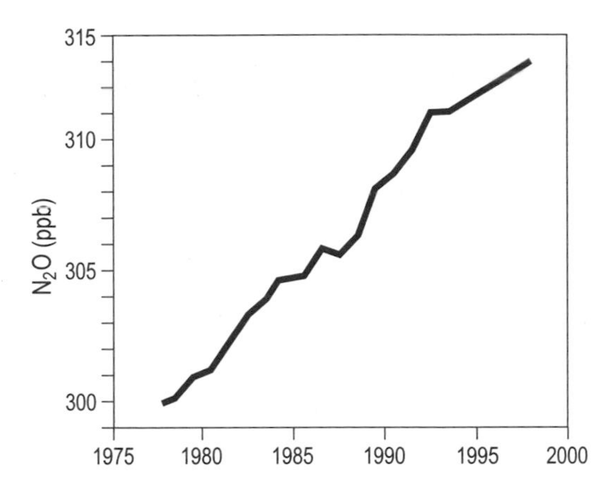

Figure 3.6
Increasing concentrations of atmospheric N_2O (Smil 1997).

310 ppb the gas is only one-one thousandth as abundant as CO_2, but because its global warming potential is roughly three hundred times as large as that of CO_2, N_2O is now responsible for about 6 percent of the anthropogenic greenhouse effect.

Possible Consequences for Food Production

Environmental changes caused by rising concentrations of greenhouse gases would affect agricultural output in many ways. A great deal of theoretical and experimental research has been recently devoted to these effects (IPCC 1996; Graves and Reavey 1996; Bazzaz and Sombroek 1996; Rosenzweig and Hillel 1998). The second round of the IPCC assessment concluded that, on the whole, the world's agricultural production would not suffer any major setbacks when the atmospheric concentrations of greenhouse gases reached an equivalent of doubled CO_2 level, although it cautioned about the possible effects of negative regional impacts, increased variability of yields, and more frequent and more damaging outbreaks of pests (IPCC 1996).

I do agree that the probability of major agricultural setbacks is very low during the next two generations. But I will try to show that the uncertainties inherent in complex and dynamic changes that may arise from rapid global warming make it difficult to decide even about the basic direction of eventual change. We know for sure that higher CO_2 levels will have no direct effect on soil organisms: soil microbes and invertebrates already live in atmosphere with up to 1.7 percent of the gas, an order of magnitude higher than in the open atmosphere. Similar, or higher, CO_2 concentrations are found in ants and termite nests, beehives, and mammalian burrows. Indirect effects are a different matter: as mineralization and nitrification rates double for every 10°C increase of temperature between 5 and 40°C, accelerated rates of these processes could lead to noticeable changes in microbial nutrient cycling.

Another incontestable reality is that the atmosphere with higher CO_2 levels should be a boon for well-fertilized and well-watered crops. The resulting higher gradient between concentrations of the gas in the ambient air and inside leaves will promote the absorption of the gas and its conversion to new photosynthate, and it will suppress photorespiration. Where water and nutrients will not limit the photosynthetic process, the yield gain should be particularly strong for all C_3 plants.

This may have many production consequences. The gap between wheat and corn yields may narrow in some regions and sugar beets might become more competitive with sugar cane. Some agronomists argue that CO_2 enrichment has already raised yields by at least 10 percent above the gains brought by new cultivars, fertilizers, and pesticides. Diminishing returns should be expected as CO_2 levels approach 1,000 ppm. Experiments have shown that some plants moved to CO_2-enriched atmosphere experience a gradual fall to the original or an intermediate level of photosynthesis after a temporary rise.

Other notable gains should be lower evapotranspiration (leaf stomata close partially in higher CO_2 levels, and this adjustment would reduce the loss of water vapor from leaves to the surrounding air), enhanced nitrogen fixation in legumes and a greater abundance of nutrient-scavenging mycorrhizal fungi (as plants will have more carbohydrates to share with symbionts); better growth in low-intensity light; better ability to withstand such environmental stresses as lower temperatures and air pollutants (largely because the partly closed stomata take up less gas from the air); and better tolerance of soil and water salinity.

For many years the experimental evidence rested on studies of individual leaves or small groups of plants in phytotrons or greenhouses.

Shortcomings of these studies were removed with the introduction of the Free-Air Carbon Dioxide Enrichment (FACE) technique pioneered by George Hendrey (Hendrey 1992). FACE microcomputers maintain a narrow range of preset CO_2 concentrations over open field plots of up to 30 m in diameter. Surprisingly, the choice of technique seems to make little difference as experiments broadly agree in their results. Doubled CO_2 should boost average yields all well-fertilized and well-watered crops. Yield gains in C_3 species should average about 30 percent, compared to just around 7 percent in C_4 plants (Kimball 1983; Cure and Acock 1986). Inasmuch twelve out of the world's fifteen leading crops are C_3 plants (including all staple cereals except corn and sorghum), their higher yields would perhaps be the most beneficial consequence of CO_2 doubling.

Both greenhouse and field experiments also confirm substantial interspecific differences. By far the highest yield gains, as much as 70 percent during several years of FACE trials at a CO_2 level of 650 ppm, were not produced by any food crop but by cotton (Kimball et al. 1993). With the same CO_2 enrichment, wheat yields increased by only about 10 percent. Soybean growth goes up with rising CO_2 level even at concentrations as high as 900 ppm, whereas rice yields do not rise above 500 ppm. There is also a great deal of uncertainty about the partitioning of additional photosynthate. If most of the additional photosynthate goes into harvestable parts of the plant, then the crop yield increase may surpass the overall growth grain—but the enrichment may actually decrease such allocation by channeling more photosynthate into leaves or roots.

A less appreciated benefit of global warming for crops would come about as the rising temperature would amplify the positive effects of elevated CO_2 on photosynthesis. A doubling of CO_2 would raise optimum temperatures for C_3 photosynthesis by about 5 to 10°C from the current 25 to 30°C—and CO_2-induced warming would help to make such higher temperatures more common. In some species the increase in productivity stimulated by higher temperatures could equal the increase caused by enhanced CO_2 uptake without any temperature change. Nichols (1997) concluded that as a result of increasing minimum temperatures Australian wheat yields have already increased by about 10–20 percent since the early 1950s, a rise accounting for nearly half of the total gain (the rest being attributed to new cultivars and better management).

Higher temperatures would also expand the area suitable for double-cropping, extend growing seasons in the mid-latitudes and open up for cultivation regions that are now too cold to cultivate. Growing winter wheat in Canada, planting crops requiring more than 120 days to mature in the same regions (the vegetation period is now as short as ninety days on Canadian prairies), and extending permanent agriculture poleward into northern parts of European Russia and in Alberta and Saskatchewan are perhaps the best examples of such changes. Every increase of 1°C in average annual temperature could mean a poleward shift of 100–200 km for major cropping regions (Carter et al. 1991). A simulation using regional temperature changes from four different GCM models resulted in net cropland gains equivalent to 10–15 percent of today's arable land (Darwin et al. 1995). Higher temperatures may have also a positive effect on some pest-predator balances (Skirvin et al. 1997).

Scores of experiments have found that doubling of CO_2 can lower transpiration losses by as much as 30 to 60 percent in some plants (Smil 1997a). Consequently, C_3 crops grown in doubled CO_2 have averaged about a 30 percent gain in water use efficiency, and up to a 10 percent gain in the case of C_4 species. Crops grown with adequate nutrients and water should thus yield more while using less water, while crops grown in arid regions could still match, or perhaps even surpass, their current productivity even with slightly lower precipitation. With intensified water cycle, the overall availability of water for irrigation should increase. Global increases in water supply with CO_2 doubling are forecast to range between 5 and 12 percent, but regional distribution of this effect is very uncertain: the highest gains are forecast in East Asia (but not in Japan), Australia, and parts of Latin America. Models differ, however, even as to the sign of the change.

As with any biospheric change, higher CO_2 levels and rising tropospheric temperatures would also have negative effects. Gains resulting from higher photosynthetic rates and from lowered photorespiration could be negated by higher temperatures: respiration rates are clearly temperature-dependent, and yield may drop once the temperatures exceed photosynthetic optima. The same inverted-U relationship appears to hold for nitrogen fixation by *Rhizobium* symbionts: its rates improve significantly as temperature rises from 15 to 25°C but fall again as it approaches 35°C (Montanez et al. 1995). Higher soil temperature and moisture would also accelerate decomposition of organic matter and cycling of essential nutrients; drier conditions would do the reverse.

Temporal distribution of temperature increases is also critical: their effects would be very different if they were spread more or less evenly throughout the year or if the higher annual mean would come about mostly from relatively large rises in a few summer months. Consequences for productivity would be particularly serious during such sensitive periods of crop development as flowering in soybeans, tasseling in corn, and grain filling in wheat. Conversely, wetter weather during harvest time could interfere with bringing in otherwise excellent crops. Excessive precipitation leading to prolonged waterlogging that can delay planting or harvesting, or to higher soil erosion, would be also unwelcome.

Net outcomes of known counteracting effects would be obviously site- and species-specific: enhanced photosynthesis and higher water use efficiency could offset most, if not all, yield losses caused by higher heat and moisture stresses—and although higher temperatures may accelerate the growth of plant canopy and increase overall productivity, a more rapid grain maturation may reduce yields. Warmer climates could thus reduce the yields of staple cereals grown today the principal agricultural regions of the northern hemisphere. Somewhat warmer and drier climate in the American Midwest would lower corn yield, causing a shift toward more drought tolerant sorghum and winter wheat. Northward displacement of the Corn Belt climate to northern Wisconsin and Michigan would lower the yield because of poorer glacial soils and often poor drainage (Follett 1993). Socioeconomic costs of such a shift would be high.

Warming could also reduce soil moisture in the subtropics, and the shorter growing season and more frequent dry spells could lower the yields. Resulting restrictions on irrigation with water from deep aquifers could exacerbate the situation. Poleward extension of farmland will be greatly limited by poor soil quality, as well as by the willingness of people to migrate to these areas. Even a relatively low ocean rise can bring intrusions of salt water to estuaries and coastal aquifers, affecting crop production, above all rice cultivation, in many low-lying areas (Egypt, Bangladesh, and China's Jiangsu province would be particularly affected). And even regions with increased precipitation may see some yield losses because of a higher frequency of damaging rains arising from stronger convection: lodging of nearly mature crops can be a major problem. Yet another set of concerns encompasses enlarged breeding areas of many pests and their faster maturation in warmer climates.

A mixture of benefits and concerns would also apply in the case of animal husbandry. Cattle production could cope with higher temperatures by relying more on relatively heat- and insect-resistant Brahman breeds or their crosses. Benefits for cattle grazing would include longer time spent on the range, reduced feed requirements during winter months, and better survival of newborns. Reduced feeding due to lowered metabolism would apply to all other free-ranging animals, while structures for housed species might need less winter heating. Larger ranges and faster life cycles of parasites would be unwelcome, as would be losses of some rangelands due to summer droughts, and heavier snows on northern grasslands. Higher C/N ratio of plants would mean lower nutritional quality of some grasses, and animals would have to consume more forage to sustain expected meat production.

But the eventual consequence of global climate change should not be imagined as a net impact of the just described gains and losses: adaptations will go long way toward eliminating negative effects of gradual climate change as economic, institutional, agronomic and technical adjustments will take place (Smit et al. 1996). In this respect, both the magnitude and the rate of future temperature change will make crucial difference. Earlier predictions of up to 4°C rise with CO_2 doubling that was to occur within half a century were obviously much more worrisome than the latest consensus of 2°C rise by the year 2000.

Magnitude of climate variability will also matter a great deal as effective adaptations are more likely to occur in response to extremes rather than to a gradual change in growing conditions. In most of the temperate zone, year-to-year variations in temperature during the last century have been on the order of 2–5°C, far greater than changes typically attributed to global climate change (on the order of 1°C). Moreover, were the global warming to unfold over a period of many generations, other, as yet completely unanticipated, environmental and socioeconomic changes might claim most of the policymaking attention.

Some of the earliest estimates of eventual impacts of global warming on agriculture neglected to consider a variety of adaptive steps farmers can take to cope with gradual climate change. The most effective measures include the introduction of new cultivars adapted to warmer conditions, changing the mix of cultivated species, advancing or delaying planting and harvesting dates, and using better irrigation and pest control techniques (figure 3.7).

Even relatively pessimistic assessments conclude that doubling of CO_2 should bring about only a small decrease of global crop production. They are exemplified by Rosenzweig and Parry's (1994) model considering two levels of adaptation on the global scale. The first one included choosing already available cultivars more suited to new conditions, changing planting dates, and applying less additional water on currently irrigated crops. The second one entailed using newly developed plant varieties, planting dates changed by more than one month, and installation of new irrigation. Depending on the GCM used with CO_2 doubling, these scenarios resulted in appreciable increases of cereal production in affluent

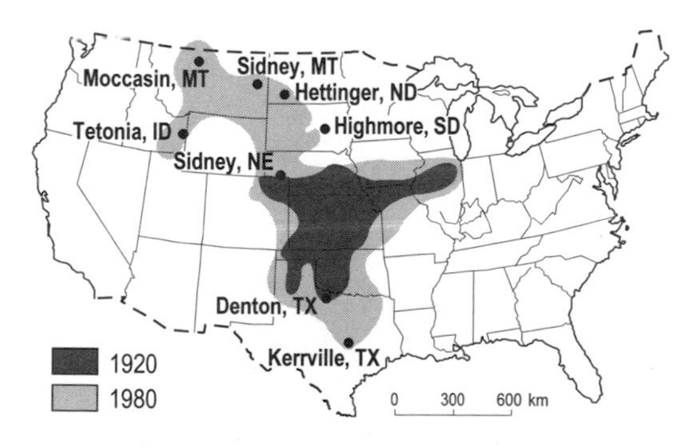

Figure 3.7
Areas of hard red winter wheat production in the United States, 1920–1980 (Rosenberg 1982).

countries (as much as 10–14 percent) and smaller losses (6–10 percent) elsewhere.

More realistic comprehensive model taking into account positive effects and adaptive responses revealed that in the world with doubled CO_2 the global output of some nongrain crops could decline, but that of grain and livestock would most likely increase (Darwin et al. 1995). Canada and Russia would be the main beneficiaries, some subtropical and tropical regions the main losers. An integrated agronomic/economic model prepared by Mount and Li (1994) for effects on Midwestern corn, soybeans, and winter wheat ended up mostly with gains for all crops under three different GCMs simulating doubled CO_2. Other models considering multiple adaptations have also ended up with either only mildly negative or slightly positive effects (Schimmelpfennig et al. 1996).

Uncertainties inherent in model assessments are perhaps best illustrated by comparing effects predicted for Iowa soybeans by Rosenzweig et al. (1994) with those of Mount and Li (1994). A scenario that shares an almost 5°C temperature increase accompanied by a nearly 40 percent decline in precipitation ends up with a 26 percent yield decline in the former case, and a 17 percent gain in the latter one. Effects of higher frequency of short-term variability are even more uncertain. We know that unusually low temperatures cause stomatal dysfunction during the following day, suboptimal temperatures affect plants particularly during flowering, and prematurely high temperatures break the dormancy of perennials. The first studies modeling the consequences of higher temperature variability have yielded equivocal results, ranging from negligible impact to appreciable yield declines (Riha et al. 1996; Semenov and Porter 1995; Mearns et al. 1996).

Major uncertainties also surround all of the current assessments of regional impacts. Will regions such as southern Africa—already repeatedly affected by severe drought associated with El Niño—be particularly hard hit? Given the region's already tight water supply situation, high degree of reliance on hydroelectric generation, fast rates of population growth, and increasing urbanization, climate change could have multiple negative impacts (Magadza 1994). In contrast, Hulme et al. (1999) showed that for some regions the impacts of anthropogenic climate change will be undetectable relative to those caused by natural multi-decadal climate variability.

Moreover, rising concentrations of greenhouse gases will be also accompanied by higher emissions of sulfur and nitrogen oxides, and the fate of these gases may have considerable effect on both the intensity and the biospheric consequences of possible global warming. Sulfates formed by the oxidation of SO_x are usually the most abundant aerosols in the atmosphere. They lower its temperature by scattering some of the incoming radiation back to space, and indirectly by supplying cloud condensation nuclei; the resulting clouds reflect more radiation back to space. These effects have been most notable in polluted, humid atmospheres of eastern North America, western Europe, and East Asia.

Their combined cooling effect has been estimated to reduce the warming effect of greenhouse gases by at least 10–20 percent, or it may virtually eliminate it in the most polluted regions. We can only speculate how the two opposite processes—further decarbonization of energy supply in Europe and North America and continuing expansion of fossil fuel combustion in Asia—will affect tropospheric temperatures by the middle of the twenty-first century.

4

Opportunities for Higher Cropping Efficiencies

During the next two generations the quest for higher food output will be dominated by intensification of cropping. Between 1995 and 2010 about 80 percent of the total increase in crop production will come from higher yields on existing farmland (Alexandratos 1995), and that share will be even higher during the following decades. In this chapter I will argue that this worldwide intensification should not be driven primarily by mobilization of new inputs. Instead, we should endeavor to derive the bulk of new needs by increasing the efficiency of existing uses. This effort should be particularly intense as far as the two key inputs—fertilizers and water—are concerned.

Analogy with post-1973 developments in energy conservation provides a perfect illustration of potential gains. Preoccupation with increased energy supplies survived the first round of OPEC's sharp crude oil price rises (Fall 1973–Spring 1974), but gradually it became clear that the reduction of inefficiencies in both production and conversion processes is a source of energy preferable to the discovery of new deposits of traditional fuels, building of more nuclear or hydro stations, or development of novel conversions. The second round of price rises triggered by the fall of the Iranian monarchy in 1979 accelerated that change. Unfortunately, this tightening of the slack in the global energy system was short-lived, as the collapse of the world crude oil prices in 1985 (from nearly $40 per barrel in 1980 to less than $15) removed the greatest incentive for more efficient energy use.

Still, the intervening gains were impressive. This largely price-driven change made it possible not only to curb the rate of energy imports

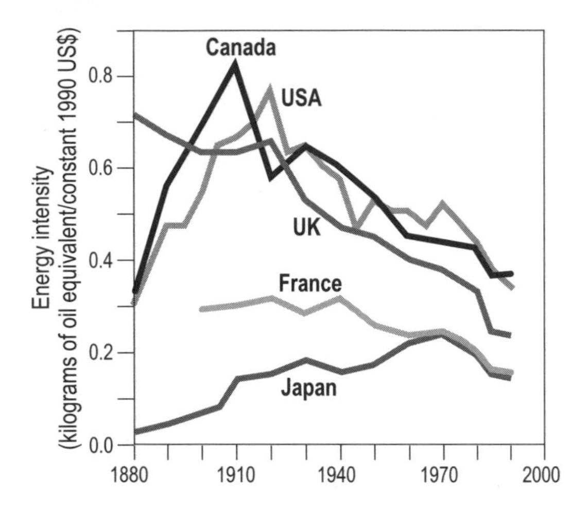

Figure 4.1
Long-term trends of energy intensity of the world's major economies (Smil 1997).

among all countries highly dependent on the Middle Eastern crude oil but also, in an unprecedented shift, to decouple economic growth from increases in energy consumption. Between 1973 and 1985 all affluent countries had only marginal increases in their energy use while their GDPs continued to grow. Even the most profligate energy user followed this trend: in 1985 U.S. primary energy consumption was less than 3 percent above the 1973 level, but the country's inflation-adjusted GDP was one third larger (figure 4.1). The experiences of highly energy-intensive industries have been even more rewarding: the eight largest OECD economies cut their energy intensities in iron and steel industry and in chemical syntheses by, respectively, 27 and 37 percent between the early 1970s and the late 1980s (Schipper and Meyers 1992).

Given their low per capita energy use, poor countries kept on increasing their absolute energy consumption, but they too have followed the trend of relative energy savings with steadily declining energy/GDP ratios. China's example is particularly impressive. In 1980 its nationwide energy/GDP ratio was about 0.7 kilograms of coal equivalent (kgce) per one yuan of inflation-adjusted GDP; by 1990 the rate declined to

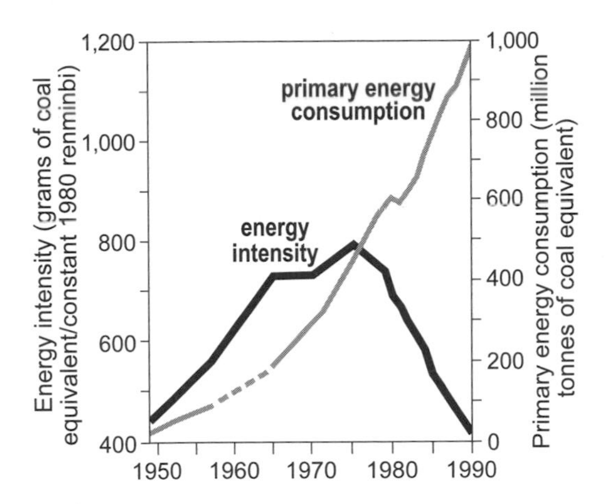

Figure 4.2
Declining energy intensity of China's economy, 1980–1995.

0.42 kgce/1980 yuan, and in 1995 it was slightly below 0.35 kgce, or a bit less than half the value fifteen years earlier (figure 4.2) (Smil 1998). Such a rate of decline has been unmatched by any other major modernizing economy. If the country's energy intensity had remained at the 1980 level, China would be now burning more than twice as much fuel to produce every yuan of its GDP—and this combustion would generate a correspondingly larger amount of environmental impacts, particularly air pollution from the burning of coal.

Similar opportunities for higher production and consumption efficiencies lie ahead in producing food. What Shepard (1991) concluded about energy efficiency is no less true about the opportunities for more efficient farming: "We've only just begun to save. Tougher standards plus stronger financial incentives will yield enormous payoffs." During the next two generations the rich nations—with their relatively stable populations and already excessive average food intakes—should produce all of their needs as well as be able to cover somewhat larger export demand, with decreased, or no worse than stabilized, agricultural inputs. Poor countries, with their still growing population and improving standards of living, will have to increase their total fertilizer and water use—but they can tap the existing performance slack in order to lower significantly the relative consumption of these resources.

As with rational energy management (a broader term I prefer to the usual "energy conservation"), there are no shortcuts toward achieving higher efficiencies in agriculture: no single action or adoption of a particular technique, even if it is highly effective, and even if it transforms a sector using a relatively large share of total energy, can save more than a few percent of the aggregate use, but a combination of similar approaches results in major and long-lasting gains. In terms of higher agricultural efficiency this means, for example, that even a nearly universal adherence to optimized timing of fertilizer applications may save no more than 5 percent of all nitrogen, but a combination of several measures—more careful timing, appropriate placement of fertilizer compounds, recycling of all crop residues, rotations including leguminous crops, and reduced tillage minimizing soil erosion—may bring a 20 to 30 percent gain in overall nitrogen recovery by crops.

My primary concern in detailing these possible gains is to make sure that all reviewed options conform strictly to biophysical realities—and that all assumptions I make will be fairly conservative. This combination will assure that all of the outlined savings could be seen as realistic goals that could be surpassed by more widespread adoption of such measures or by more assiduous application of available technical fixes and managerial improvements.

More Efficient Fertilization

After decades of steady and substantial growth (the total use had grown more than twentyfold between the late 1940s and the late 1980s) the global production of fertilizers reached a peak in 1988–1989, and then it kept declining for the next five years, with cumulative reductions of more than 7 percent for nitrogen, nearly 25 percent for phosphate, and 35 percent for potash (figure 4.3) (FAO 1999). This decline stemmed from a variety of causes. The collapse of the Soviet empire followed by rationalization of collectivized farming in former European Communist countries, whose heavy subsidies led to excessive use of fertilizers, was a major reason (the USSR was the largest producer of nitrogen fertilizers before 1989). So was the long-overdue reduction of generous farm subsidies in the European Union, a temporary stagna-

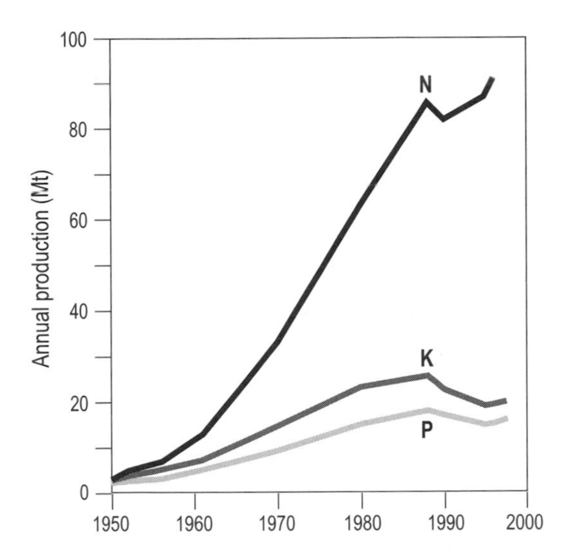

Figure 4.3
Global production of inorganic fertilizers, 1950–2000.

tion of grain production in China, and generally low world prices for cereal grains.

As the global grain market tightened, a recovery began in 1994, and by 1998 the output of nitrogen fertilizers was back to above 80 Mt N/year. Global production of these fertilizers is now almost exactly split between affluent and low-income countries, but the latter use almost three-fifth of the applied total (FAO 1999)—and they will account for more than 90 percent of the future additional demand, which is forecast to grow at 2 percent a year (and 3 percent a year for P and K). The need for expanded fertilizer applications in poor countries is obvious, as the gap between their average crop yields and those in affluent countries has actually grown during the past generation (FAO 1999).

The most urgent need is to increase fertilizer use in sub-Saharan Africa, where insufficient supply of nutrients results in the continuing decline of soil fertility. So far the most comprehensive assessment of soil degradation concluded that a mere 15 percent of potassium, less than 30 percent of nitrogen, and less than 40 percent of phosphorus needed by crops are actually replaced by fertilizers (Oldeman et al. 1990). In order to stop this degradation, the region should quintuple its currently negligible use of nitrogen, averaging a mere 10 kg/ha. If this were to be done within one decade, as Larson and Frisvold (1996) argue it should be in order to assure the region's food security, the fertilizer use would have to grow by nearly 20 percent a year—but its consumption has been actually declining.

In contrast to Africa, Asia's food production has expanded considerably: its successes in feeding larger population have been based on steadily increasing fertilization. This development has made possible the adoption of high-yielding rice and wheat varieties. It has also boosted the production of residual phytomass, which can be used to improve organic content of nutrient-poor tropical soils, and hence indirectly further improve the yields. Given the fact that most of the world's future population growth will be concentrated in a handful of Asian countries (see chapter 1), the continent will have to see further substantial increases in fertilizer use.

Total applications of synthetic fertilizers will have to increase if the more intensive cropping practices are to be expanded and sustained beyond the next century. But, at the same time, relative application rates will have to be limited in many regions or locales in order to prevent excessive environmental impacts. Possible advances in genetic engineering may eventually impart nitrogen-fixing capability to nonleguminous crops, but there are no signs that such a monumental breakthrough is on the horizon. Increased production and more efficient application of synthetic nitrogen compounds will thus remain two key determinants of larger harvest for the foreseeable future. And because nitrogen is usually the nutrient needed in the highest amount by staple crops (see chapter 2), the following survey will focus mostly, though not solely, on its losses and uses.

Nutrient Leakage

As I have already emphasized (in chapter 2), a significant portion of nitrogen available in farm soils is never assimilated by growing crops. Leaching removes readily soluble nitrates into both surface waters and aquifers, and the element in both inorganic and organic solids is taken away in eroded soils. Volatilization transfers ammonia into the atmosphere, and bacterial denitrification, the closing arm of nitrogen cycle, returns the element to the air as dinitrogen, together with some N_2O.

Leaching transfers any soluble nutrients, above all nitrates and phosphates, to ponds, lakes, streams, estuaries, and shallow seas where they can cause serious eutrophication (see chapter 3). Leaching rates are determined above all by levels of fertilization, compounds used (ammonia fertilizers leach very little in comparison to readily soluble nitrates), soil permeability, and the amount of precipitation. Extremes for nitrogen leaching range from negligible losses in arid and semiarid areas (for example, in the interior grain fields of North America) to 20–30 kg N/ha in such intensively cultivated temperate fields as the U.S. Corn Belt, and to over 50 kg N/ha in the most heavily fertilized crop fields of northwestern Europe. Wet winters in Mediterranean Europe also cause high nitrogen leaching losses (Carvalho and Basch 1996).

The World Health Organization (WHO) sets a limit of 50 mg of nitrates per liter in drinking water, and this level of contamination is now present in many wells and aquifers in intensively farmed areas. Most notably, extensive surveys of groundwaters had shown that the WHO limit was surpassed during the early 1990s on 95 percent of Dutch farms, with mean levels of 134 mg/L and 243 mg/L on intensive dairy farms (Jongbloed and Henkens 1996). Approximate relative losses due to leaching would be about 10 percent of applied nitrogen with less intensive fertilization in rainy regions, and 20 percent with higher applications. In contrast, low solubility of phosphorus compounds causes leaching losses no higher than one kg P/ha, or rarely more than a few percent of total applications.

Because erosion rates vary widely even within a single field (chapter 3), any large-scale generalizations are merely order-of-magnitude indicators. Annual losses of 10 kg N/ha for gently sloping farmland not subjected to high wind erosion, and 50 kg N/ha for sloping and windy areas, should be seen as minima; respective maxima would be 50 and 100 kg N/ha. Analogical phosphorus losses in eroded soil would be, respectively, 1 and 5, and 5 and 10 kg/ha.

Volatilization of ammonia is responsible for large nitrogen losses from both animal manures and from all ammonia fertilizers. Its rates generally increase with higher application and shallower incorporation of NH_3 fertilizers. Solid compounds spread on the surface lose commonly 5–20 percent of their nitrogen, mostly within just a week after application. Field conditions increasing the losses are the low buffering capacity of soils, high pH of soil solutions (volatilization is minimal at pH below 7), higher temperature (seasonal maxima in summer, daily ones around noon), and soil water content at or near field capacity.

Crop residues left on the surface boost the volatilization rate by slowing down the fertilizer diffusion into soils and by contributing more urease for faster urea hydrolysis. A conservative estimate would put nitrogen losses at at least 10–15 percent of applied ammonia fertilizers. Average losses from animal manures are even high, usually around 30–40 percent, and a substantial share of the initially voided nitrogen (easily 10–20 percent), is volatilized during the farmyard storage. Loss of nitrogen from dairy operations before and during the manure application to fields can reach 50–75 percent (van Horn et al. 1994). In Europe perhaps as much as 80 percent of all airborne ammonia may originate from livestock wastes, and the highest annual emission densities in the Netherlands have been estimated at between 40 and 65 kg N/ha (Sutton et al. 1993).

Because of its great temporal and spatial variability, denitrification can be responsible for just a fraction of 1 percent or for up to three-quarters of all postapplication losses. For basic estimates, Bouwman (1996) used background annual emission of one kg N_2O-N/ha and fertilizer-induced emission equal to 1.25 percent of the nitrogen application. Values between 5 and 20 kg N/ha would represent typical encountered losses from fertilized crop fields, and rates equal to 10–20 percent of applied fertilizers are perhaps most common.

Aggregate losses add up commonly to more than half of all nitrogen applied in synthetic fertilizers. Detailed analyses of thirty different representative agroecosystems investigated during the 1970s showed that two-thirds of them had nitrogen recovery efficiencies below 50 percent, and that the most efficient cropping sequences absorbed nearly 70 percent of the applied nutrient (Frissel and Kolenbrander 1978). Where fertilization rates were below 150 kg N/ha uptake efficiencies were up to 60–65 percent, but with higher applications they were scattered around 50 percent (both P and K recoveries were typically well below 50). Utilization of nitrogen is particularly low—between 20 and 40 percent—in irrigated rice (de Datta and Buresh 1989; Cassman et al. 1996).

Unlike nitrogen, phosphorus does not undergo any rapid global cycling: because the element is not volatile and it does not form any stable gaseous compounds it has no significant atmospheric link from ocean to land. On a civilizational time-scale, the element's path in the biosphere is just a one-way flow from land to ocean with interruptions, and only slow geotectonic processes close the cycle after 10^7 to 10^9 years. With no volatilization and with hardly any leaching losses, erosion and runoff are by far the most important causes of the nutrient's unwanted transfer from agricultural soils.

Arguments about phosphorus uptake rates—what portion of the nutrient is irreversibly fixed in the soil soon after the application and how much remains in utilizable form—have been among the most controversial topics in modern soil science. Traditional interpretation saw phosphorus fertilization largely as an expensive mass transfer of the nutrient from mined and treated rocks to insoluble, and hence unavailable, soil compounds, and this misconception led to decades of excessive P applications. In reality, phosphorus use by crops is at least as high, and commonly even higher, as that of fertilizer nitrogen (Karlovsky 1981).

This means that problems with phosphorus in waters have not resulted primarily from inferior nutrient utilization. They are the consequence of a very sensitive phytoplanktonic response to the presence of the available nutrient in water, but phosphorus from organic wastes, rather than the nutrient in fertilizers, is almost always dominant in the eutrophication process: approximate global ratio of P contamination from fertilizers and urban sewage is 15/85 per unit area of intensively farmed and densely settled land.

Because there can be no direct field measurements of large-scale nutrient flows, and because the estimates of nutrient losses are particularly uncertain, it is very difficult to offer any firm conclusions concerning field balances of the three macronutrients. Many of the best estimates available for particular cropping systems, or for individual fields, show that both N an P soil pools are either steady or are slightly increasing (Frissel 1978). This conclusion is not surprising given the often high rates of fertilizer applications—but without synthetic fertilizers, more than half of those agroecosystems would have had net nutrient losses. A good approximation would be then to assume that no more than 45–50 percent of all nitrogen applied in synthetic fertilizers in rainfed cropping, and between 30 and 40 percent used in irrigated farming, is absorbed by the world's crops.

Reducing Fertilizer Losses

There are many proven ways to cut nutrient losses and improve fertilizer efficiency (FAO 1980a; Hargrove 1988; Cooke 1988; Munson and Runge 1990; Fragoso 1993; Dudal and Roy 1995; Prasad and Power 1997). Soil testing, choice of appropriate fertilizing compounds, maintenance of proper nutrient ratios, and attention to the timing and placement of fertilizers are the most important direct measures of universal applicability. Indirect approaches that can either reduce the need for synthetic fertilizers or increase the efficiency of their use rely primarily on greater contributions by biofixation (accomplished by more frequent planting of leguminous crops or by optimizing conditions for other diazotrophs) and on good agronomic practices embracing crop rotations, conservation tillage, and weed control.

Periodic soil testing is a highly effective way to improve fertilizer uptake: it makes it possible to give much better recommendations for appropriate macronutrient applications, and it can uncover growth-limiting micronutrient deficiencies (Havlin et al. 1994). Periodic testing for major macronutrients has been common in rich nations for decades, but testing for micronutrient deficiencies (ranging from boron and copper to molybdenum and cobalt) has been much less frequent. In many areas of the poor world even occasional basic soil tests are rare or absent, and too many farmers simply follow general recommendations for fertilizer applications rather than modifying them according to their particular field needs. Efficient farming in regions with increasing cropping intensification is predicated on good soil tests, which can prevent either insufficient or excessive application of nutrients.

There is also place for additional soil testing in well-established intensive farming. For example, the amount of mineral nitrogen present in soil is usually not taken into consideration in making recommendations for spring fertilization of winter wheat. Scharf and Alley (1995) demonstrated that in the humid climate of eastern North America levels of mineral nitrogen persisting until spring can be so high that there is no additional response to spring applications. Similarly, Schlegel and Havlin (1995) found that a model used to calculate recommendations for N and P in continuous corn cultivation overestimated the actual rates by 30 to 60 percent.

Because of the wide range of residual nitrogen, only field-specific recommendations can maximize nitrogen recovery. Large improvements in nitrogen recovery are also possible in flooded rice fields when farmers adjust their application rates to the highly variable nitrogen supply capacity of fertilized soils (Cassman et al. 1993). When this is done, nitrogen recovery by rice can surpass 50 percent even with high applications needed to support yields in excess of 9 t/ha.

Not all synthetic nitrogenous compounds have been always considered equal in their efficacy. For example, some earlier experiments showing that urea may be less effective than other sources of nitrogen limited its use in a number of European countries, as did the traditional European preference for mineral fertilizers. Yet there is no doubt that urea, now the world's leading source of fertilizer nitrogen, is as effective for cereal fertilization as ammonium nitrate (Webb et al. 1997). Many experiments have shown that application methods make usually much greater difference than the choice of compounds. Elimination of one particular fertilizer, however, would greatly improve the efficiency of nitrogen use in the world's largest producer of fertilizers.

About one-third of the fixed nitrogen in China is produced in small, coal-based plants as ammonium carbonate, an unstable, highly volatile compound that should be vacuum-packed for distribution and incorporated deep into soil to eliminate large losses, recently estimated to be as high as 20 percent even before application during transportation and storage (Smil 1993). Gradual replacement of many remaining small- and medium-sized plant synthesizing ammonium bicarbonate by modern ammonia-urea facilities would thus greatly boost the overall efficiency of China's fertilization.

Importance of balanced nutrient supply has been appreciated for more than 150 years: Liebig's realization that the essential nutrient in the shortest supply will limit the yield—and hence that only properly balanced applications of macro- and micronutrients can maximize it—stands at the very beginning of modern agronomic science (Liebig 1840). Since that time greenhouse and field trials showing the interdependence of nutrient applications have been among the most common topics of plant and soil science research. To cite just one impressive recent example, apparent fertilizer nitrogen recovery in the grain in a ten-year sequence of corn response to nitrogen and phosphorus fertilization was twice as high with phosphorus as without it (Schlegel and Havlin 1995).

Still, unbalanced applications of synthetic fertilizers are far from exceptional. Until about a generation ago, they were characterized by an excess of phosphorus. The already noted traditional misconception—that the major part of applied nutrient is rather quickly tied up by the soil, piling up in unavailable forms—led to irrational use of fertilizers: the worldwide N:P ratio passed 2.0 only in the mid-1950s, by 1970 it was about 3.5, and now it is close to 6.0.

Currently excessive use of nitrogen is the most common sign of unbalanced macronutrient applications. The practice diminishes the efficiency of the element's applications and promotes unnecessarily high losses of the nutrient, particularly when phosphorus is deficient. This has been, unfortunately, a chronic problem in China. While the worldwide N:P:K mean is now about 100:18:22, Chinese applications have been highly deficient in both P and K, with the nationwide ratio of 100:14:8, and with much higher imbalances in many intensively cultivated regions. Inappropriate application ratios are common in many low-income countries. For example, El-Fouly and Fawzi (1996) concluded that proper N:P:K ratios based on soil testing and plant analysis and adjusted to the prevailing cropping sequence could raise typical Egyptian yields by 20 percent without using more nitrogen.

Numerous field tests have confirmed that proper timing of applications is a highly effective way of saving fertilizers: applications should coincide, as much as practicable, with periods of the highest nutrient need. Uptakes are relatively low during early growth, and they are generally highest during rapid vegetative growth stage before crops shift to reproductive growth. For example, the highest rates of nitrogen assimilation occur in corn when the plants are near the tasseling stage, and in rice during the period of primordial initiation. Unfortunately, a highly transient nature of available nitrogen in flooded paddies makes it particularly difficult to synchronize the nutrient's supply with crop needs.

Applications many weeks ahead of planting, and particularly fall applications for spring crops, usually risk substantial losses of nitrogen. Applying fertilizer at the time of seedbed preparation and planting may be convenient, and, depending on many environmental and agronomic factors, these practices may not be too wasteful. But, generally, the closer to the time of the peak requirement the applications will be the higher the rate of recovery. Split applications—the initial one before or at the time of planting, the second, or third, nearer the time of maximum uptake—are thus a better choice.

An interesting case of specific timing is the addition of nitrogen fertilizer to irrigated spring wheat late in the growing season. This is done in order to raise the grain quality needed to meet bread wheat standards. Experiments in California, using ^{15}N-labeled fertilizer, showed that the recovery of nitrogen applied at planting time ranged between 30 and 55 percent, but that 55–80 percent of the nutrient applied at anthesis were assimilated (Wuest and Cassman 1992).

Placement of fertilizers must reconcile some conflicting needs: it should promote high uptake efficiency and hence should deposit nutrients as close to the position of future, or existing, plant roots as possible—but it should not cause injury to plants and it should be doable in a practical and speedy manner. Surface applications—simple broadcasting or side dressing—are almost always less efficient than subsurface incorporation (broadcasting followed by plowing or discing, row application with seed, deep banding apart from seed).

In sideband row placement nutrients are inserted to the side and slightly below the seed at planting time. Spoked-wheel placement, penetrating the soil and releasing predetermined amount of nutrients as it turns, provides an even greater application accuracy than continuous banding. Although ammonia injection into soils, as opposed to simple broadcasting of solid fertilizers, requires specialized storage, distribution, and application techniques, these higher costs are repaid by better crop uptakes.

Recent studies in the highly erosible loess soil of China's Shaanxi province show the advantages of subsurface application. Nitrogen fertilizer uptakes were just 18 percent for surface application to corn and 25 percent to wheat, but the rates rose to between 33 and 36 percent for mixing treatments and banded applications to wheat (Rees et al. 1997). Deep placement of nutrients, however, is also wasteful: studies with labeled nitrogen show that the use of nutrients decreases appreciably with depth. Comprehensive consideration of soil conditions (its physical, microbial, and microclimatic status), fertilizer properties (nitrates are readily soluble, urea must be first enzymatically hydrolyzed), and plant requirements would determine the best placement.

A three-year study in Missouri demonstrated that, as long as adequate water supply was assured, placement of nitrogen in alternate furrows produced harvests at least equivalent or even higher than those resulting from fertilizing every furrow (Hefner and Tracy 1995). On the other hand, overwatering, frequently associated with traditional surface furrow irrigation, can also reduce the efficiency of nitrogen fertilizer.

Unfortunately, even well-timed applications of properly emplaced ammonia fertilizers may result in appreciable nitrogen losses because of relatively high volatilization rates and fairly rapid nitrification, the bacterial conversion of ammonia to highly soluble nitrates. Slowing down the rate of nitrification should reduce the losses of waterborne nitrogen, and reduced rates of nitrification also reduce potential losses by denitrification, increase the extent of nitrogen immobilization by soil microorganisms, and reduce the nitrate content of groundwater.

This option became commercially possible with the development of nitrapyrin, the first nitrification inhibitor, in the early 1960s. About twenty of these chemicals are now on the market, with less than half widely tested in the field (Prasad and Power 1995). They include some natural products, most notably those from *Azadirachta indica*, or neem tree, widely grown in India, and a variety of synthetic chemicals ranging from ammonium polyphosphate to KCl. The two most common inhibitors, nitrapyrin and dicyandiamide, were found to be highly, but not always equally, effective. Inhibition rates twenty-five days after treatment range from less than to 10 more than 90 percent. Reported yield increases with the use of inhibitors have been mostly between 5 and 20 percent for major field crops—but there have been also reports of no yield gains under field conditions. Two recent studies exemplify this range of responses.

Indian field experiments with dicyandiamide, previously shown to be effective both in European and Asian conditions, confirmed that wheat yields went up compared to those achieved with prilled urea alone; moreover, residual nitrogen left after corn harvest increased substantially the yield of the succeeding wheat crop (Sharma and Prasad 1996). In contrast, Scharf and Alley (1995) found that none of the five tested inhibitors increased yields of wheat in the eastern United States. In general, low content of soil organic matter and relatively low temperatures promote inhibition.

Controlling nitrogen losses from animal manures is generally more difficult than limiting the escape of the nutrient from synthetic compounds, largely because of high rates of volatilization. Modeling of the cost of nitrogen management on Dutch dairy farms has shown that reducing ammonia emissions is much more expensive than cutting down the same amount of nitrogen losses from leaching and runoff (Berentsen and Giesen 1994). One effective way to reduce substantially volatilization from cattle slurry surface applied to cropland is to treat it with HNO_3 in order to acidify it to pH between 4.5 (during high temperatures and high water evaporation) and 6.0 (during low temperature and evaporation). Dutch trials showed that typical 60 percent (29–98 percent range) loss from untreated slurry can be reduced by 55–85 percent regardless of grassland soil types and application levels (Bussink et al. 1994).

Biofixation

Greater reliance on biofixation in all areas where principal grain, oil, or sugar crops can be rotated with leguminous species—be they grown as food legumes (soybeans, peas, beans), fodder crops (alfalfa, clover), or green manures (any leguminous cover crop plowed in after thirty to one hundred days rather than harvested)—should have the dual advantage of sparing the soil and creating a store of nitrogen. Leguminous species should require very little of the accumulated soil nitrogen, and they should leave varying amounts of the nutrient for the subsequent crop. In reality, these gains vary widely, ranging from no net benefits to substantial enrichment.

Symbiotic fixation may provide much more nitrogen than needed by a growing leguminous crop, but it may also cover only a fraction of its nitrogen need. These differences are both inter- and intraspecific. Comparisons of different cultivars of *Phaseolus* beans show that the plants derive less than 25 and as much as nearly 65 percent of its nitrogen from symbiotic fixation, or less than 25 to more than 150 kg N/ha (Hera 1995).

How much nitrogen is available to nonleguminous crops that follow in the rotation is highly dependent on the share of the fixed nutrient taken up by the harvested legume seeds. This so-called nitrogen harvest index may be as low as 30 percent for beans and as high as 80 percent for soybeans. Plowing bean plant residues into the soil thus returns most of the fixed nitrogen, but soybeans, despite their reputation as prolific fixers of nitrogen, may not be even able to fix all the nutrient they need (Heichel 1987). Only when the soybeans can derive more than about 80 percent of all nitrogen from symbiosis, or when they are planted as green manures, can they add up to 90 kg N/ha, the enrichment surpassed only by plowing-in good stands of clover or alfalfa. Nitrogen recovery from green manures is generally much higher (commonly around 70 percent and up to 90 percent) than from synthetic fertilizers (de Datta 1995).

Plowing in such green manure cover crops as alfalfa, vetch, or clover can incorporate more than 300, or even 500, kg N/ha—but no more than 5–10 percent of this total may be mineralized during the growth of the following crop. Green manures are thus an outstanding component of long-term management of soil quality, but not a means of boosting short-term nitrogen supply for higher harvests of cereals. In spite of their potentially large contributions to soil nitrogen stores, green manures are being planted less frequently than in the past.

In China, where in many regions cropping used to be highly dependent on it, green manure cultivation fell from a peak of 9.9 Mha in 1975 to only about 4 Mha by 1989, with more than 20 percent of their total area concentrated in a single province (Smil 1993). The reason for abandoning green manure cultivation is clearly the higher pressure to produce more food on limited land. If Chinese peasants were to return to the plantings of green manures that prevailed in the mid-1970s they would have to forego grain harvest on some 6 Mha, or roughly 20 Mt of food cereals, equivalent to almost 5 percent of China's total output and sufficient to feed some 75 million people. Much like China, other countries with a limited amount of arable land cannot afford to use that space for green manure whose growth would preempt the cultivation of a food crop, and must turn to synthetic nitrogen fertilizers.

Even those legumes that do not usually produce a net nutrient loss, such as peanuts or chickpeas, may be more valuable in crop rotations more because of their sparing effect on soil than because they fix substantial quantities of nitrogen available for the following crop. Some field experiments suggest that even forage legumes are credited with contributing more nitrogen to the soil-plant system than they actually do. Six years of rotations in Minnesota demonstrated that corn yields were as high when the crop followed wheat as when it followed soybeans, or wheat and alfalfa.

Moreover, many soils do not have sufficient densities of natural bacterial populations needed to establish effective nodulation. In such cases annual biofixation amounts to only a small fraction of potential production—but seed inoculation with particular *Rhizobium* strain has been a very effective remedy. Commercial inoculants have been applied for several decades by using sticking agents to coat seeds with a peat-based carrier as they passes through a grain auger. Peat provides an excellent medium for rhizobia, but the practice is time consuming and impractical for sowing large areas. Liquid inoculants are now available: they are easy to apply, provide uniform coverage and result in yields equal to or better than with peat-based inoculants (Hynes et al. 1995).

Selection of the most suitable cultivars is also an effective option. When Brazil embarked on its ambitious soybean cultivation program, the selection of genotypes was done with the view of minimizing nitrogen fertilizer applications. Today the country is the world's second largest produce of the crop, with some 20 Mt harvested annually, and unlike the United States, where up to a fifth of the area planted to soybeans receives an average of about 50 kg N/ha, Brazil uses no nitrogenous fertilizers on its soybeans.

Symbiotic *Rhizobium* provides virtually all of the nitrogen needed by soybeans. But Brazilian crops also benefit from nitrogen contributed by bacteria living symbiotically inside roots, stems, and leaves of several grass, tuber, and oil palm species, rather than being attached, as rhizobia, to their roots. The first of these endophytic diazotrophs, *Herbaspirillum seropedicae*, was identified in 1986, and another species of the same genus, long known for causing mottled stripe disease in sugar cane, was found to have the same ability a few years later (Baldani et al. 1986; Gillis et al. 1991). Substantial reduction, or even elimination, of nitrogen fertilizers for Brazilian sugar cane is a key reason for a positive energy balance of the country's fuel alcohol program.

Inside sugar cane, the two species are commonly joined by *Acetobacter diazotrophicus*, an endophyte also found in sweet potatoes. *Herbaspirillum* was subsequently found also in rice, corn, sorghum, and some forage grasses where its contributions to plant nitrogen balance are not as high as in sugar cane. In 1996 yet another new endophytic species was identified in rice, banana, pineapple, and oil palms.

This surprisingly rich, and previously unsuspected, endophytic colonization of many crops grown in the tropics, including rice, explains why the region's biofixation rates are substantially higher than in temperate climates. Obviously, adoption of sugar cane cultivars requiring very low, or no, nitrogen fertilization should be encouraged in all appropriate tropical locations—and a better understanding of endophytic organisms should eventually increase biofixation in a number of other tropical crops. Yet another effective option is to increase the natural presence of free-living, nitrogen-fixing bacteria. Inoculation of rice fields with N-fixing cyanobacteria, developed in India, can raise grain yields by anywhere between 15 and 29 percent (Hamdi 1995).

Agronomic Contributions

The quest for high efficiencies of nutrient use must go beyond optimizing fertilizer applications and legume rotations: improvements to a variety of agronomic practices, some of them not directly connected with fertilization, can add up to significant gains. Overall nutrient uptakes can be improved by timely planting of N-responsive modern varieties, by maximized recycling of organic wastes (particularly needed to raise the commonly low organic matter content of tropical and subtropical soils) and integrating the uses of organic and synthetic fertilizers, by minimizing soil erosion (be it through contour cultivation or reduced tillage), by assuring adequate moisture supply and by controlling pests (FAO 1980a; Dudal and Roy 1995).

This integrated approach is exemplified by codes of best agricultural practices recently adopted in several countries of the European Union. They recommend that applications not exceed crop needs (after the contributions from organic sources are taken properly into account), ask that soils not be let bare during rainy periods, and urge that nitrates present in soil between crops be limited through planting of trap crops.

Certainly the most difficult environments in which to achieve higher efficiencies of fertilizer use are summer rainfed areas in tropical and subtropical belts. In those regions, accounting for a significant share of global food output, limited spells of good water supply coincide with exceptionally high evapotranspiration, and interannual moisture fluctuations can be quite large. But a large body of Indian field experiments and commercial practices confirms that higher rates of fertilization can bring substantial economic returns and that nitrogen use efficiencies better than the traditional rates of as little as 20 percent, and rarely more than 50 percent, are possible (De 1988). Appropriate application of nitrogen fertilizers also increases water-use efficiency as plants can use soil moisture from deeper horizons.

A long-term comparison of the agronomic efficiency and residual benefits of organic and inorganic nitrogen sources for irrigated rice cultivation brought some valuable practical conclusions for optimizing the nutrient's management in the tropics (Cassman et al. 1996). Results based on nine to eleven years of studies showed that the nitrogen uptake, its agronomic efficiency and grain yield were highest with applications of relatively expensive and more labor-demanding urea super granules, and that these measures were only slightly lower when using prilled urea, the most common solid nitrogenous fertilizer. Similar agronomic efficiencies were obtained when organic nitrogen from azolla and sesbania was used as the only source of the nutrient or when it was combined with urea.

However, these green manures are more expensive sources of the nutrient than prilled urea, and nitrogen uptake from both sesbania and azolla was always lower than from the inorganic fertilizer. Rice straw was an even poorer source of nitrogen when used alone—but its combination with prilled urea resulted in agronomic efficiency just 15 percent lower than for the inorganic nitrogen alone. This slight disadvantage is offset by several compensating factors: rice straw provided greater residual benefit than other organic sources of nitrogen and, given its high C/N ratio, it was a larger source of carbon, and it increased bacterial biofixation.

And, unlike green manures, which require additional labor and preempt temporarily cultivation of food or feed crops, rice straw is readily available. Given typical grain/straw ratio of around one for modern rice cultivars, Cassman et al. (1996) conclude that well-managed rice crops of tropical lowland rice could derive annually as much 75 kg N/ha. Efficient use of this valuable resource would be promoted by optimized timing of fertilizer applications, by better incorporation of the recycled straw into soil, and eventually by mechanical harvesters that leave straw in the field.

As beneficial as it is, recycling of crop residues also requires attention to particular negative impacts. Inherently high C/N ratio of straws and stalks (commonly in excess of 100, easily ten times higher than soil humus) is perhaps the most common problem when recycling large amounts of residues. In such cases a source of readily available nitrogen (synthetic fertilizer, manure) should be added at the same time in order to prevent temporary immobilization of relatively large amounts of nitrogen by decomposer bacteria. Large volumes of residues may also produce various phytotoxic substances, and oil seed cakes, which have relatively high levels of plant nutrients, also have high concentrations of alkaloids that inhibit nitrification.

The combined effect of improved fertilizer use, first in rich countries, later throughout the poor world, should be impressive. Gains in fertilizer efficiency can be already seen in the United States, in Japan, and in many European countries where the total applications of nitrogen (and also of P and K) have been either declining or stagnating, while average yields of major crops, and hence the amount of nitrogen they incorporate, have continued to increase. Naturally, the gains will be uneven, but with careful agronomic practices it should be possible to raise the average nitrogen use efficiency by at least 25–30 percent during the next two generations, that is, to average uptakes of at least 50–55 percent in modernizing countries, and to around 65–70 percent in affluent nations.

Even if the utilization of nitrogen from other source remained constant, such higher fertilizer efficiencies would use 10–12 Mt of the currently wasted nitrogen applications. In reality, effective supply of nitrogen from organic wastes, biofixation, mineralization, and atmospheric deposition should also increase because of reduced nutrient losses in soil erosion and because of more frequent rotations and more vigorous recycling.

Again, relatively modest improvements would translate into impressive total gains: reducing erosion losses by 20 percent would save roughly 5 Mt N from nonfertilizer sources, and expanding biofixation (largely through proper rotations with legumes) and waste recycling by just 10 percent would add another 5 Mt N. Cumulative effect of adopting well-proven and mostly low-cost measures aimed at increasing efficiency of nutrient uptake would be then equal to expanding effective nitrogen supply to crops by 20–22 Mt a year.

Given the fact that nitrogen is almost always the nutrient with highest field losses, similar relative efficiency improvements should be possible for the other two macronutrients. This nutrient gain would be sufficient, even with a much lower crop response, to produce additional harvests equivalent to about 500 Mt of grain. No less beneficial would be the moderation of environmental stresses from reduced nutrient loss, and lower demand for energy needed to synthesize and apply fertilizers.

Better Use of Water

Analogy with rational use of energy is particularly relevant for better use of water: both commodities have been priced far below their real cost, insufficient attention has been given to environmental consequences of their excessive use, and wasteful demand has been ignored while aggressive quest for enlarged supply has been the typical response to any tightening of availability. And, as with energy conversions, efficiency of water uses can be improved not only with technical fixes but also with moderated demand. But, unlike in the case of thermodynamically dictated dissipation of useful energy, used or wasted water can be also reused after its quality is restored by suitable treatments.

Not surprisingly, efforts to increase those water supplies that feed generally very wasteful uses continue around the world. Perhaps its two grandest, and dubious, instances are the Chinese plans to use the ancient Grand Canal as the conduit for diverting the Chang Jiang's waters to the lower part of the Huang He basin, and the projected New Delta Canal, which would take water from the Aswan dam lake and carry it to Dakhla and Farafra oases deep in Egypt's Western Desert. Yet, at the same time, both countries continue to provide irrigation water to farmers at a fraction of its actual delivered cost—and they have also done too little to lower industrial and household water consumption.

The Indian situation is similarly unsatisfactory: the country's agriculture consumes a third of all electricity, but there is no tariff and no metering of electricity used for water pumping. Payments are simply by pump size (mere 2–3 cents a month per installed horsepower), but a great deal of consumption is actually free as illegal connections abound (Sathaye and Gadgil 1992).

Arguing the need for higher efficiencies does not mean dismissing all efforts aimed at securing larger supply. Such projects are particularly needed in arid zones where two approaches—water harvesting and the use of saline waters—can be particularly useful. Water harvesting can put to use a significant fraction of very limited rainfall that would otherwise run off during erratic, but often intense, precipitation events. Improvements to many traditional water harvesting methods—including a variety of contour bunds and ridges, terraces, crescent-form microbasins, and shallow wells—can make a great deal of difference for local field crop or orchard production (van Dijk and Ahmed 1993; FAO 1994). Additional water gains in drier cold environments can come from such a simple adjustment as leaving higher stubble to catch more snow. Winter wheat grown in the northern part of the Great Plains can yield

30–50 percent more with 20-cm stubble than when the straw is cut just 5 cm above the ground (Black and Bauer 1990).

Saline waters contain relatively high concentrations of major inorganic ions, but proper selection, planting, and irrigation procedures make it possible to use some of them in crop production (Rhoades et al. 1992). Among field crops, barley, semidwarf wheat, sorghum, sugar beets, and cotton have the highest salt tolerance. In fact, barley's tolerance is surpassed only by that of tall wheat grass among forages (many of which are moderately salt-tolerant) and of date palm among fruit trees (which are generally salt-intolerant). Experiments have also shown that some wild salt-tolerant halophytes—above all shrubby *Salicornia* (glasswort)—can be irrigated only by sea water, have yields comparable to those of alfalfa, and used as good livestock feeds (Glenn et al. 1998).

Irrigation Efficiencies

Efforts to improve low water-use efficiencies should dominate water management in both arid and humid areas. Perhaps the most notable example is that of China, the country that irrigates more land than any other. In March 1997, at Forum Engelberg in Switzerland, Jian Song, the chairman of China's State Science and Technology Commission, claimed that China's agricultural water-use efficiency averages a mere 10 percent. This is surely an underestimate, but the real figure is difficult to estimate because the concept of water-use efficiency is a more complex matter than the assessment of fertilizer-use efficiency.

Typical irrigation efficiency rates more commonly cited for China are between 25 and 30 percent, very similar to the rate of roughly one-third usually given for the efficiency of U.S. surface irrigation. Such figures are derived from estimates for individual irrigation systems rather than from basinwide appraisals. If water wasted by an upstream irrigation system becomes completely available for downstream use, then its initial conservation would represent only theoretical savings. Only if that wasted water were to totally evaporate before it could be tapped again, or if it were to flow directly to the ocean or to aquifers too deep to tap, would its initial conservation represent real efficiency gain. Overall efficiencies for whole river basins can be thus appreciably higher than the usually cited nationwide totals ranging from just around 25 percent for the least efficient water users to more than 80 percent in Israel.

Still, the potential for real savings is considerable because cropping is invariably by far the largest consumer of water in all areas with extensive irrigation. A great deal of water is lost even before it can reach the fields. Many of the poor world's irrigation systems are old, designed without much, if any, concern about water-use efficiency. For example, China's Dujiang Yan on the Chengdu Plain of western Sichuan, which waters more than half a million hectares, dates back to 230 B.C. Such old and extensive irrigation systems, where water travels from reservoirs to fields through often unlined canals with both high bottom and side seepage and considerable evaporation losses, have low conveyance efficiencies resulting in rather high ratios between the volumes leaving the storage and those actually available for field irrigation.

There are two fundamentally different measures of efficiency once the water reaches crop fields (Stanhill 1986). Hydrological approach is a straightforward ratio of the water reaching the root zone of irrigated crops to the total volume of water applied to the field. This irrigation efficiency (the term commonly used by irrigation engineers) is quite different from the physiological measure which is expressed as the total yield (or in its equivalents as total biomass or total assimilated carbon) to total water inputs (Sinclair et al. 1984; Stanhill 1985). This ratio is commonly referred to as water-use efficiency, and it can be improved by planting more water-efficient crops even when the field water use (measured by hydrological ratio) is at its practical maximum. As we have seen (chapter 2), water use efficiency of C_4 plants is about twice the rate for C_3 species. On the other hand, C_3 plants will have a greater response to higher CO_2 levels (chapter 3).

Assuming that a quarter of photosynthate is respired and that only a third of the net dry phytomass produced is harvested as a yield gives theoretical minima of evapotranspiration ratio (water use efficiency in g of water for a g of dry weight yield) of between 320 for harvests in temperate climates and 800 in arid regions (Stanhill 1986). In contrast, even the highly water-efficient Israeli cropping averages about 1,000, and the global mean for annual and permanent crops is above 5,000. Consequently, there is roughly an order-of-magnitude difference between the minima and average worldwide performance, and a severalfold difference between this mean and the best national averages.

Improving Water Use

To begin with, these gaps can be narrowed by better distribution of water to plants. The two traditional methods of water distribution that still dominate irrigation in most countries—flood irrigation covering the whole field with a layer of water, and furrow irrigation chaneling water from ditches to crops along slightly inclined parallel rows—can have irrigation efficiencies as high as 50 percent, but rates between 25 and 40 percent are more common. Major improvements of these inherently inefficient performances are possible with conversion to pressurized systems, including a variety of portable sprinklers, center pivots, lateral lines, and drip irrigation (Finkel 1982; Stanhill 1985; Nakayama and Bucks 1986; McKnight 1990; Postel 1999). The greatest global impact would come from efficiency improvements in a handful of countries with the largest irrigated areas, in China, India, and Pakistan, which have nearly half of the world's irrigated land.

Center pivot sprinklers, first introduced in the United States during the early 1950s, spray water from numerous emitters placed along interconnected aluminum pipes that are elevated two to three meters above ground in order to pass over tall-growing crops (Splinter 1976). The pipes are mounted on A-frame wheeled towers to distribute well water over circular areas. Linear (or lateral) move sprinklers look much like center pivots, but they are not anchored anywhere; they pump water from a ditch in the middle or side of the field and distribute it over rectangular patterns. Like pivoted sprinklers, they, too, can travel over gently rising or undulating ground. Both of these systems can irrigate very large fields with minimal water losses. Center pivot pipes are commonly up to 400 m long, covering a circle of 50 ha, lateral distributors up to 800 m long, irrigating rectangular fields of up to 150 ha.

Delivery can be programmed for optimum amounts of water, scheduled during night hours with low temperatures and low winds (and hence with the lowest evapotranspiration), and fertilizers and pesticides can be accurately distributed in the spray. Irrigation efficiencies with downward spraying from pivoted or lateral pipes should not be lower than 70 percent, rates of 75–80 percent are a good standard performance, and the best efficiencies may be around 90 percent.

Drip irrigation, distributing small volumes of water through plastic pipes right to the roots of plants, has its genesis in the Israeli quest to eliminate water waste associated with traditional field flooding or furrow irrigation techniques (Hillel 1994). This technique—substituting infrequent, high-volume applications wetting whole fields by very frequent, low volume applications wetting small areas—was made possible only with the development of suitable inexpensive plastics after World War II.

By 1990 about 60 percent of Israeli irrigation was done with drip systems (Stanhill 1990), and diffusion of this highly efficient (up to 95 percent) method has been the main reason for the country's more than 30 percent drop of average water use rate and for the doubling of crop yield per unit of water application. Drip irrigation and other microirrigation techniques (ranging from ancient pitcher to modern subsurface irrigation)—do not only conserve water and reduce water stress on plants but are also less likely to cause waterlogging and salinization (Batchelor et al. 1996; Hillel 1997).

The cost of pressurized irrigation systems may not be such a tall barrier for their adoption as is often maintained: prices quoted for irrigation equipment delivered today in Texas are a misleading proxy for future costs in sub-Saharan Africa or China, and an appropriate accounting should reveal more favorable cost/benefit ratios. In the long run, inefficient irrigation will almost certainly result either in waterlogging or salinization, and eventual costs of land rehabilitation may surpass the cost of introducing more efficient irrigation.

Still, there is no doubt about relatively high energy cost of water saving by such irrigation equipment. Stanhill (1986) calculated that each cubic meter of water saved by replacing traditional surface irrigation by a drip system (with irrigation efficiency rising from 55 to 85 percent) costs roughly one liter of oil. Only when using the saved water for the production of very high value crops does such an exchange rate become economical.

But outstanding efficiencies do not necessarily require the substitution of surface irrigation by pressurized water distribution. Where the conditions are suitable, a combination of recent technical advances ranging from laser-guided land-leveling to runoff recovery systems makes it possible to raise the efficiencies of gravity systems close to those of pressurized techniques with a fraction of cost required for the installation and operation of pivots, laterals, or drips.

As in the case of fertilizer applications, optimized timing can greatly increase the efficiency of field irrigation. Optimal irrigation schedules should take into account not only such variables as wind speeds and soil moisture but also the substantial differences in the sensitivity of various crops to water stress at different stages of their growth (Frederick 1988). In affluent countries such appraisals can be now done rather inexpensively with relevant data fed into appropriately programmed microcomputers. Even farmers in poor countries can now afford this service on a collective basis for a larger area with fairly homogeneous soils and with only a few dominant cropping patterns.

And a surprisingly simple American device makes an almost perfect scheduling available to any farmer willing to invest only a very modest amount in an auger, a couple of dozen gypsum blocks, and an AC resistance meter. Cheap gypsum blocks containing two electrodes are buried at several locations and depths in root zones (Richardson et al. 1989). Because the blocks absorb and lose moisture at a rate very similar to that of the surrounding soil, regular measurements of changing current flow with a pocket-size impedance meter give reliable indications of soil's moisture status. Proven benefits have included considerable savings for reduced water purchase and pumping and higher crop yields.

Other inexpensive measures include irrigating every other furrow (this saves about one-third of water with only modest decline of crop yields) and a better matching of crops with natural moisture supply. Inappropriate cultivation is now widespread, ranging from alfalfa irrigated in California's desert climate to Florida sugar cane draining the Everglades, and from a steady extension of Chinese rice cultivation into the arid North to the cultivation of center-pivot irrigated wheat in deserts of Saudi Arabia. In contrast, appropriate crop choices can result in appreciable water savings: replacing irrigated corn with sorghum can lower the water need by 10–15 percent, and planting sunflowers instead of soybeans as an oil crop can save easily 20–25 percent of water.

Where climate and market conditions are right, the savings can come not only from reduced water use but also from a higher profit from new crops. Such savings were realized by many Chinese farmers after the privatization of the early 1980s when they planted more oil and fiber crops and concentrated the water-intensive grain production on better soils (Smil 1993). As I will show in the next chapter, a considerable amount of water could also be saved by reducing feedlot-based beef production.

Again, as in the case of higher fertilizer efficiency, improvements in overall agronomic management can improve water use in both irrigated and rainfed fields (Davis 1994; Sharma and Datta 1994; Bhagat et al. 1996). Concerns about the future of irrigated farming should not obscure the basic fact that most of the food comes, and will continue to come, from rainfed regions. Even in monsoonal Asia only about a third of all rice is actually irrigated. Better water management in those fields is thus critical for higher harvests because many of these areas experience considerable variability of precipitation. Key agronomic contributions to better water use include planting of appropriate cultivars, water-sparing tillage, and careful management of crop residues. Planting improved, drought-resistant varieties of winter wheat in China's arid Gansu province has helped to increase the yield by 14–17 percent compared to local cultivars (Song et al. 1997).

In dryland farming, appropriate degree of tillage helps to control runoff and erosion, maintain soil porosity, and increase surface area of roots or rooting density; it can also help to control weeds whose evapotranspiration reduces overall water-use efficiency. In rice farming, proper puddling of soil drastically reduces water percolation losses. Incorporation of crop residues can be particularly beneficial as they reduce water runoff and erosion and evaporation losses and increase soil organic matter content and thus improve water infiltration and water availability in the root zone: water absorption capacity, in kilograms of water per kilogram of raw material, is no more than 0.25 for sand—but it ranges from 2.5 to 2.9 for chopped straws and shredded corn stover.

The quest for better use of water also requires attention to the performance of motors and pumps. Electric motors and internal combustion engines powering the pumps have life expectancies of twenty to thirty years, and pumps are usually good for twenty to twenty-five years (Finkel 1982). As power irrigation expanded rapidly since the 1960s, a large share of irrigation machinery is now ripe for replacement. This massive upgrading is not needed only because of the growing breakdowns and unreliability of old motors and pumps: these machines are also highly inefficient, wasting large amounts of electricity and refined fuels. For example, whereas a new electric motor (95 percent efficient) and a new pump (75 percent efficient) can offer an overall 65 percent efficiency of delivering water to a field, a combination of an aging motor (70 percent efficient) and an old pump (40 percent efficient) will do the same task with less than 30 percent efficiency, requiring more than twice electricity.

There is also no shortage of accounts demonstrating that the institutional side of irrigation is no less important than technical means or water prices. Effectiveness of cooperative management, eliminating friction about water scheduling between upstream and downstream farmers, is perhaps best demonstrated by ancient Balinese arrangements that reflect ecosystemic variability of cropping and irrigation processes (Lansing 1991). An excellent illustration of the need for cooperation is provided by repeated severe damages inflicted by farmers to the Gandak canal system in the eastern part of the Indo-Gangetic Plain as they desperately try to save their crops during wet season drought (Burns 1993).

Finally, the best way to use water more efficiently may mean not using it for cropping at all. Unlike seeds, agricultural machinery, or farm chemicals, non-drinking water is not easily fungible on an intercontinental basis, and for many countries in arid regions with inherently high evapotranspiration and very limited water resources it can be advantageous to import water in the form of grain and to spare local supplies for urban uses and for specialty crops grown very efficiently with drip irrigation or in greenhouses. International trade in staple food commodities—above all in grains, oils, and sugar—may be thus seen as an effective strategy of overcoming national water shortages.

This strategy already amounts to considerable water savings for low-income countries. Assuming that one tonne of traded grain required at least 1,200 tonnes of water to produce, about 150 Mt of cereals and cereal products that are now being imported annually by low-income countries in Africa and Asia required transpiration of at least 150 km^3 of water, a volume equal to about one-tenth of all irrigation water used annually in those nations. Imports of water disguised as grain will almost certainly increase throughout the Middle East, and may become very important in North China. But not every country has been following this strategy: Saudi Arabia and Libya have used their oil-derived wealth to subsidize extremely inefficient cultivation of cereal crops in extreme desert climates by tapping deep aquifers and desalinating sea water.

As I already noted (in chapter 3), estimates of the total volume of additional water needed to produce food during the coming generations are highly dependent on chains of assumptions. Approximations are thus better calculated by taking a composite mean based on assumed average food intakes. I will assume that to grow crops supporting a nutritious, but low-meat, diet (for details see chapter 8) would take annually between 1,800 and 2,000 m^3/capita.

Similarly, cumulative assumptions are unavoidable when trying to estimate potential water savings brought by more efficient distribution and application techniques and arrangements. Consequently, the following figures should be seen merely as realistic indicators of potential gains. Inasmuch as most of the poor world's irrigation water is delivered by traditional methods, improvements of these practices can bring relatively large rewards. Overall distribution, seepage, and evaporation losses in the poor world's field irrigation add up typically to between 60 and 70 percent of initial water flow. The best available techniques can improve large scale water use efficiencies by 70–80 percent. As a result, current irrigation efficiencies could be nearly doubled—but because of its enormous capital demands such an achievement is pointed out here merely as a theoretical possibility.

Much more modest, and clearly achievable, gains can still make a great deal of difference. Raising the overall water use efficiency in low-income countries by 25–30 percent should be possible through a combination of relatively simple techniques and application adjustments reviewed in this section. Even then the prevailing rates would be much below the level achieved by today's good practices, leaving further room for improvement.

But using irrigation water with efficiencies of 40–50 percent would result in an annual gain of up to 200 km^3, enough to feed more than 100 million people. As an example of such gains on a national scale, a 25 percent increase in irrigation efficiency in China would release up to 50 km^3 of water, about twice the total annual residential and industrial water use in China's fifty largest cities during the mid-1990s (State Statistical Bureau 1999). Costlier approaches—including substitution of gravity irrigation by various sprinklers and higher industrial water recycling—could eventually raise the performance of inefficient traditional irrigation to averages around 60 percent, bringing annual savings of up to 350 km^3 compared to today's use, enough water to produce balanced, low-meat diets for additional 200 million people.

Precision Farming

Combination of progressively cheaper microcomputers and commercial availability of navigation systems guided by signals from global positioning satellites (GPS) made it possible to begin the development of site-specific application techniques, adjusting the amount of agricultural chemicals not only according to field-to-field variations in soil quality but also to often substantial differences of nutrient content within a single field (figure 4.4). Desirability of such discriminating applications is based on a great deal of evidence showing that crop yields vary within fields, and that variations can occur within soil types as much as across soil types. Advantages of such precise dispensation of fertilizers or pesticides seem to be obvious: yields should be maximized and variable applications should greatly reduce the losses of farm chemicals, and hence the degree of environmental pollution.

I prefer to see the images of huge tractors with GPS-linked onboard microcomputers as only one extreme of a very broad continuum of precision farming. Far behind that cutting edge are the gypsum blocks mentioned earlier in this chapter—but when their soil moisture reading

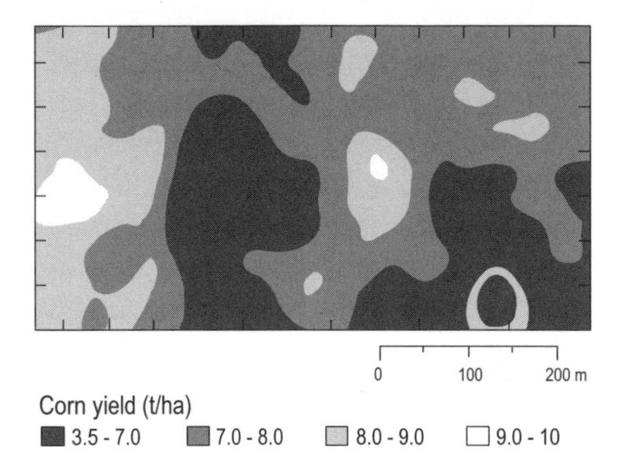

Figure 4.4
Variations of corn yield within a field near Beltsville, Maryland (Lu et al. 1997).

becomes an input for a microcomputer program that takes into account key cropping and weather variables, a farmer can get an accurate forecast of his irrigation needs for a week ahead. At the other extreme is an ancient Middle Eastern technique using simple ceramic jars buried in soil: their porous walls release water to nearby tree or crop roots almost without any waste (Hillel 1997).

High-Tech Possibilities

Variable rate applications have been made possible by an integration of several advanced information techniques (Lu et al. 1997). When equipped with differential adjustments using geodetically surveyed locations, GPS using the Department of Defense NavStar satellites can pinpoint real-time positions with an accuracy of 1–3 m. Easily manipulable information stored in geographic information systems contains a variety of digitized measurements characterizing the site's physical features, microclimatic conditions, and variability of soils and yields.

Some of this information must be collected by grid sampling in a field. This may require repeated testing done at appropriate densities, obviously an expensive process (Franzen and Peck 1995). Spatial distribution of other variables can be determined from satellite imagery, aerial photographs, and existing soil maps. Crop yield monitors that continuously measure and display yields and moisture content of the harvested grain are now commercially available, together with such real-time sensors of soil quality as near-infrared reflectance devices determining soil organic matter and soil moisture. After appropriate processing in microcomputers this information can be coupled with GPS to control machinery—liquid sprayers, air sprayers and spreaders, granular applicators, and seed drills and planters—dispensing variable amounts of agricultural chemicals or seeds and improving input efficiencies through high-precision farming (Munson and Runge 1990).

All components of the precision farming are now readily available, but the costs of many of these advanced technique are high, and the rewards may be underwhelming. Obviously, the expense of testing and variable application will not be rewarded where factors determining crop yield vary only marginally within particular fields, or where the variation is incorrectly evaluated. But Robertson et al. (1997) found that even in a field with a remarkable degree of spatial variability there was surprisingly little correlation between crop (soybean) productivity and static soil properties. If such findings prove to be more widespread, then site-specific management should not be based solely on detailed analysis of soil properties.

Some experiments with variable fertilizer applications did not result in higher yields; others yielded more, but the conventional method gave better net return because of relatively high cost of grid soil sampling (Wibawa et al. 1993; Sawyer 1994). A review of eleven field studies of precise farming found the practice to be unprofitable in five cases, and the results were inconclusive in four other instances (Lowenberg-DeBoer and Swinton 1995).

Continuing refinement of this potentially highly valuable technique will lower its costs and make it more widely accessible. An important consideration in favor of the technique is that custom applications can make it profitable even for smaller operators. Whatever the future of the GPS-guided variable-rate applications will be, we should not associate the term *precision farming* solely with this high-tech method.

More Precise Farming

The greatest task for agricultural research of the next generation is not, as it may appear on the basis of developments during the 1990s, to make every plant transgenic, but rather to come up with effective solutions for more precise farming. Field-specific delivery of inputs on time and without excess is a key ingredient of this approach (Cassman 1999). Optimization of essential agroecosystemic services that sustain high yields will also be essential. Both tasks present an enormous challenge for small-scale farming in low-income countries, and environmental stochasticities will always prevent field farming from being as precise as manufacturing. They do not, however, preclude an appreciable increase in the overall performance of cropping.

To fulfill this potential, many desirable measures making up the practice of precision farming will have to deployed concurrently, or in appropriate sequences, and farmers will have to know much more about the total agroecosystem. The last point is well illustrated by recent field studies of nitrogen efficiency in irrigated rice (Cassman et al. 1993). They suggest that a large improvement is possible by adjusting nitrogen rates to the nitrogen-supplying capacity of soil, a dynamic variable whose knowledge will require repeated assessments. Precision farming will thus demand a much more informed management, moving agriculture along the path already traversed by industrial production.

Field-specific application of nutrients, water, and pesticides is the cornerstone of this approach. Timeliness of operations will be rewarded by substantial yield increases. Early planting is particularly rewarding. A two-week difference in planting Manitoba spring wheat may translate to harvests up to 40 percent higher, and 10–35 percent gains can be also realized with earlier planting of corn and soybeans. But in all regions with rains preceding or coinciding with the rainy period timeliness cannot be hoped for without good field drainage: every additional day of waiting for fields to dry means a mounting loss of harvest. Timeliness is also important in weed control and in harvesting. With timely cultivation and rotary hoeing, losses of soybeans can be cut by half (Gonsolus 1990)—but they mount with every day beyond the harvest optimum.

Proper seeding rates magnify the effect of timely planting. Greater planting densities, closer spacing of plants within the row, and narrower row widths generally boost the yield. Planting of improved crop varieties can boost efficiencies of water and fertilizer use and minimize applications of herbicides and insecticides. Reduced tillage, leaving anywhere between a quarter and a third of the soil surface covered with crop residues, is very effective for controlling surface runoff and soil erosion. Some form of conservation tillage is now practiced by roughly a quarter of Canadian producers and by almost 20 percent of U.S. farmers.

Nitrogen leaching can be also reduced by no-till farming, which generally increases moisture infiltration into soil but prevents further downward movement of the nutrient. As I already noted (chapter 2), crop rotations including legumes can enrich soil's nitrogen stores, increase cereal yields, and reduce losses due to pests. Besides recycling nutrients and renewing soil organic matter, careful management of crop residues also enhances nitrogen fixation, improves water absorption capacity, reduces soil compaction, diminishes soil erosion, and, in cold areas, captures more snow and prevents deep soil freeze.

5

Rationalizing Animal Food Production

We do not have to eat any animal products in order to lead healthy and active lives and to look forward to generous life-spans. But only a very small fraction of humanity adheres to strictly vegetarian diets. Arguments about the desirability of a wider adoption of at least some modified versions of vegetarianism (lacto- or lactoovovegetarianism being the most common choices) have been offered as one of the ways to accommodate larger populations. These arguments have been disparaged as misleading countercultural calls undermining the viability of major sectors of modern agriculture. I have no wish to adjudicate the cultural dimension of these arguments, but resolving the factual question about the comparative merits of vegetarianism and omnivory is easy once one takes energetic and evolutionary perspectives.

The food-chain argument in favor of vegetarianism rests on a straightforward interpretation of an undeniable energetic imperative: eating closer to the Sun will support a larger number of people inasmuch as the interposition of another link in the food chain has to be paid by large metabolic losses associated with animal reproduction, growth, maintenance, and activity. Phytomass used for animal feed would be converted more efficiently if it were consumed directly by humans. This argument would be universally valid if our species had developed multiple stomachs or if our guts carried bacterial communities capable of digesting phytomass containing high shares of hemicellulose and cellulose.

Because none of this is the case, not all domesticated animals compete with people for edible harvests. When cattle, sheep, and goats are grazing on land that should be never converted to crop fields—on semiarid grasslands, slope lands, and mountain meadows—they are not even in any theoretical competition with us, in that the highly cellulosic phytomass of these ecosystems is indigestible by humans. In addition, grassed surfaces provide valuable environmental services by conserving soil moisture and preventing soil erosion. In these cases any objections to harvests of meat and milk based on considerations of human carrying capacity are inappropriate—as long as care is taken to prevent overgrazing and to assure the survival of herds in often highly stressed environments.

Resource competition is also absent, or minimal, in the case of animals grazing on crop residues, on grasses or leguminous cover crops planted in rotation with food species, and fed residues from crop processing and a variety of organic wastes. Again, we could not or would not wish to digest this phytomass, whereas animals metabolize it readily to produce meat, eggs, and milk and wastes whose recycling maintains, or even improves, soil quality.

The carrying capacity argument based on fundamental energetic considerations thus does not exclude appreciable shares of animal food in human diets. The evolutionary argument in favor of omnivory is even stronger. We now know that hunting for meat has an important place—nutritionally and socially—in the lives of both chimpanzee species (*Pan troglodytes* and *Pan paniscus*), and hence also in the evolution of Pliocene hominids (Stanford 1996, 1998). Diet made up primarily of plant foods but supplemented, especially seasonally, by meat is our evolutionary heritage, and strict herbivory is a culturally induced adaptation. The expensive-tissue hypothesis and considerations of practical satisfaction of protein requirements strengthen this conclusion.

Although human brains are much larger than the primate ones, the total mass of our metabolically expensive tissues (internal organs and muscles) is very much as expected for a primate of our size. Aiello and Wheeler (1995) argue that the only way to accommodate larger brains without raising the average metabolic rate was by reducing the size of another metabolically expensive organ. This option is highly constrained in the cases of liver, heart, and kidneys, leaving the gastrointestinal tract as the only metabolically expensive tissue that could vary considerably in size. Indeed, our guts are about 40 percent smaller than they should be in a similarly-sized primate. The primary reason for this difference is that—unlike in the case of herbivores consuming large amounts of low-energy density leaves, roots, and fruits—our diets evolved to include smaller quantities of energy-dense, and easily digestible, foods, including seeds, nuts, and meats.

In those environments where nuts and seeds, which also have relatively high protein content, were readily available, preagricultural foragers could obtain adequate diets by remaining overwhelmingly vegetarian. But in environments where fruits and leaves with low energy density and low protein content made up most of the accessible edible phytomass, satisfaction of basic food energy requirements would have required daily intakes of more than 3 kg of the former and over 10 kg of the latter. Even then this impractically bulky plant diet would have supplied no more than about half of the needed protein (Southgate 1991).

Evidence for human omnivory is provided by numerous anthropogenic studies that have found that the provision and eating of animal foods—be they scavenged (from kills by large carnivores), collected (shellfish, eggs), hunted (from tropical birds to aquatic mammals), or produced eventually by domesticated species—has been a universal ingredient of human behavior with obviously valuable nutritional and social implications. In addition, a relatively recent adaptation—domestication of milking animals—further extended the human consumption of animal foods to include milk and dairy products from at least half a dozen mammalian species (Zeuner 1963; Clutton-Brock 1989).

An obvious conclusion is that our normal diets should contain a variety of animal foods, including meat. Minimum intakes of these foodstuffs can be estimated on the basis of evolutionary considerations. Average consumption of meat among the studied chimpanzee groups (they eat mostly colobus monkeys and also a few other smaller animals) has been between 4 and 11 kg a year (Stanford 1996, 1998). In proportion to body mass, this intake is equivalent to about 6–17 kg a year for humans—a range clearly overlapping with typical per capita meat consumption in preindustrial societies. Per capita rates near the lower end of the range (5–10 kg of meat a year) were common in most peasant societies where meat was eaten no more frequently than once a week and relatively larger amounts were consumed only during some festive occasions. Somewhat higher intakes were found in some pastoral societies and among better-off groups in richer temperate regions with mixed farming.

There would seem to be a good evolutionary argument for the annual presence of at least 10–20 kg of meat in average diets. Looking at traditional per capita intakes in milking societies (leaving such extremes as East African pastoralists, where milk supplied three-fifths to two-thirds of all food energy, aside) suggests annual consumption of dairy products equivalent to around 100 liters of fluid milk. Recent per capita global average for milk supply has been almost 80 liters a year, and meat availability in the mid-1990s averaged about 35 kg.

Both of these levels have been greatly surpassed by all affluent Western societies. During the 1990s population of affluent nations added up to only a fifth of the global total—but these countries produced one-third of hen eggs, two-fifths of all meat, and three-fifths of all poultry and cow milk. Their annual per capita milk supply surpasses 300 kg, and their meat eating ranges from 70 to 110 kg. These rates are up to an order of magnitude higher than in many poor countries—and extending even the current global means to an additional three to four billion people would call for a substantial expansion of animal husbandry.

Better management of grasslands, as well as their mostly regrettable but inevitable extension due to continuing deforestation in Latin America, Africa, and Asia, will provide some of this additional need. Even so, an increasing share of animal foods will have to come from feeding of phytomass grown in direct competition with food crops. This trend has been most obvious in the case of cereals. In 1900 just over 10 percent of the world's grain harvest was fed to animals; by 1950 the share surpassed 20 percent, and it was about 45 percent in the late 1990s. National shares of grain fed to animals now range from just over 60 percent in the United States to less than 5 percent in India. The continuing rise in global demand for meat means that even a larger share of cereals will be fed to animals.

By far the most important means of making diets with a decent share of animal foodstuffs available to billions of additional people would be then to maximize feeding efficiencies, a quest that will at the same time help to lower the claims of animal husbandry on land and water, and moderate its rate of waste metabolism and environmental impacts. A fundamental evaluation of long-term prospects of animal food consumption should set out first accurate comparisons of efficiencies with which animal foods can be produced, and it should also appraise their specific claims on key natural resources. Given a high degree of substitutability among animal foods, such knowledge should help us to introduce more rational ways of their production that would secure the nutritional needs of larger populations with minimal environmental impacts.

Feeding Efficiencies and Resource Claims

Production of milk, meat, and eggs based on feeding crops harvested on arable land inevitably entails a loss of potential food output: edible crops cultivated in place of feed crops would always yield more digestible energy, as well as more protein, than the animal foodstuffs produced from the feed. (However, the plant protein's quality would be inferior to meat, milk and egg proteins; see chapter 7.) Both energy and protein losses caused by inherent inefficiencies of animal growth and metabolism differ substantially among domesticated species—but making accurate comparisons is not as simple as repeating frequently quoted feed/gain ratios.

When they are, as is usually the case, presented without detailed explanation of how they were derived and what they actually represent, these measures are quite misleading. Moreover, the most frequent way of expressing the ratio—units of feed per unit of live weight gain—is not the best means of appreciating energy and protein costs of producing animal foods. I will first calculate these ratios in this common manner—in units of air-dry feed (that is, the phytomass as is usually fed, with only 10–15 percent of moisture) per unit of gain (live body weight or gross product output)—but then I will adjust them to reflect feeding costs per unit of edible product, and express them in terms of energy and protein efficiencies. All these calculations are based on a variety of widely accepted equations and recommendations predicting feed intake of food-producing animals (National Research Council 1987, 1988a, 1988b, 1989, 1994, 1996b).

Protein conversion efficiencies will be given simply as percentages (grams of edible protein per gram of feed protein). But because high-quality protein is the most valuable nutrient in animal foods, I will also express its cost in terms of gross feed energy. Energy conversions will be expressed both in terms of gross energy (GE) and metabolizable energy (ME) of the feed. The former term expresses the energy content of plant or animal-derived feed, and its knowledge is necessary in order to find the burden animal feeding puts on agricultural resources. The second term is the energy in the feed less energy lost in animal feces, urine and gases (Subcommittee on Biological Energy 1981).

In the global ranking of animal foods by mass, cow milk comes first: its annual production is now approaching 500 Mt. Pork, with annual output of about 80 Mt, is by far the most important meat and its output continues to rise. In 1995 the world production of nearly 55 Mt of poultry surpassed the combined beef and veal output, and it will continue to rise steadily while the consumption of the two bovine meats will increase only slowly. Consumption of hen eggs is now at more than 40 Mt a year, and recent rapid growth of aquaculture (a near tripling in fifteen years) has put the combined output of finfish, molluscs, and crustaceans ahead of mutton.

Before following this order of animal food production in discussing specific feeding efficiencies, I must note that all unqualified single figures are misleading. Diversity of breeds, environments, and feeding practices and the necessity of making simplifying assumptions (whose even relatively small uncertainties are magnified by successive multiplications) means that a definite rate is valid only for a particular animal, herd, or flock. These problems are further compounded by wide ranges of average energy densities that could be used in efficiency calculations. For example, average energy densities of pig carcasses range between 3,100 kcal/kg for very lean pigs to 5,500 kcal/kg for lardy animals. To avoid an excessive amount of data, I will calculate values for the best current commercial practices and then point out the lags between these rates and standard performances.

Rates and Trends for Major Animal Foods

Mammalian milk production is an inherently efficient energy conversion process. Feed/milk ratio for the most efficient dairy cows is less than 0.6, which means that, depending on the energy density of their diet, between 55 and 67 percent of gross energy in the feed can end up as food energy in milk. Fat accumulated in cow's bodies can be metabolized to produce milk with even higher efficiency of up to around 80 percent.

Typical performances are obviously less efficient. Feed/milk ratios for most animals on well-balanced diets are between 0.7 and 0.8, and the U.S. nationwide mean is now below 0.8, a 25 percent improvement since the early 1960s (figure 5.1) (U.S. Department of Agriculture 1910–1999). When comparing feed/milk ratios with feed/gain rates for animal foodstuffs produced by nonruminant animals, it must be kept in mind that normal ruminant diets must include adequate amounts and proper composition of roughage that is either indigestible by nonruminant species or can be converted by them with only very low efficiencies.

A general rule with milking cows is that at least one-third of all dry matter feed, or roughly 1.5 percent of the animal's body weight, should be long hay or an equivalent amount of other coarse roughage or silage

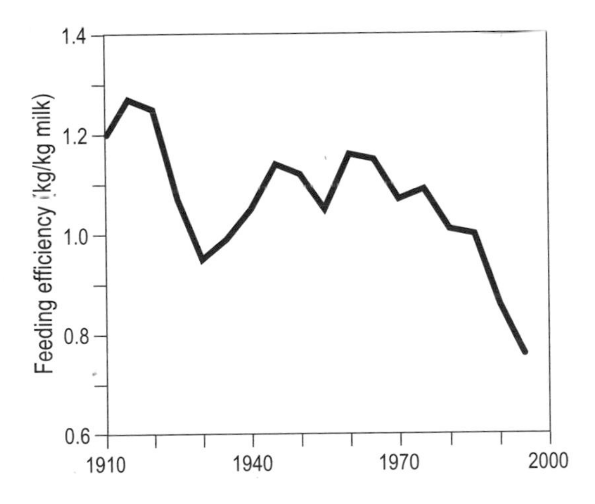

Figure 5.1
Feeding efficiency in U.S. milk production, 1910–1998; plotted from data in U.S. Department of Agriculture (1910–1999).

(National Research Council 1988a). Milk production is thus even more efficient in terms of its claim on resources that could be used as good feed by other animals or directly as food. On the other hand, the overall cost of milk production should be adjusted to include the feed needed to maintain the whole dairy herd.

Depending on the breed, average herd life (from the first parturition to removal) of dairy cattle is between thirty-five and thirty-nine months, equivalent to 3.1–3.5 parities, and to about 1.3 surviving potential female replacements. This means that at least 75 percent of heifers would have to be retained to maintain constant herd size. The retained share is usually higher in order to make possible culling based on first and second parity milk yields and to make up for often considerable mortality, which may exceed 15 percent in some herds.

Moreover, the average age at first calving is twenty-nine to thirty months, rather than the possible twenty-two to twenty-four months. All of this adds up considerably to the feeding cost of dairy herd—but only part of these costs could be attributed to milk production, inasmuch as most of the male calves from dairy herds are now fattened for beef and at least one-fifth of dairy cows is culled annually for meat slaughter.

The most fundamental bioenergetic reason for the high efficiency of pork production is the animal's inherently low basal metabolism. Kleiber's equation, expressing the basal metabolism of mammals as the function of their body mass raised to the power of 0.75, translates into a straight log-log mouse-to-elephant line (Kleiber 1961). Pigs need as much as 40 percent less energy than expected for their weight, whereas basal metabolic rates for cattle lie between the value expected from Kleiber's equation and levels up to 15 percent above it (figure 5.2). As a result, pigs at the midpoint of their growth will channel almost two-thirds of their metabolized energy into weight gain, whereas the share for a 300-kg steer is only around 45 percent, and between 50 and 60 percent for chickens (Miller et al. 1991).

Bioenergetic advantages that make pigs highly desirable producers of meat do not end with low metabolic rates (Pond et al. 1991; Miller et al. 1991; Whittemore 1993). The animals also have a short gestation time and high reproduction rate and grow rapidly. Reproduction usually

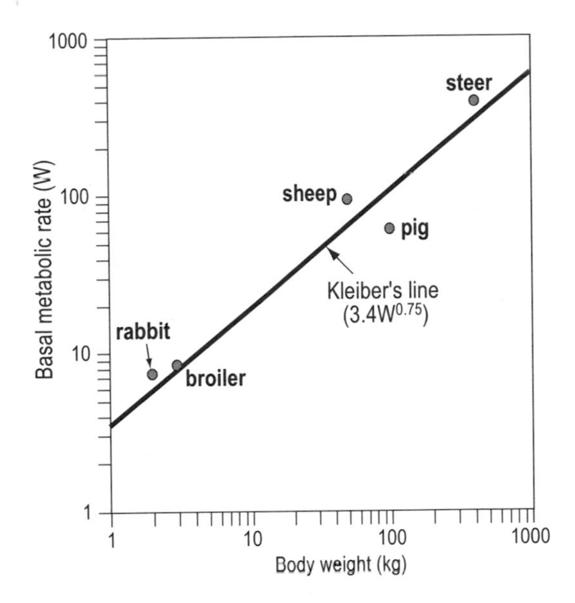

Figure 5.2
Comparison of basal metabolic rates of domestic animals (Smil 1998).

starts at four to eight months of age, pregnancy averages 114 days, and litter sizes range from eight to eighteen. Because the pig's period of gestation is short relative to its life span, its birthweight is a much smaller fraction of its mature weight (a mere 1/300) than in most mammals (Hollis 1993).

As a result, the pig has a very rapid postnatal growth. Piglets are weaned after three to five weeks, and extreme slaughter ages encompass a range from suckling pigs to huge animals used in making Parma hams. Typical slaughter weights in commercial production of modern breeds (90 to 100 kg) are reached 100–160 days after weaning. This means that many pigs are now marketed in less than five months after they are born. Takeoff rate—the number of pigs slaughtered in a year divided by the total number of pigs on farms—measures production efficiency. During the 1990s the rate was in excess of 1.5 in North America and in the European Union (implying an average slaughter age of less than eight months), but just 1.0 in Mexico, 0.8 in China, and only 0.5 in Brazil.

As in other animals, the pig's energy conversion efficiencies decline with age: weaned piglets need only 1.25 kg of feed for a kg of gain, and then the conversion rate of feed to live weight in growing pigs, by definition those between weaning and 70 kg of body weight, rises from less than 1.5 to about 3.0. The rate for finishing pigs, that is, those weighing more than 70 kg but not yet heavy enough for slaughter (at 90–120 kg), is between three and four (National Research Council 1988b; Miller et al. 1991).

With *ad libitum* intake of feed, overall feed/gain rate for North American pigs from weaning to slaughter ranges between 2.5 and 3.5. Rates around 3.0 would be good standard performance with feed whose ME averages about 3,200 kcal/kg and some 15 percent of protein (National Research Council 1988b). With 55–57 percent of the pig's live weight in edible tissues, adjustment of the numerator from the total live weight to edible energy raises the ratio from 3.2 to about 5.4. The rate would be proportionately lower for the best American operations, which now have feed-to-live weight conversion ratios of about 2.5 (Smith 1997).

Addition of feeding costs of the breeding stock (of its reproduction and maintenance, and of fetal growth and subsequent lactation periods) and adjustments for environmental stresses, disease, and premature mortality can raise the overall feed/gain rates quite significantly. Perhaps the best long-term record at the national level has been kept by the U.S. Department of Agriculture since 1910 (USDA 1910–1999). They are expressed in terms of corn feeding units (GE of 3,670 kcal/kg) per unit of cattle, pig, and broiler live weight and per unit of produced eggs and milk.

Nationwide feed/live weight gain ratio for pigs was about 6.7 in 1910. After an initial decline it has fluctuated between 5 and 6.5 ever since (figure 5.3). The main reason why the trend of continuous improvements in feeding has not been reflected in the national mean has been the quest for less lardy pigs. Leaner animals are inherently more costly to produce: efficiency of metabolizable energy conversion to protein in pigs peaks at about 45 percent , while conversion to fat can be as much as 75 percent efficient.

Birds have inherently higher metabolism than mammals: the difference for identically massive creatures (for example, a rabbit and a hen) will

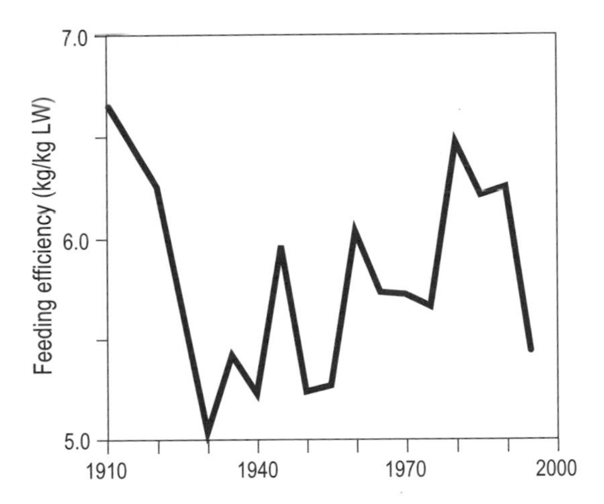

Figure 5.3
Feeding efficiency in U.S. pork production, 1910–1998; plotted from data in U.S. Department of Agriculture (1910–1999).

be up to 10 percent. The higher body temperature of nonpasserine birds—39.5°C vs. 38°C for eutherian mammals—explains most of the difference. Passerine birds have even higher metabolic rates, but no songbirds are reared as meat animals, although plenty of them are caught and eaten in some countries, particularly during their annual migrations. Chickens grown for meat are marketed at ages of four to nine weeks when their body weights range between 0.9 and 3 kg. In the United States the average time needed to produce a broiler was cut from seventy-two days in 1960 to forty-eight days in 1995, while the bird's average slaughter weight rose from 1.8 to 2.2 kg and the feed/gain rate fell by about 15 percent (Rinehart 1996).

When fed well-balanced diet (ME of 3,200 kcal/kg, containing about 21 percent protein) cumulative feed/gain ratios are as low as 1.5–1.8 for lighter birds slaughtered after four to six weeks, and between 1.8 and 2.0 for the birds in the most common 2.0–2.5 kg range (National Research Council 1994). Feed/gain ratios for other common poultry species are somewhat higher, ranging between 2.5 and 2.9 for ducks and between 2.5 and 3.2 for turkeys. Feed requirements of breeder hens and cockerels and feed wasted on birds that die before reaching maturity raise the mean by at least 10 percent. Ratios between 2.0 and 2.2—or between

3.3 and 3.6 for the edible portion—represent the standard of recent good performance.

USDA's (1930–1997) long-term record of feeding efficiency ratios for chickens is available only since the mid-1930s, when the value stood above five, identical to that of pigs. A continuous subsequent decline had halved that rate by the mid-1980s; this has been the only case of a steady improvement of a USDA-tabulated national mean of feeding efficiency among U.S. meat animals (figure 5.4).

Calculating comprehensive feed/gain efficiency ratios for beef is a task greatly complicated by a variety of arrangements under which the meat production takes place (Orskov 1990; Jarrige and Beranger 1992). The ratios for purely grass-fed beef are of interest to sustainable grassland management, but such animals do not compete for feed resources with other domesticated species and have no impact on field crop production. Cattle raised without any grazing on commercial feeds (including the minimal share of roughage) are the other extreme of the beef-producing spectrum. After weaning, calves are moved to feedlots, where they are fed a diet dominated by concentrates combined with feed additives, growth promoters, and disease preventers. These animals

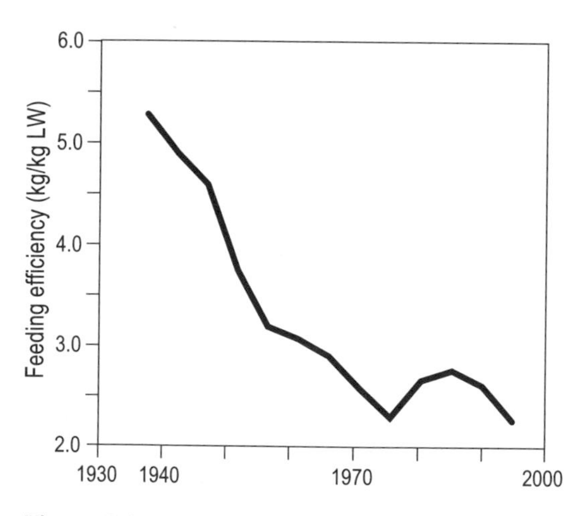

Figure 5.4
Feeding efficiency in U.S. broiler production, 1935–1998; plotted from data in U.S. Department of Agriculture (1935–1998).

gain between 1 and 1.3 kg a day, growing much faster than grazing animals whose daily gains, even on good pasture, average no more than 0.5 kg.

Animals spend commonly between 120 and 170 days in feedlots before reaching the market weight of 450–500 kg, but many of them are now fed in lots for more than 200 days. Feed/meat ratios for growing and finishing calves and yearlings in this kind of operation are usually cited when comparing beef with other animal foods. Naturally, they will represent the most demanding, that is, the least energy-efficient, alternative.

Several bioenergetic realities make cattle less than ideal convertors of feed to meat. As already noted, basal metabolism in cattle is appreciably higher than in pigs. In addition, large body mass and long gestation and lactation periods mean that feed requirements of breeding females in cattle herds claim at least 50 percent more energy than for pigs, and almost three times as much as in chickens. For growing and finishing steer and heifers (calves and yearlings) North American and European feed/gain ratios range between 7 and 9.

With 8 as a common mean, and with roughly 40 percent of live weight in edible biomass, feed energy gets converted to beef with efficiencies between 4 and 5 percent, and protein conversion efficiency is around 8 or 9 percent. Adjusting these rates for the costs of reproduction and growth and maintenance of sire and dam animals raises the feed/gain ratio of herds over 10. The USDA's historic feed/meat data for all of the nation's cattle and calves show an undulating pattern rising and falling between lows of about 9 and highs of 14 (figure 5.5). These rates would mean that as little as 2.5 percent of gross feed energy are converted into food, and that protein conversion efficiency may be lower than 5 percent. The United States does more of this highly inefficient feeding than any other country in the world. It is, with about a quarter of total consumption, also the world's largest importer of beef.

Performance of modern egg production has improved quite impressively during the latter half of the twentieth century: in the best operations the number of eggs laid per hen has doubled, while the amount of feed needed to produce an egg fell by half. When calculating the

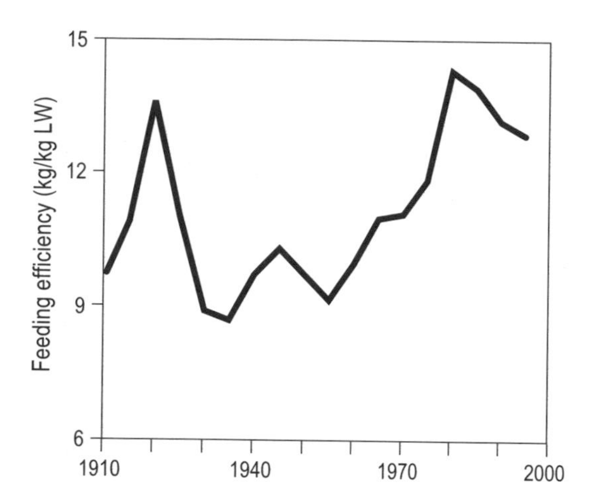

Figure 5.5
Feeding efficiency in U.S. beef production, 1910–1998; plotted from data in U.S. Department of Agriculture (1910–1999).

efficiency ratio for eggs, only the maintenance of the mature laying hen and the nutrition needed to produce egg should be added; requirements of the growing bird before the laying begins should be charged against its eventual use as a low-quality meat source. Feeding rates depend on the hen's body weight (larger birds will, obviously, require more feed for maintenance) and on the rate of egg production (the higher the rate the lower the share of maintenance in egg's feed cost).

Feed/egg mass ratios (based on ME of about 2,900 kcal/kg, with 15 percent of protein) range between 1.8 and 4.1 (Subcommitee on Poultry Nutrition 1994). The most common ratios for Leghorn-type laying hens fed 100–120 g of feed a day and producing eggs nine times every ten days (average weight of 60 g) are between 1.9 and 2.2. The U.S. national ratio has been fairly steady, fluctuating between 2.3 and 2.9 with no discernible trend (figure 5.6).

No food animals are as efficient in converting feed to body tissues as fish (Cowey et al. 1985; Halver 1989; Hepher 1988; Steffens 1989; Barnabe 1994; Parker 1995). There are four major reasons for this primacy. As ectotherms, fish do not have to divert energy to maintain steady bodily temperature. Their low maintenance energy requirements mean that more of their metabolizable feed intake can be diverted to

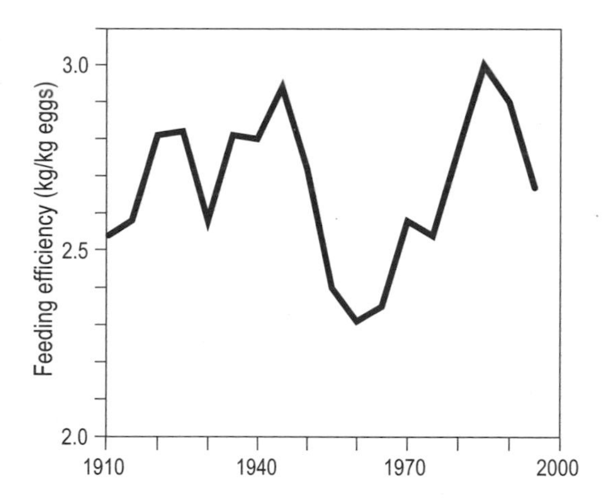

Figure 5.6
Feeding efficiency in U.S. hen egg production, 1910–1998; plotted from data in U.S. Department of Agriculture (1910–1999).

growth than among mammals and birds. Energy costs of locomotion are low inasmuch as life in a buoyant medium does not require large muscles to overcome gravity and inasmuch as streamlined body shapes lower resistance during swimming. Finally, unlike mammals and birds, which spend energy on converting ammonia to urea and uric acid, fish simply excrete most of their waste nitrogen as ammonia through their gills.

Fish need 35–45 percent of protein in their diets, considerably more than either poultry or mammals (Hepher 1988). But their energy needs per unit gain are much lower than in terrestrial species. Conversion ratios for semi-intensively bred carp in warm waters are between 1.4 and 1.8, for catfish between 1.4 and 1.6. Danish statistics show that the average for all intensive freshwater fish farms is almost exactly 1.0 (Ministry of Environment and Energy 1997). Feed/gain ratio for Norway's Atlantic salmon declined from 2.3 in 1972 to 0.9 in 1994 (Blakstad 1995).

Because this reduction was achieved not only by improved breeding and better management, but also by feeding more energy-dense meals (4,700 kcal/kg in 1995, compared to 3,500 kcal/kg in 1972), the actual efficiency gain was somewhat smaller. Still, in terms of edible salmon, the conversion ratio almost doubled, from about 19 percent to about 36 percent.

Comparisons of Efficiencies and Resource Claims

No domesticated animal can produce edible energy and high-quality protein with higher efficiency than a dairy cow (figure 5.7). Highly productive animals convert at least 20 percent of their gross, and more than 30 percent of metabolizable, feed energy to edible energy in milk's lipids and carbohydrates, and between 30 and 40 percent of feed protein is converted to milk protein. Hen eggs from the most efficient layers take an overall second place. Comparisons of feeding efficiencies indicate that pig is the most efficient domesticated animal for converting feed energy to meat (table 5.1): with the best rates of more than 20 percent of metabolizable energy deposited in edible tissues, its performance is superior to that of cattle, and it is well ahead of the rate achieved by poultry.

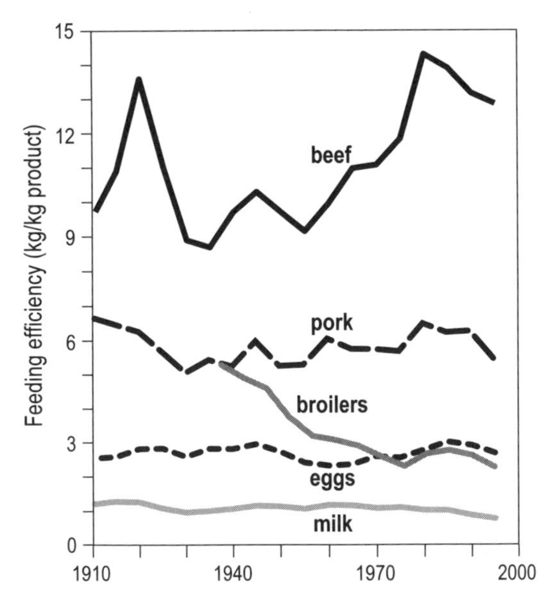

Figure 5.7
Comparison of feeding efficiencies in U.S. milk, meat, and egg production, 1910–1998 (based on figures 5.1 and 5.3–5.6).

Table 5.1
Efficiencies of Animal Food Production

	Milk	Eggs	Chicken	Pork	Beef	Carp	Salmon
Feed (kg/kg LW)	1.0	2.5	2.5	4.0	8.0	1.5	0.9
Edible weight (% of LW)	95	90	55	55	40	65	65
Feed (kg/kg EW)	1.1	2.8	4.5	7.3	20.0	2.3	1.4
Food energy (kcal/kg)	650	1,600	1,800	3,100	3,000	1,200	2,200
Protein content (% EW)	3.5	13	20	14	15	18	22
Energy conversion efficiency (% GE)	20 25	15 20	10 15	15 20	4 5	10 15	30 35
Energy conversion efficiency (% ME)	30 40	20 25	15 20	20 25	6 7	15 20	35 40
Protein conversion efficiency (%)	30 40	30 40	20 30	10 15	5 8	20 25	40 45
Protein conversion efficiency (MJ/g)	0.3	0.3	0.3	0.6	2	0.15	0.1

But we have never produced meat simply for its food energy, but rather for its palatability, attributable to incorporated lipids, and for its protein content. Again, pig is the most efficient converter of feed energy to edible lipids—but in terms of feed energy costs per unit of edible protein, chicken leads. The most efficient broilers need less than 1.2 Mcal/kg of protein, while pigs require more than twice, and beef cattle about seven times as much. The same ranking prevails when the efficiency is expressed in terms of edible protein per unit of gross feed protein.

Animal husbandry's overall claim on agricultural resources is illustrated most obviously by the already noted rising share of feed grains in the world's cereal production. Total mass of cereal and leguminous grain now eaten annually by animals (more than 700 Mt during the late 1990s) contains enough energy to feed more than three billion people—but only if they were willing to eat a largely vegetarian diet with corn, barley, sorghum, and soybeans as staples providing most of their food energy.

A more realistic comparison, assuming that the area now devoted to feed crops were planted to a mixture of food crops, and only their milling residues were used for feeding, would lower this estimate to about one billion people on predominantly vegetarian diets.

Direct land claims of farm animals—for their barns, sties, coops, yards, and enclosures—are only a small fraction of indirect requirements necessitated by production of feeds. For example, an optimum allotment of space for growing and finishing pigs is about one m^2/head (National Research Council 1987); the two animals that occupy sequentially that area during one year will consume at least 600 kg of feed, which, assuming that the pigs are raised on a mixture of concentrate feed, will need on the order of 1,000 m^2 of arable land to grow. Even if we quadruple the amount of space per housed pig in order to account for corridors, accesses, stores, and yards, the difference of two orders of magnitude remains unchanged.

Farmland needed to grow feed for animals is not simply proportional to specific conversion efficiencies. To begin with, a significant share of feeds comes from by-products generated by processing (above all oilseed pressing and grain milling) of food crops, and this contribution should be subtracted from the feed total because its provision does not take away any arable land potentially usable for food crops.

An additional adjustment for a noncompetitive share of feeds must be made in the case of ruminants in order to account for their necessary consumption of straw and grass. Given a great variety of concentrate feed mixtures, differences and fluctuations in feed crop yields, and variability of feed shares supplied by by-products and ruminant roughages, it is not surprising that the values of competitive footprints of animal feeds can range two- or even threefold.

To calculate representative North American means, I have used a weighted average for typical yields of concentrate feed crops, assumed a common share of 20 percent of the total coming from by-products and the minimum 15 percent share of ruminant roughage, and applied these factors to the previously derived efficiency comparison. Once again, milk ends up first, with as little as one m^2 of farmland needed to produce one Mcal of food; eggs are about 50 percent more demanding and pigs are not far behind (table 5.2). Production of chicken meat needs twice as much land as milk, and beef demands at least six times as much.

But chickens need the least amount of space to produce a unit of protein (of course, it helps that, carnivorous fish aside, their meat has the highest protein content of the all animal foods), closely followed by eggs and milk. When comparing on the basis of lean meat, cattle may require more than 20 m^2/Mcal when fed with grain grown in drier regions with lower average crop yields, and they may need up to twenty times as much as space as chickens to produce the same amount of protein. But pasture-fed cattle that receives only a small amount of feed protein supplement may demand no more arable land per unit of edible energy than do broilers or pigs.

As in the case of land requirements, total volumes of water needed to grow feed crops are orders of magnitude higher than the direct consumption by animals—but the presence of thousands of heads of

Table 5.2
Comparison of Land Requirements of Animal Foods*

	Milk	Eggs	Chicken	Pork	Beef
Space per unit of food energy (m^2/Mcal)	1–1.5	1.5–2	2.5–3	2–2.5	6–10
Space per unit of food protein (m^2/kg)	19–28	19–25	13–15	80–100	180–310

* Assuming average feed crop yield of 6 t/ha.

large animals and tens of thousands of birds in feedlots and barns requiring *ad libitum* water supply adds up to considerable quantities. These needs vary greatly with prevailing diet, productivity, and ambient temperature.

For example, a small cow producing less than 20 kg of milk a day will need less than 100 kg of water, whereas a 600 kg animal producing 30 liters of milk will need about 150 kg of water in spring, but more than 300 kg in summer. Poultry has by far the lowest water requirement: a growing broiler needs just 1–2 liters of water a week, egg-laying hens 500–800 ml (National Research Council 1994). On the other hand, water needs of dairy cattle are higher in relation to their size than for any domestic animal. For nonlactating animals they equal three times their dry matter feed intake in cold temperatures, and the multiple rises to about five when the temperature is around 25°C; afterward there is an even faster increase to the multiple of 15 at 35°C. When lactating, the animals will require additional 2.2 liters for every kg of milk produced at 10°C, and over three liters at 30°C (National Research Council 1988a).

I have calculated ranges of typical water requirements per unit of edible foodstuff for animals in temperate climates and relative differences are very similar in both energy and protein terms. Beef requires roughly three times as much water as milk, which, in turn, is two-three times as demanding as pork; water cost of edible energy and protein in eggs are a mere one-twentieth to one-fiftieth of those for beef (table 5.3).

Table 5.3
Comparison of Water Requirements of Animal Foods

	Milk	Eggs	Chicken	Pork	Beef
Water per unit of food energy (g/kcal)	10–15	1.5	6	5	25–35
Water per unit of food protein (g/g protein)	200–300	15	50	150–200	700–800

Obviously, aquacultural water requirements are comparatively large. Those for shrimp cultivation are particularly high, between 50,000 and 60,000 tonnes per tonne of shrimp. In ponds with intensive production about a third of water has to be changed daily, and about half of it is fresh water needed to obtain the optimum salinity level. Not surprisingly, groundwater levels of several coastal regions dotted with shrimp ponds have been dropping rapidly. Large-volume pumping of freshwater and seawater also effects the biodiversity of affected areas.

Finally, a resource constraint peculiar to animal farming is its generation of voluminous wastes, which is already resulting in introductions of legal limits on their field applications and which is bound to cause even greater environmental stresses in the future. Domestic animals are prodigious producers of organic wastes, but generalizations about their manure output are not easy, inasmuch as the rates differ with animal breeds, sizes, feed quality, and health.

In relative terms (per kg of live weight) dairy cows are by far the largest producers of feces and urine, followed by beef cattle, poultry, and pigs. Annual waste production from a dairy farm with 1,000 animals will amount to around 20,000 t of feces and urine, with solids accounting for about 14 percent of this total. Animals are also particularly inefficient users of nitrogen. Even such good protein convertors as young pigs will excrete 70 percent of all ingested nitrogen. Dutch dairy production now uses no more than 12–16 percent of total nitrogen input (Steverink et al. 1994). Similarly, Bleken and Bakken (1997) calculated average nitrogen retention in animal foods in Norway at just about 20 percent.

The worsening environmental impact of manure production stems from a fundamental shift in the structure of animal husbandry, from the still continuing separation of livestock production from field agriculture. In preindustrial agricultures wastes from small-scale animal production using farm-produced feeds were a valued resource critical for renewing soil fertility. Recycling of animal, and often also human, wastes in traditional mixed farming, as well as lower harvest indices of unimproved cultivars (which resulted in a larger share of assimilated nutrients retained in crop residues) kept a substantial share of N, P, and K excreted by animals circulating within agroecosystems.

Manures, when properly stored and applied, are a very valuable ingredient of rational food production. Their application results in similar or higher crop and pasture yields as those obtained through the use of inorganic fertilizers (Choudhury et al. 1996). In areas with concentrated poultry or pig production (prime U.S. examples are the Delmarva peninsula and western Arkansas for poultry) manure could supply all nutrients needed by all crops. But the combination of intensive production of large numbers of animals in confinement and of low-cost synthetic fertilizers turned those wastes from assets to liabilities.

In terms of dry solids the global production of animal manures amounted to more than 2 Gt during the mid-1990s and, assuming average nitrogen content of about 5 percent, it contained about 100 Mt of nitrogen. This is an impressive total, but not all of these wastes are available for collection and recycling: most of the world's cattle, camel, horse, sheep, and goat manure is not produced in confinement. And because the relative nutrient content of fresh wastes is low—mostly between 0.5 and 1.5 percent N and 0.1 and 0.2 percent P—its handling, transportation, and application costs are high in comparison to much more concentrated synthetic compounds. In most instances costs of manure transportation usually limit the distribution of wastes to radii of a few km (Sims and Wolf 1994).

As is the case with housing the animals, space to store wastes is a small fraction of the area claimed by cropland, but manure applications to fields face growing limitations. Because of the continuing concentration of animal production—for example, six Midwestern states produce about two-thirds of all U.S. pork, and some four-fifths of all U.S. pigs are now fed on farms selling a thousand or more animals a year—cropland in some regions, and particularly in the vicinity of the largest enterprises, may become rapidly saturated with organic wastes.

Inevitably, waste generated by modern animal husbandry has become a major source of not just local but also regional environmental pollution. Volatilization of ammonia is the source of objectionable odors from large-scale operations, particularly dairy farms and piggeries; the gas also contributes to both eutrophication and acidification of terrestrial ecosystems. Leaching of nitrates, contaminating and eutrophying waters, has been given perhaps most of the attention, but accumulation of phosphorus and heavy metals—copper, zinc, and cadmium originating in fertilizers used to grow feed crops and in compounds added to animal diet—is also a serious problem.

Copper and zinc levels are usually highest, and cadmium levels are typically two orders of magnitude lower. Unlike the first two elements, cadmium is not an essential micronutrient; it is highly toxic and its accumulation in plant tissues is a clear health hazard. In addition, pesticides used to control insects in poultry houses and antibiotics used in all forms of animal husbandry can be found in manures. We know little about the fate of these chemicals, or their residues, after manure applications.

After decades of warnings about the impossibility of sustaining environmental burdens of intensive manuring, some countries have begun legislated limits on the practice. Most notably, the Netherlands enacted limits based on manure's phosphorus content and prescribed better methods of application (Archer and Marks 1997).

Wastes that are not recycled to fields could be used *in situ* as substrates for methane generation or microbial or protein synthesis, or turned into feed ingredients. The first two options, technically well proven, are economically unappealing. The third one has potentially by far the highest economic return because of the relatively high nutrient content of some wastes, particularly poultry litter and pig excreta: dry-matter crude protein content of these wastes is as high as 25–30 percent, and they have also high concentrations of many minerals (Fontenot et al. 1996).

Numerous experiments have shown that feeding of wastes does not adversely affect either the quality or taste of meat, or milk composition or flavor. Safe utilization, however, requires that wastes be free of drug residues and that they be properly processed (by heat treatment or ensiling) in order to destroy potential pathogens. Ruminants are the best animals to use high-fiber wastes of other species. Environmental benefits of this practice can be high because up to 60 percent of its dry matter can be digested by cattle.

Opportunities in Milk and Meat Production

Given the variety of livestock systems around the world, different concerns will dominate future development. The two extremes are highly intensive production systems prevalent in affluent countries and in parts of Asia, and low-intensity production in the humid and subhumid tropics (Sere and Steinfeld 1996). Intensive systems, now producing more than half of all meat by raising mainly pigs and poultry on well-balanced mixtures of prepared commercial feeds, have severed the traditional integration with field cropping: they do not engage either in feed production or in the disposal of manure on farmland. At the same time, they are vulnerable to fluctuating costs of commercial feeds and are handicapped by concentrated volumes of waste that they generate. Producing in saturated (or declining) markets, their primary goal is to use existing inputs with higher efficiency and to deal effectively with environmental effects of their operations.

In contrast, management in the humid and subhumid tropics and subtropics has to concentrate on higher production rates and on improved feeding efficiencies in order to meet the demand of growing populations. Its most important challenge is the adaptation of traditionally proven systems where animal husbandry has been integrated, through feeding of diverse phytomass and recycling of organic wastes, with field farming. Its greatest opportunities are in increasing the output of the three most efficiently produced animal foodstuffs, milk, pork, and chicken. A particular Asian concern is the transformation of water buffaloes from working animals to valuable meat and dairy species.

More Efficient Feeding

There is no shortage of means to improve animal feeding. The U.S. Office of Technology Assessment (1992) identified fourty-one potentially available techniques that can improve feed, reproductive and production efficiency in beef and dairy cattle, pigs, and poultry. Its forecasts predicted gains in feed efficiency rising at an annual rate of 0.39 percent for dairy cattle and up to 1.63 percent for pigs during the 1990s. There is no reason why somewhat lower gains should not follow in the new century. Universally applicable routes toward higher feeding efficiency include such basic improvements as better processing of both concentrate and roughage feeds and such advanced measures as the use of additives ranging from supplementary amino acids to compounds raising conversion efficiencies.

Proper processing of concentrates is universally helpful: it requires not only requisite capital investment but also considerable expenditures of energy for moisturizing, steam flaking, exploding, gelatinization, roasting, grinding, and pelleting of grains and oil meals. These investments are repaid by increased efficiency with which the concentrates are converted to animal foodstuffs. Most low-income countries could achieve substantial efficiency gains by pursuing this course.

At the same time, feeding every suitable organic waste will obviously maximize the overall efficiency of the entire food chain. There is a large assortment of these materials, and as commercial processing of crops becomes more widespread they are continuously, or seasonally, available in considerable quantities. Using these wastes requires additional management and storage, but in return the producers in the proximity of processing facilities can get good quality residual feeds (peanut skins, rice bran, distillers grains) or inexpensive roughages (citrus pulp, cottonseed hulls).

Dairy production everywhere can benefit from better harvesting, storage, and feeding of roughages. This phytomass still supplies most of feed energy in poor countries, and although concentrates now provide as much as 70 percent of all dairy feed in rich countries, forages remain indispensable in milk production. Their quality matters as they provide large shares of feed energy and minerals, and as their inherently lower digestibility has a great influence on overall productivity. In order to ensure high nutritive value, forage cultivars should be selected for high digestibility (low lignin content), harvested at appropriate time, and stored in ways maximizing nutrient recovery.

This requires close attention both to harvesting and preserving of the feed. Attention to such variables as dry matter content of the forage, stage of kernel maturity in grain feeds, and leaf-to-stem ratio in alfalfa is critical to harvest high-quality forage. Plant and microbial enzymes will take their toll on stored forages, and even good ensiling and haying practices will end up with losses of up to 20 percent of the original dry matter.

Correct moisture for ensiling, quick filling and tight packing of silos, and careful sealing of ensiled phytomass will minimize the losses by limiting the period of high microbial activity. Fairly tight seals, assuring anaerobic conditions, are relatively simple to provide, and they are particularly rewarding in smaller storages: because most of the damage due to feed oxidation takes place in the uppermost meter of silo, the resulting loss of dry matter will be relatively much larger in smaller storages. Rewards for tight seals are high: young Holsteins could gain between 100 and 200 g more a day when fed alfalfa from a tightly covered silage as opposed to feed coming from uncovered silos (Staples 1992).

Appreciable loss of dry matter (anywhere between 5 and 15 percent) is inevitable as hay drying reduces the moisture from about 80 to less than 20 percent. Solar radiation, ambient temperature, and air and soil moisture are critical drying factors beyond a producer's control. Still, a few simple measures—such as harvesting smaller batches of peak-quality forage rather than waiting for an assured period of prolonged dry weather, or mowing early in the day to maximize the available drying time—can make a big difference.

Improper hay storage may undo all the effort of good harvesting. Baling of wet (more than 20 percent moisture) forage encourages mould growth and damage to plant proteins from excess heat production (even spontaneous combustion is possible). Rewards of proper storage are high: for example, alfalfa baled at more than 20 percent moisture can lose nearly twice as much dry matter as the crop stored with moisture below that critical level (Mader et al. 1991). As with farm-produced forages, a proper storage and timely feeding of commercial feeds can prevent considerable losses, particularly with such moist matter as brewers grains (60–80 percent water content).

Enhancing the digestibility and palatability of roughages offers yet another route toward higher efficiency of feeding. Common chemical treatments of straw include hydrolysis using sodium hydroxide, ammonia or urea, or oxidation. Digestibility of crop residues can be also greatly improved by delignification (done through solid-state fermentation with white-rot fungi), treatment with purified enzymes (degrading cellulose and hemicellulose), or bacterial inoculants (Fahey et al. 1996).

Considerable efficiency gains can be realized as dairy farmers outside affluent countries move away from relying solely, or largely, on self-produced forages and raw concentrates and begin feeding premixed commercial feeds formulated to deliver accurately balanced nutrition. Consistency of these feeds is another advantage; although bovine rumen helps to even out daily fluctuations in nutrient intakes, even ruminant productivity suffers as a result of inconsistent rations. This shift toward a higher share of commercial feeds should help to narrow the gap between the rich and the poor world's typical dairy productivities. Average milk yield in rich countries (on the order of 4,500 kg a year per animal) is now more than four times the global mean and more than ten times the average for sub-Saharan Africa.

With pork accounting for two-fifths of all meat consumed worldwide, improvements in feeding represent a particularly rewarding investment in research. Taking advantage of the pig's proverbial omnivory is an important approach, above all in land-scarce countries. Pigs now consume about half of the U.S. corn harvest. They gain well on sorghum when the grain is incorporated into a mixed feed; in northern latitudes they are often fed barley (a feed inferior to corn) and potatoes, whereas in the tropics a combination of cassava and a protein supplement works as well as the corn-soybean meal (Pond et al. 1991).

Pigs find both wheat and rice brans highly palatable, and consume a variety of distillery and brewery by-products. Other good carbohydrate feeds include sweet potatoes, bananas and plantains, cane sugar, and molasses. Both animal fats and many oilseed meals make good feeds, as do blood and fish meals. Properly treated (thoroughly boiled) food wastes are a common pig feed. Naturally, young pigs reared only on garbage, a feed of low energy density with low protein content, grow slowly compared to animals fed good concentrate feeds: where garbage constitutes a large share of pig feed, market weight may not be reached in fewer than eight to ten months. The pig's omnivory presents tremendous opportunities for tapping currently underutilized or wasted feed resources, ranging from unmarketable bananas to leucaena leaf meal and from cocoyams to seaweeds (Thacker and Kirkwood 1990).

These alternative feeds are available particularly in tropical countries, many of them being high-yielding crops of lower nutrient density than grains, but produced with higher photosynthetic efficiency (sugar cane, water plants, tree crops). For example, bananas can be a substantial source of carbohydrates, for 10–50 percent of the total crop in major exporting countries are rejected for human consumption (Ravindran 1990). When combined with a protein supplement, ripe waste bananas can provide up to three-quarters of all dry matter intake in growing and finishing pigs, while dried green fruit can substitute up to half of all grain in usual diets.

Sophisticated manipulation of individual nutrients, or of their constituents, can result not only in significant conversion gains but also in substantially reduced waste. Genetic manipulation of crops or adjustments of protein quality could lower or eliminate the presence of nonessential amino acids and produce feeds with more fitting proteins that would generate less waste nitrogen. For example, every gain of 0.25 percent in feeding efficiency of pigs reduced their nitrogen excretions by 5–10 percent. Formulation of diets with synthetic amino acids in order to provide requisite levels of protein while avoiding overfeeding can reduce total nitrogen excretions in urine by as much as 25–30 percent. Lysine and methionine are already common feed additives; threonine and tryptophan will follow.

Nitrogen utilization in ruminants can be improved by using beta-agonist drugs and a growth hormone (bovine somatotropin, BST) whose intakes, singly or combined, cause appreciable increase in nitrogen retention (Sillence 1996). BST is produced by bacterial fermentation, and its implantation or injection to lactating cows increases their milk production by up to 40 percent. Higher feed and water requirements are more than compensated by higher milk output and by higher feed efficiencies.

BST also improves feed conversion in beef feedlots by about 9 percent and increases carcass lean content by the same amount. The use of BST has been highly controversial (Fallert et al. 1987; Jarvis 1996), but its future large-scale production could lead to worldwide use, as well as to its extension to sheep, goats, and water buffaloes (the hormone boosts production in hot climates). Feed efficiencies in ruminants can be also increased by addition of antibioticlike compounds produced by *Streptomycetes* fungi, which alter the metabolism of several ruminant bacteria and decrease energy loss as methane, saving about 10 percent of feed (Bent 1993).

Because about two-thirds of plant phosphorus is locked in phytic acid, a compound almost indigestible to monogastric animals, increased availability of phytase, the requisite hydrolytic enzyme, will reduce phosphorus excretion. Only wheat has plenty of the enzyme, while corn, sorghum, and most other feedstuffs are relatively phytase-deficient, and hence additions of microbial phytase, commercially available in the Netherlands since 1991, is very helpful inasmuch as it can increase phosphorus utilization by 20–30 percent (Jongbloed and Henkens 1996).

Better Management

Improved feeding efficiencies are far from being the only goal of more rational production of animal foods. Conversion of feed to lean meat is less efficient than the conversion to fat, but leaner meat is nutritionally much more desirable. The relative ease with which we have manipulated the lean/fat ratio in pigs is an excellent example of better management pursuing the goal of healthier nutrition.

When lard was a popular fat pigs were bred accordingly; as plant oils have virtually eliminated lard as a kitchen fat, and as more health-conscious consumers began demanding less fatty meat, leaner pigs came to dominate the market. The pig's genetic plasticity made it possible to achieve many desired changes fairly rapidly, over the course of twelve to twenty generations. For example, U.S. breeders reduced the average lard weight from about 14 percent of the carcass in 1960 to less than 5 percent by 1983. Total fat was reduced from 30 percent of carcass in the early 1960s to less than 15 percent by the early 1990s, with the subcutaneous layer of fat declining by 0.5 mm a year to as little as 12 mm in mature animals (Pond et al. 1991; Whittemore 1993). Choice, prime-grade, boneless cuts of pork have between 25 and 35 percent less fat than similar cuts of beef.

Pigs have other advantages for universally better management. Unlike cattle—whose production suffers in the tropics due to persistent parasites and in large parts of tropical Africa due to the presence of the tsetse fly—pigs tolerate a wide range of environments and can be reared in climates ranging from the subarctic to the equatorial. Also unlike cattle, whose best temperate breeds do not readily adjust to hot environments, pigs bred for temperate climates adapt easily in the tropics, and hence the selection of the most productive animals is merely a matter of transfer rather than lengthy breeding requiring incorporation of indigenous genes.

Other measures that can result in higher dairy production range from very simple environmental manipulations to novel genetic interventions. Extending the exposure of cows to light is a surprisingly effective way to improve lactation, growth, and reproduction. Lactating cows living with sixteen to eighteen hours of light a day produce 5–16 percent more milk than the animals exposed to twelve to thirteen hours of light. Supplemental light boosts winter production in all latitudes beyond the tropics. Fluorescent lights should be used in facilities with low ceilings, and high-pressure sodium lamps can be installed in buildings with high ceilings or in open corrals.

Advances in dairy research also promise many less controversial improvements than the use of BST. As many cow embryos are currently not carried to term, improvement in reproduction rates brought by better understanding of uterine and embryonic synchrony will be an important source of increased efficiency. Dairy production will be also enhanced by better vaccines against common pathogens and by genetically engineered protein used in treatment of diseases.

Throughout monsoonal Asia, and elsewhere in humid tropics, much greater research and breeding attention should be paid to water buffaloes (*Bos bubalis*). These docile animals are well adapted for tropical and subtropical climates. Because of the higher count of cellulose-breaking bacteria and protozoa in their guts, they also use low-grade roughages more efficiently than both *Bos taurus* and *Bos indicus* cattle (Cockrill 1974; Rao and Nagarcenkar 1977). As a result, their overall feed/gain ratios, typically ranging between 5 and 7, tend to be lower than in cattle. In addition, buffalo milk is richer in protein and fat than cow milk, but both average milk and meat yields are far behind the means for temperate-climate cattle.

Potentially significant changes requiring better management of domestic animals may come because of rising concerns about animal welfare. These concerns should not be dismissed merely as the emotional outbursts of vegetarian activists. After all, all domesticated species used for food production are social animals with well-developed group organizations, and modern farming obviously disrupts these arrangements in ways ranging from overcrowding to complete isolation (Mench and van Tienhoven 1986). Resulting stresses, as well as density-promoted disease, contribute to their discomfort.

Crowding is most obvious in poultry and hen production. Broilers and laying hens reared by groups of many thousands in tightly spaced cages can have as little as 450–500 cm^2 per bird. In contrast, free-range hens may have as much as 25 m^2 of grass per bird, or five hundred times as much space—but because of their higher metabolism they will consume up to 20 percent more feed than their caged counterparts (Appleby et al. 1992). Free-range hens will also produce about 10 percent fewer eggs, and hence their feed/egg ratio is 25–30 percent higher than for the caged birds.

Aquacultural Possibilities

Biospheric imperatives—above all the scarcity of macronutrients in surface waters—limit the annual primary production of the open ocean to just between 50 and 100 g C/m^2, resulting in fish yields of well below 100, and commonly less than 10, kg/km^2. Fish yield on continental shelves is, on the average, two orders of magnitude higher, and that of ponds is ten times higher still. Annual global catch has averaged about 100 Mt, of which about 30 percent are reduced into fish meals, oils, and other industrial products for nonfood uses.

During the late 1990s the world's average per capita supply of some 14 kg of fish, molluscs, and crustaceans contained only a few percent of all available food energy, but it supplied about one-sixth of all animal protein. The latter share is much higher regionally: aquatic species provide more than a third of animal protein to at least two hundred million people, mostly in East and Southeast Asia. Yet the natural foundations of this valuable harvest have become seriously endangered (Safina 1995; FAO 1997).

After several years of slight decline and stagnation, the total marine catch rose once again in the mid-1990s, but further substantial increases are unlikely. Indeed, a conservative assessment of the global marine potential concluded that as of the year 1996 the world ocean is being fully fished (FAO 1997). This means that a large share—about 60 percent of some two hundred major marine fish resources—has been either overexploited or is at the peak level of its sustainable harvest.

The Atlantic Ocean has been fully fished since 1980. The Pacific reached that threshold before the year 2000. Only in the Indian Ocean are there still opportunities for increased catch. That is why more optimistic assessments of global marine potential foresee a possibility of substantial increase (on the order of 15–20 Mt/year). But these forecasts rest on weak foundations, and even relatively major gains in the Indian Ocean may be more than negated by the sudden collapse of mature or senescent fisheries. As far as North Sea cod stocks are concerned, even a regimen close to the maximum sustainable yield may be prone to risk (Cook et al. 1997).

The only prudent course is then to assume that long-term marine catches should not be boosted above the recent rate of 80–85 Mt a year. Even if this level of fishing could be maintained during the first half of the next century, population growth would cut per capita fish supply by up to 50 percent (Population Action International 1995). This brings an obvious question: can expanded aquaculture fill most, if not all, of the rising demand for finfish, molluscs, and crustaceans?

Potential Gains

Unlike with our thousands years of experience with land animals, relatively large-scale and widespread breeding of aquatic species, and particularly of ocean finfish, is a very recent phenomenon outside the areas of traditional Asian aquaculture. This means that the recent trends of rapid expansion characterizing aquacultural developments in the United States and Canada, in both the Atlantic and Mediterranean Europe, and in parts of Latin America are poor predictors of future developments.

These trends may subside in the near future, or they may persist, with inevitable downturns, to make aquaculture a major contributor to global nutrition of the twenty-first century. FAO's projections are optimistic: they see aquaculture supplying as much as 20 Mt of fish and crustaceans a year by the year 2010 after expanding by about half compared to the early 1990s. The first of the three basic modes of aquacultural production—extensive practices relying entirely on feed biomass produced naturally within the pond or other confinement—has inevitably low productivity and will make little difference to global fish supply. Yields in extensive aquaculture arc as low as 100–300 kg/ha for carp raised without any fertilization, converting to no more than about 50 kg of protein per hectare.

Semi-intensive practices augment natural production of feed biomass by applying organic wastes or synthetic fertilizers and by supplementation with a variety of feeds, ranging from live invertebrates to farm-made feed mixtures and from aquatic weeds to commercially produced and nutritionally well-balanced pellets. Semi-intensive production in freshwater ponds is the world's dominant type of aquaculture.

Given China's long tradition of such practices and its continuing primacy in freshwater finfish production, it is not surprising that herbivorous and omnivorous carp now accounts for about three-quarters of the total finfish output, and that silver, common, grass, and bighead carp are by far the most abundant cultured species. Depending on the intensity of fertilization and supplementary feeding, these semi-intensive polycultures yield anywhere between 2 and 4 t/ha (up to about 700 kg/ha of protein) and the most productive ponds may yield over 5 t/ha (figure 5.8).

Intensive aquaculture produces finfish and crustaceans at high stocking densities by feeding the animals solely with commercially available or farm-prepared feeds (and in some places with locally caught low-grade fish). Carnivorous finfish (salmon, trout, yellowtail, seabream) and crustaceans (shrimps, prawns) reared in intensive monocultures within tanks, floating cages or ponds can yield over 100 t/ha. Traditional compounding of sea water containing wild shrimp spawn, and the harvesting of mature crustaceans five to six months later followed by return of the land to fields or pasture, yielded between 100 and 500 kg/ha. In contrast, two to three shrimp crops grown in ponds yield 1,000–10,000 kg of shrimp per hectare—but only by using commercial feeds.

The only case of highly intensive aquaculture not totally dependent on commercial feeds is the dike-pond culture that has evolved in parts of South China, most notably in the Zhujiang (Pearl River) Delta of

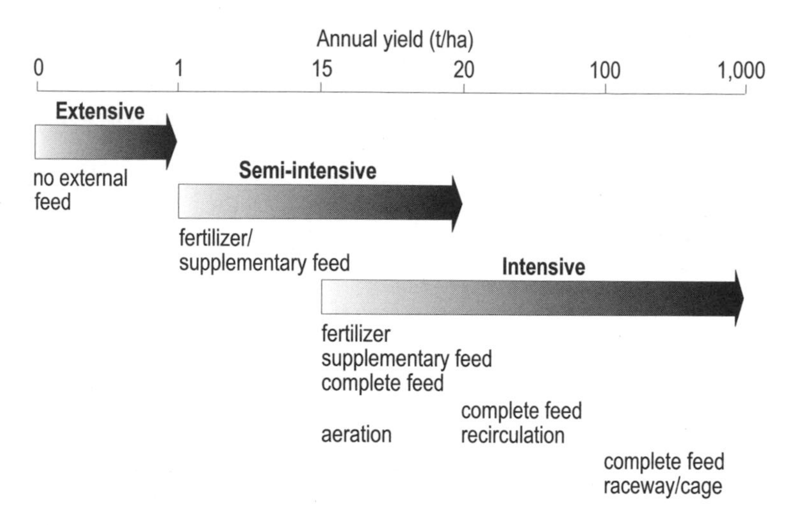

Figure 5.8
Yields and feeding options of different types of aquaculture; based on Tacon (1996).

Guangdong province, where its greatest extent reached about 800 km^2 (Ruddle and Zhong 1988; Korn 1996). This arrangement, integrating agriculture and aquaculture, is more productive than any other traditional agroecosystem. Individual units are small, just 0.2–0.5 ha; on land they comprise domestic animals (cows, pigs), intensively cultivated vegetables, sugar cane, and mulberries; in ponds they always include ducks and fish polyculture. Dike-pond units can be managed as purely organic systems, using animal wastes to fertilize crops and ponds, or they can supplement biomass recycling with externally produced feeds.

Typical annual yields are 10–15 t/ha of live fish, about 80 t/ha of vegetables or sugar cane, and 20–30 t/ha of fruits; typical annual harvests are thus between 30 and 40 t/ha of fish and crop biomass. While the management of the unit requires labor inputs equivalent to three full-time jobs, per hectare annual harvest can feed more than fifty people a year; in contrast, intensive rice farming in the same region can support no more than eleven people per hectare.

Stocking with finfish species with different feeding niches is a major reason for high aquatic productivity. Silver carp (*Hypophthalmichthys molitrix*) lives near the surface, feeding on phytoplankton. Omnivorous plankton-filtering bighead carp (*Aristichthys nobilis*) occupies middle water layers. Grass carp (*Ctenopharyngodon idella*) is a herbivore preferring aquatic macrophytes and feeding readily on organic wastes; the low energy content of its coarse feed makes for poor feed/meat conversion efficiency, but its feces are used, with other settled debris, by bottom dwelling detritivore and omnivore species, most often by common carp (*Cyprinus carpio*). Carp polycultures are also very efficient in using water, fit well into irrigated agroecosystems, and provide many opportunities for year-round employment.

Aquaculture's importance has grown steadily since the early 1980s to claim an increasingly larger share of the global harvest (FAO 1995a; FAO 1997). Cultured species now supply nearly 20 percent of all fishery production, with the shares of nearly two-thirds for inland production, and about 7 percent for marine output. Ocean aquaculture makes the greatest contribution in the production of molluscs—by the mid-1990s four-fifths of the world's mollusc harvest was cultured, with China, Japan, and South Korea being the largest producers—and crustaceans (with about one-fifth of all shrimp and prawns, mostly from China, Thailand, and Indonesia). And although aquaculture's contribution is still tiny as far as most ocean species are concerned, its share has now risen to over a third of the global supply for salmonids.

This recent expansion has been remarkable both because of its rate and its spatial pattern (FAO 1995a). Between 1980 and 1995 harvests of finfish, molluscs, and crustaceans nearly tripled. Since the late 1980s cultured finfish production has yielded more meat than mutton and lamb and is now equivalent to nearly a third of the world's chicken meat output (the two foodstuffs are comparable in terms of their total energy content and protein share).

Thanks to China's extraordinarily rapid growth of pond farming (from less than one Mt in 1980 to 9.4 Mt by 1995), aquacultural products are among the few foodstuffs whose production comes largely from low-income countries: in the mid-1990s about four-fifths of all finfish, and almost three-fifths of all molluscs and crustaceans, came from such countries, while affluent nations (led by the United States, France, Spain, Italy, and Norway) harvested only about one-seventh of the total mass (leaving farmed aquatic plants aside: three-quarters of them were also produced by low-income countries).

About three-fifths of cultured finfish come from inland regions, with herbivorous and omnivorous carp species accounting for more than four-fifths of all output, with carnivorous catfishes being a distant second. Carnivorous salmonids and yellowtails dominate among cultured ocean fishes. Thanks largely to rapid expansion of shrimp production, coastal aquaculture has been expanding more rapidly than inland ponds, and it now accounts for more than a quarter of total product value.

Aquaculture has several obvious nutritional and socioeconomic advantages, particularly in populous, low-income countries where—besides enriching diets with locally produced, affordable and tasty meat and high-quality protein—it can employ a relatively large number of people and be a good source of cash income. Some fish cultures can be also integrated with staple grain production: rearing fish in rice fields is a highly desirable form of agroecosystemic management that produces affordable animal protein for local consumption, provides additional labor opportunities, and boosts income of families (Choudhury 1995). Even a modest diffusion (no more than 5 percent of all paddies) and low yields (just 300 kg/ha) would produce close to 3 Mt of fish annually.

Production of nontraditional species could make a significant long-term difference. Omnivorous, mild-tasting, white-fleshed tilapia has a particularly great potential; it can be grown in warm climates in both low-intensity settings of a poor country's ponds and by intensive means in cages. Nor are the environmental benefits absent: extensive aquaculture can prevent eutrophication of waters by withdrawing suspended nutrients, and it may benefit human health by reducing certain disease vectors. But high productivities, high feeding efficiencies, and some environmental advantages are not enough to assure aquaculture's long-term future. Its practices will have to rest on a demonstrably sustainable basis, and its overall impact will have to be environmentally benign.

Problems with Aquaculture

Today's aquaculture has no greater species variety than large-scale field agriculture. About 10 percent of some 25,000 species of finfish are harvested by subsistence and commercial fisheries for food, but only some hundred species of fish—with 70 percent coming from fresh and brackish waters—are reared in aquacultures (Williams 1996). Moreover, only a handful of carp species account for nearly four-fifths of all freshwater finfish production, while salmonids provide nearly half of all carnivorous marine harvest (FAO 1995a). Narrow genetic basis is a challenge to aquaculture: the practice is dominated by a handful of freshwater herbivores and the contribution from marine species rests on even fewer carnivores. Transgenic fishes may offer many advantages by growing faster and by tolerating cooler or warmer waters (salmon with an antifreeze protein gene from the winter flounder is already available).

Availability of suitable freshwater environments, or of options for converting farmland to ponds, will be the most common constraint on expansion on land. Concerns about effects on wild aquatic species and on natural ecosystems will be a major constraint on coastal marine cultures. Environmental impacts of predatory aquaculture have become particularly common in Southeast Asia's coastal regions, and they led to a sweeping ban on virtually all commercial shrimp farming in five coastal states in India. Pollution from shrimp ponds has contaminated drinking and irrigation water, seeped into aquifers, and affected coastal fisheries (Masood 1997). Asia's shrimp aquaculture has been also responsible for destruction of mangroves and wetlands and, after the ponds are abandoned (sometimes in just five years), for creation of infertile land (Gujja and Finger-Stich 1996).

Both freshwater and marine fish farms generate pollution from the uneaten feed and fish feces made up of both suspended and dissolved organic solids, including nitrogenous and phosphorous compounds. The impact of these wastes can be particularly damaging in locations with limited water exchange where the decomposition of organic matter consumes dissolved oxygen, and nutrients contained in wastes contribute to local eutrophication, stimulate growth of bacteria and fungi, and often shift planktonic balance in favor of a few, and possibly undesirable, species. Hypoxic waters plagued by algal blooms will also affect the cultured fish, while excessive plankton can clog fish gills and its decay can produce toxins.

Deposition of uneaten feed and feces can suffocate benthic organisms and alter the chemistry of bottom sediments: The resulting anaerobic fermentation generates hydrogen sulfide, whose emanations can impair fish health and cause sudden mass mortality. The esthetic impact of manmade structures, particularly in previously untouched remote coastal regions, cannot be underestimated. Escape of the cultured species may become a more worrisome problem as aquaculture expands, especially if the transgenic organisms would be involved.

Dense stocking of most cultured species obviously promotes diffusion of pathogens; this may be aggravated by stress-induced immunosuppression. The difference between wild and farmed species is most obvious in the case of salmon: wild salmon feeds unrestrained in huge volumes of cool and clear deep-ocean waters, aquafarmed fish is confined in nearshore cages whose coastal water has higher concentrations of plankton, parasites, and pathogens. Expectedly, epizootic diseases exact considerable toll. Estimates from China's Jiangsu province indicate that losses to diseases amount to at least 20 percent of total annual finfish production, to as much as 40 percent of shrimp, and 50 percent of mollusc output (FAO 1995a). During the early 1990s epizootics caused a collapse of shrimp farming in China, a more than 70 percent reduction of total output between 1991 (when the country was the world's largest producer) and 1994.

Chemicals used to prevent and limit these diseases naturally contribute to selection for more resistant strains of bacteria and may contaminate the edible product as well as sediments and other aquatic species. Controls of serious epizootics by antibiotics or other chemicals may also pollute confined waters; leave residues in the fish, in wild organisms, and in other farm products; and help to develop drug-resistant pathogens. As a result, consumer acceptance of aquacultural products may decline. Some of these concerns are relatively easy to address; others will persist. Feeding waste can be much reduced by using highly digestible compounds dispensed at rates maximizing intake. Waste recycling is practical in intensive tank cultures as is the effective waste removal (sedimentation, filtering) in flow-through arrangements—but difficult or very costly in larger ponds and lagoons and from beneath submersed cages.

Securing appropriate feed for carnivorous species will be an increasing challenge. While in some carnivorous fishes (catfish and eels) protein digestibility differs little among high-quality feeds of animal and plant origin, other species, including rainbow trout and salmon, prefer feeds of animal origin to meals derived from soybean or cottonseed (Steffens 1989). Moreover, fish meal and fish oil are the only available source of highly unsaturated fatty acids that are both essential nutrients for all carnivorous finfish and a key reason for health appeal of these foods. These high-protein and high-fat feeds are often derived from marine products: fish meals and fish oils usually make up about 70 percent of feeds for carnivorous finfish and (supplemented by shrimp and squid meals) about half for crustaceans (Chamberlain 1993).

All carnivorous finfish and shrimp species reared in an semi-intensive and intensive manner are thus net protein consumers rather than producers: depending on the arrangements and particular feeds, their needs for fish proteins exceed their protein output two to five times (Tacon 1995). And the culturing of carnivorous species contributes to depletion of marine stocks: the rapid expansion of Thai shrimp farming was made possible by inexpensive fish meal supplied by the trawl fishery in the Gulf of Thailand.

Aquaculture is still an insignificant consumer of commercial animal feeds, but its use of fish-derived feeds is already substantial, accounting for some 15 percent of all fish meals and fish oils in 1995 (Tacon 1994). Most of these fish-derived feeds are still used in chicken and pig production, but the rising output of carnivorous finfish (now accounting for about one-eighth of all farmed fish) would increase the share of high-quality feeds. Although it is unlikely that shortages of fish meal and fish oil will restrict aquaculture in short or even medium term, the use of fish biomass in fish production obviously represents a long-term limit on the output of carnivorous species. Rising feed prices may eventually limit the use of fish-derived feeds and crustaceans, and mesopelagic fish may be used for feed.

Affluent countries can afford carnivorous aquaculture. Japan's output is dominated by carnivorous yellowtail, seabream, and eel. In contrast, in China, the world's leading aquacultural power, about 98 percent of

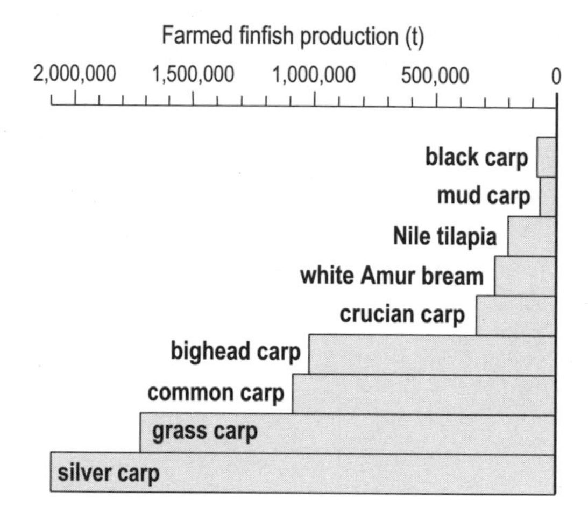

Figure 5.9
China's aquacultural finfish production of the mid-1990s was dominated by herbivorous species; the most important fish, silver carp, contributed about 2 Mt/year. Based on Tacon (1996).

cultured species are herbivores and omnivores (figure 5.9). If aquaculture is to make substantial long-term nutritional difference on the global level, then energetic imperatives will favor the production of herbivorous and omnivorous species.

6

Consuming the Harvests

If this book had followed the standard pattern of assessing agricultural prospects, then the closing chapter would occupy this space. There is a curious, and deep, divide between the inquiries into demands, improvements, limits, and modes of agricultural production and the understanding of food chains beyond the farm gate. In accord with increasing specialization, studies of individual crops, particular cropping practices, and specific problems in animal husbandry abound. Although much less common, integrative works looking at whole farming systems are also available.

But the works on food supply futures have one thing in common: they almost never go beyond the production stage. Studies detailing a specific farming practice as well as those tracing the food chain between the Sun and marketable grain in a bin or a fattened pig in a sty belong to a very different intellectual and research tradition than those trying to comprehend what happens to those crops and animals once they leave the farm. What share of staple grain or a fattened animal does end up as food? How much of that retailed food do we actually eat? And, most fundamentally, how much nutrition, and of what kind, do we really need?

In this chapter I will look at the first three aspects of this separate universe. I will deal first with a very important, and inexplicably under-researched, matter of postharvest food losses; then I will look at the limits of standard methods used to account for food supply; and, finally, I will turn to our surprisingly uncertain understanding of actual food intakes.

Harvests and Postharvest Losses

Summing up the harvests of principal crops and the supplies of meats, fish and dairy foods appears to be a fairly straightforward task, if not a boring chore devoid of intellectual challenge. In reality, neither is the case: assembling the record is difficult and tricky. The work is done regularly by national agricultural and statistical institutions, which often publish specialized statistical yearbooks. On the global level, it is one of the principal duties of the FAO. FAO's *Production Yearbooks*, and the organization's electronic database, list the areas planted to some two dozen crops, total harvests, and average yields, as well as annual output of all major kinds of meat, eggs, and milk.

These figures are widely quoted and reprinted, and they are used in computer models of future supply and demand. Perhaps most importantly, they are the inputs for the preparation of FAO's food balance sheets, which in turn are used for assessing the extent of national and global malnutrition. The informational, analytical, and policymaking impact of these figures is thus far-reaching. But although this information is neatly arrayed in statistical yearbooks and accessible on the Web, its accuracy is a different matter.

Even an uninitiated observer might wonder how we came to know the area planted to millet in the midst of Somalia's murderous tribal wars, or who had been monitoring peanut harvests in Sudan or counted cattle herds in Angola during a generation of armed conflicts. A more astute observer might question the reliability of data in most poor countries where a high level of traditional subsistence farming, whose products rarely reach any markets, makes it much more difficult to come up with reliable nationwide totals of agricultural production.

Uncertainties do not end here. I have been always puzzled by an incredible lack of attention paid to food losses during and after harvests. This neglect is particularly astonishing given the stakes involved. On one hand, there is the continuing quest for higher yields so that the annual growth of crop harvests and livestock productivities would at least match the population growth rates of low-income countries. On the other hand, in most of those countries harvest and postharvest losses of cereals commonly surpass 10 percent. Losses for tubers, fruits, and vegetables may be higher than 25 percent, and spoilage can reduce fish catches by an even higher share (James 1986; FAO 1989, 1997; Hanley 1991).

Common reasoning concerning the world's food prospects sees larger outputs as practically the only way of satisfying higher demand. It pays hardly any attention to the postharvest fate of foodstuffs and their consumption. This neglect makes no sense as there are enormous opportunities for increased, and improved, food supply that could result from reducing postharvest food losses and from rationalizing human nutrition. Once again, energy analogy is very apposite: we should look for more food not only in fields, pastures, barns, and ponds but also in storage bins and sheds, in warehouses and supermarkets, in food service and in household pantries and refrigerators. And, no less importantly, in our eating habits—and in scientific understanding of our nutritional needs.

Appraising the Harvests

Gathering reliable food production statistics is not easy even in affluent countries. Automated monitoring and accounting setups that either check physically every outgoing item or use reliable sampling procedures are standard in modern industrial production. In contrast, for crops planted on large, and fluctuating, areas and producing highly variable yields it is very costly to do direct physical measurements even on minimum samples to assure reasonable accuracy of a region's or a country's harvest totals. Rich countries can afford some of these expenses. For example, every June some 1,100 USDA enumerators survey crop areas to gather direct data from nearly 17,000 sampling units encompassing about 0.6 percent of total land area. French have determined crop yields since 1956 by using aerial photographs to select an equal-probability sample of 7,200 points per *département* for direct field survey.

High-resolution satellite monitoring can provide accurate information on the extent of particular crops and fairly reliable knowledge of standing phytomass and anticipated yield, but its costs are beyond the means of most countries, and it obviously cannot capture often considerable harvest losses. Most of the time even the affluent societies prefer to collect their crop statistics by less expensive means, relying on interviews, and mail surveys. Even in most European countries only about one-tenth of all agricultural data is based on direct physical measurements, and more than half is estimated in offices from secondary sources.

Naturally, in poor countries the indirect assessments—at best the estimates of supposedly knowledgeable observers in the field, often just second- or third-hand guesses—are dominant. At least two-thirds of all countries appear to base their crop data on such estimates, and the shares are similar for published statistics on livestock numbers and meat and dairy production. Even direct physical measurements of yields are not error-free. Substantial overestimates (10–40 percent) have been well documented owing to the choice of plots that are too small but, naturally, cheaper to sample. If the whole process of sample cutting, threshing, and drying is not done in ways closely comparable to the prevailing practices the measurements result in further overestimates.

There are many other opportunities for statistical lapses. Uncertainties concerning the actually cropped areas (see chapter 2) matter a great deal when estimates of typical harvests are multiplied by totals of arable land. Lesser discrepancies arise from accounting for land in ditches, hedgerows, and shelter belts, from exclusion of small, patchy plantings or, conversely, inclusion of output from unreported fields. Errors in conversions from traditional volume or weight measures to metric units are also common. Other errors arise from reporting the output of continuous crops (some vegetables and fruits) and of incomplete harvests (such as cassava, the leading tropical tuber, which is often harvested only according to immediate need). Understatements of crop harvests are common in countries whose statistical services are still under development: Poleman (1996) cites examples that in early periods of harvest estimates errors can be 30–40 percent below the real value.

Consequently, it is not at all surprising that estimates of standing crop phytomass may easily differ by 5–10 percent from actual field totals, and that for some crops in some countries these errors may be sometimes as large as 20–25 percent. Similarly, head counts of domestic animals in countries that never had any reliable agricultural census (that is, in at least two-thirds of the world's nations) are far from accurate.

Field, Storage, and Processing Losses

We have large-scale national and international institutions devoted to plant and animal breeding for higher yields—complex and costly activities that are usually able to add on the order of 1 percent of additional production per year—but we do not have even a single organization devoted to the worldwide problem of food losses. The academic record is no better: search for "post-harvest losses" in *Agricola*, the largest agricultural electronic data base, uncovers fewer than twenty items published during the 1990s, when several thousand articles were published on wheat, rice, and corn yields.

By far the best explanation of this extreme dichotomy is provided by an already noted analogy with the contrast between household energy conservation and increases of centralized energy production. Reduction of postharvest losses throughout a country's food system requires, much like the conservation of electricity or natural gas in millions of homes, myriads of individual decisions and repeated actions. Even those individual changes that are relatively quite impressive—be it harvesting a cereal crop at the stage of kernel maturity that minimizes combining losses or substituting a fluorescent light for an incandescent bulb—are insignificant in absolute aggregate terms, and they make a substantial difference only because of their eventual combined impact.

Although there is no shortage of figures on the extent of harvesting and postharvest losses, especially as far as food grains and vegetables are concerned, the overwhelming majority of these values are just approximations and guesstimates. Sometimes they are based on detailed understanding of local or regional situation, but often they are just recycled figures of dated, and unattributable, origins. A handful of more reliable recent studies confirms that typical losses continue to be unacceptably high. A recent five-year survey of grain losses in leading cereal-producing provinces found that about 15 percent of grain crops are lost annually during harvesting, threshing, drying, storage, transport, and processing (Liang et al. 1993). For rice, China's detailed multiprovincial survey came up with losses of 7 percent in harvesting, 2.5 percent in threshing and drying, 2 percent in transport, and 5–11 percent in storage (four-fifths of all grain is stored by peasant households).

In the absence of representative surveys, any generalizations concerning nationwide food crop losses must be suspect as so many variables determine the outcome (FAO 1980b, 1995b). Depending on the method of harvesting (ranging from traditional cutting of individual rice stalks by knives or sickles to multirow combining of corn), variety of crops grown (from ancient local cultivars to modern short-stalked hybrids), timing of the harvest and ripeness of the grain, the losses may range from a fraction of 1 percent to a major share of the standing crop.

Two examples from carefully conducted harvesting studies illustrate wide ranges even in modern settings. Clarke (1989) found that losses in direct combining Canadian barley ranged over an order of magnitude (0.07–2.81 percent) when the harvesting was done at optimum time, but that delaying the harvest for two weeks following combine ripeness increased the typical loss to 5–6 percent. Similarly, differences in planting density, pod weight, plant height, degree of lodging, and crop cutting heights resulted in harvest losses ranging from just 0.4 to 16.5 percent of yield for three cultivars of Kentucky soybeans (Grabau and Pfeiffer 1990). And typical losses for two successive harvests can differ by an order of magnitude.

In every subsequent stage—handling, threshing, drying, storage, and milling—the losses may be as low as one and as high as 10 percent for the cumulative totals ranging from well below 10 percent to as much as 40 percent (and even higher figures were reported for some African crops). Moreover, year-to-year fluctuations may be quite considerable, but constant waste rates are commonly used in compilations of national food balance sheets. Similarly, the fractions of food grains, tubers, and sugar crops used in feeding are often highly variable, responding to availability, marketing regulations, and price levels but fairly accurate totals are available only for rich countries.

Because of their enormous spatial and temporal variability, any average figures for losses of fish harvests are seen as meaningless by some experts (James 1986). The matter is further complicated by the fact that appreciable amounts of spoiled fish are still eaten. Counterintuitively, smoking the fish may not reduce the loss: because of insect infestation losses of outdoor-smoked fish are commonly higher than those for fresh unfrozen catch, with rates raging up to more than 50 percent in some locales, and with typical shares around 25 percent.

Postharvest storage of cereals, pulses, tubers, fruits, and vegetables also causes various chemical changes, some of them affecting the quality of available nutrients. Although the total protein content of stored grains remains unchanged, availability of individual amino acids, especially of lysine, may decline quite substantially. Wheat may lose up to 8 percent of available lysine after just six weeks of typical storage (at 13 percent of moisture and 20°C) and over 40 percent after two years (FAO 1984). Prolonged storage can reduce available lysine by 40 percent in soybeans and by nearly 20 percent in rice. There may be also large losses of thiamine, carotene, and tocopherols and a smaller loss of lipids. Vegetables grown with too much nitrogen fertilizer will mature later and reach larger sizes but will be more susceptible to postharvest injury.

Qualitative changes are also common as foods are increasingly undergoing various forms of processing. All of the common ways of commercial food preparation were originally developed without any attention to chemical consequences of these processes but extensive research is now available to demonstrate a variety of undesirable changes (Richardson and Finley 1985; Davidek et al. 1990). Common changes of color, flavor, and structure may not cause any degradation of nutritional value, but decreased appeal and palatability of affected foods will increase the likelihood of waste. Sterilization gelatinizes starches, alters natural food colors, and brings out undesirable food flavors. Warmed-over flavor of precooked refrigerated meats (particularly in turkey, beef, and pork) is caused by oxidation of intramuscular lipids and it can develop in a very short time (Reineccius 1989). Breads and pizza crusts become stale, and they also undergo textural changes.

Vitamins are perhaps the most vulnerable group of nutrients damaged during food processing. Vitamins A, C, D, and E are prone to inactivation by oxidation; C and B-group of vitamins are degraded during heating as well as during freezing. Processing can also denature and degrade proteins, while lipids are subject to hydrogenation of the double bond producing nutritionally less desirable trans-fatty acids. Chill injuries degrade every kind of produce, with critical minimum temperatures ranging from 13°C for bananas to 3°C for apples. Enzymatic browning is common in many fresh fruits and vegetables, as is lignification, an enzymatically catalyzed process toughening plant tissues. Clearly, the common practice of multiplying food quantities by standard nutritional equivalents while preparing food balance sheets may seriously misrepresent the quality, and sometimes even the total amount, of nutrients actually available at the consumer level.

How Much Food Do We Have?

This is a surprisingly difficult question to answer. Uncertainties concerning the actual harvests and a variety of postharvest losses are just a part of the problem. There is little choice but to use the variety of slightly to substantially inaccurate crop production data and animal head counts as the first entry in the construction of comprehensive accounts of food availability.

But many more estimates and assumptions must be made in order to construct national food balance sheets, the summaries offering a comprehensive review of a country's food supply in a particular time period. In countries with assured food supply a calendar year is the standard choice—but in order to account for delayed use of produced food and to lower the effect of interannual fluctuations in less fortunate countries, an average of two or three years may be preferable. Fewer than fifty countries publish regularly their own FBS, but FAO prepares them regularly for all of its member states. Useful as these food balance sheets are, in the case of many low-income countries they may leave appreciable amounts of available food unaccounted for, and in all but a few cases average per capita rates resulting from these exercises cannot be equated with actual food intakes: gaps between average availability and typical consumption are often surprisingly large.

Food Balance Sheets

Procedures for the preparation of FBS are straightforward. They start with the preparation of supply/utilization accounts for nearly one hundred different crop, livestock, and aquatic commodities (FAO 1995c). Domestic production of food crops is adjusted for changes

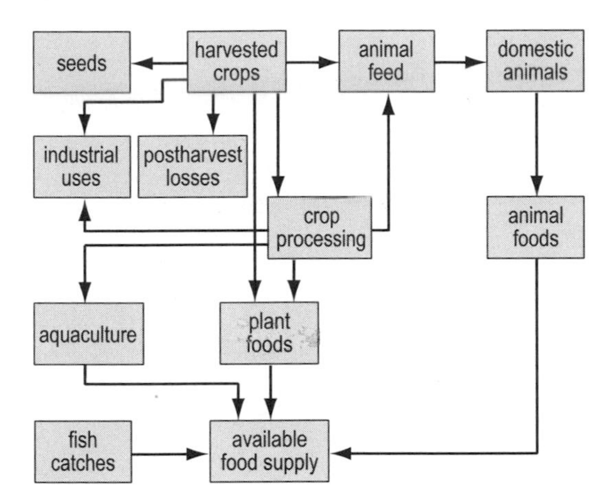

Figure 6.1
Steps in the preparation of food balance sheets.

in stocks and for the net external trade; amounts used for seed, in industrial conversions (above all for alcoholic beverages and starch), and fed to livestock and wasted during storage, processing, and transportation are subtracted from these totals to determine quantities available for use as food. Appropriate extraction rates for milling of cereals or pressing of oil seeds are then applied to derive the actual supply of individual plant foodstuffs (figure 6.1).

Livestock accounts begin with gathering data on head counts of animals and their reproduction rates, natural losses, trade, and slaughters. Meat production is expressed in terms of carcass weight, excluding slaughtering fats; edible offal is given as a separate category. Milk and egg output is either taken from available statistics or derived by multiplying the numbers of producing animals by estimates of specific productivities. Aquatic catches are adjusted for nonfood uses (fish meals and fish oils for animal and fish feeds) and processing and storage losses. Totals for individual foodstuffs are then converted to average daily per capita consumption rates and expressed both as overall food energy supply, and in terms of carbohydrate, protein, lipid, and major micronutrient availabilities. These averages of daily per capita food supply given in FAO's FBS are undoubtedly the most frequently cited and reprinted values concerning human nutrition.

Inevitably, this quantification of food supply requires many approximations and is open to numerous errors. For each crop in supply/utilization accounts FAO statisticians must enter information for the seven principal supply categories and for nine utilization elements; eight entries are used for livestock products, fifteen elements for fishery commodities. Not surprisingly, this quantity and detail of rates, shares, or multipliers are not available as reliably established pieces of primary information for most of the organization's member states. Even in rich countries with comprehensive statistical systems the requisite information may not conform to balance sheet concepts, and it may not be consistent with respect to measurement units and reference periods.

Only about one-third of requisite figures are supplied directly (and, naturally, with varying reliability) by the member states. The rest must be estimated in the organization's Rome headquarters from information originating from a variety of sources, including irregular surveys by government or private agencies, reports and experiences of extension agents, records of marketing authorities, and expertise of industrial users. Inevitably, direct incorporation of this diverse information into the FBS framework is rarely possible, and adjustments are necessary to take care of incompatible concepts, inconsistencies, and missing data. Needless to say, all these problems are more extensive and more acute in the least developed countries where even the basic output statistics for a few major commodities may not be available.

Cumulative opportunities for significant errors are obvious. As we have already seen, adjusting for postharvest losses is largely a matter of best guesses. Uncertainties regarding the choice of typical food processing multipliers may not be as large as those characterizing the assumptions about losses, but they, too, can add up to large discrepancies. For example, rice milling removes between 25 and 35 percent of the grain. Using the nationwide average milling rate of 75 rather than 67 percent translates into over 15 Mt of additional food for China's 1990 rice harvest, roughly the equivalent of total annual Japanese rice production! Qualitative differences are no less important: brown and highly milled rice do not differ much in their total energy content, but the milled grain may have only one-third of the former's niacin content. Similarly, energy contents of wheat flours differ by less than 10 percent, but their thiamine contents may differ more than fifteen-fold!

Seeding rates may be fairly uniform in the Corn Belt fields or in French wheat fields planted with a few mass-marketed varieties, but they are highly variable throughout most of the poor world, especially when expressed, as is common in construction of food balance sheets, as a fraction of total crop harvest. Depending on the cultivars and agronomic practices, seed requirements range from 2 to 7 percent of harvest for rice, 6 to 12 percent for wheat, and 8 to 20 percent for peanuts. Quantifying production of animal foodstuffs is not easy even when the available head counts of domestic animals are fairly reliable. Again, the cumulative effect of assuming a slightly higher takeoff rate and slightly heavier carcass with a slightly higher share of lipids can translate into a significant difference in total food energy supply.

Uncertainties do not end with calculating the availability totals. The final step in the preparation of food balance sheets is the conversion of all mass values into their energy equivalents and calculation of contributions by the three macronutrients. The challenge is even greater for micronutrients. Indeed, quantification of all important nutrients (about forty) in all of the available foods (currently about four thousand items in the United States, and on the order of ten thousand worldwide) would be an extremely expensive task: with just ten replications to obtain a representative set for each item, four million analyses would be required.

At the same time, preparation of major nutrient totals in nationwide food balance sheets does not rest on exhaustive and precise understanding of food compositions. By far the most common problem is the selection of a representative average from often wide ranges of well-established values. In any case, the quest for highly accurate supply totals in affluent countries is irrelevant from a nutritional point of view, inasmuch as all these nations have surfeit of food. Average per capita food energy and protein means have been very high for more than a century in North America, and they rose to very similar levels in the richest countries of Western Europe by the early 1960s. Now even the poorer Mediterranean countries are not far behind the continent's top rates of over 3,600 kcal/day, and Greece is actually right there with

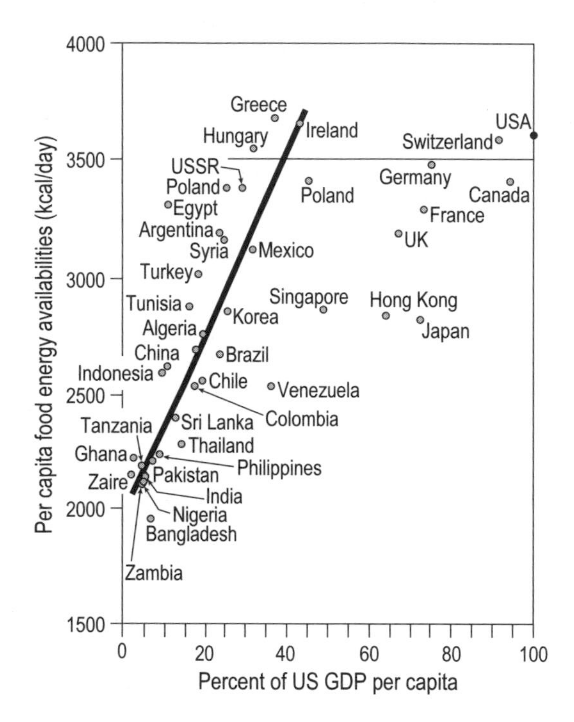

Figure 6.2
Average per capita food availability in relation to national GDP measured in terms of the purchasing power parity (Poleman and Thomas 1995).

Denmark and Belgium. The lowest rates, besides Portugal's 3,100 kcal/day, are in the United Kingdom, the Netherlands, Spain, and France, all at about 3,300 kcal/day (figure 6.2).

Supply data derived from food balance sheets are not in any way direct measures of food availability but rather outcomes of complex constructs all too frequently resting on dubious foundations. All they do is provide an equivalent of an order of magnitude estimate, an approximate picture of the overall food situation in a country. Daily per capita averages require particularly careful interpretation: they are just approximate indicators, not reflections of food actually consumed by individuals. Actual intakes vary a great deal among different groups of a given population and between individuals.

Degrees of uncertainty are perhaps best illustrated by comparing FAO's food balance sheets with itemized national accounts of food consumption available for most of the affluent nations in their own statistical summaries. Even in these cases—with good sampling of harvests, constant monitoring of crop uses, and regular reporting of relevant statistics to international organizations—food balance sheets starting with identical inputs may end up with substantial discrepancies caused by differences in assumptions and calculating procedures.

For individual foodstuffs these differences can be as large as 30–35 percent in either direction. For example, the French *Annuaire Statistique* puts per capita beef and veal consumption at about 22 kg in the early 1990s, whereas FAO's FBS have it at just over 30 kg (INSEF 1995); in contrast, they put Czech fruit supply at 50 kg, whereas the Czech statistics peg it at nearly 73 kg (Czech Statistical Office 1995). A very large gap between the red meat intakes for the United States—FAO's per capita mean is about 74 kg, USDA's average is only 52 kg—is the difference between bone-in and boneless meat.

Rates of the daily per capita food energy supply and of protein and fat availabilities may in reality be 5–10 percent different (in either direction), and the actual average supply of minerals and vitamins may be up to 25 percent higher or lower. Which data sets are the more reliable ones? Without detailed line-by-line review of all assumptions and without extensive knowledge of particular national conditions there is no easy way to answer the question.

Beyond the Standard Accounts

Disparate as they may be in some of their parts, food balance sheets of rich countries at least do not omit any major inputs. Their questionable averages of various nutrients are the inevitable result of the exercises' complexity rather than the consequence of neglecting the whole classes of inputs. The situation is different in many poor nations where the problem with commonly unreliable input totals and utilization assumptions is aggravated by the complete omission of some qualitatively essential food supplies.

Wildlife is perhaps the most important component of all the inputs that never make it into a country's FBS because, even when consumed in relatively small amounts, it may be an important source of high-quality protein. Numerous local and regional assessments attest to the ubiquity of this intake and to its truly omnivorous nature. Besides the wild meat cherished in temperate regions—mainly large ungulates and larger birds—subtropical and tropical diets have always contained nearly all locally available edible species. These lists can start with ants, maggots, and snails and proceed with lizards, small birds, and rats to hedgehogs and pythons. The largest part of these wild animal foods is secured by opportunistic collecting and hunting and hence cannot be reliably quantified on a regional or national scale—but it may provide, especially seasonally, a large share of local diet.

For example, Price (1997) found that during the rainy season wild foods, gathered mainly from paddy fields, made up about half of the total food consumed by farmer's in Thailand's central northeast. A pioneering study of urban wildlife consumption found that at least 160 t of meat were for sale in a single market in Accra, Ghana, during a period of eighteen months (Assibey 1974). Shortly afterwards a review by de Vos (1977) concluded that in many areas of Africa and Latin America wild meat can provide at least 20 percent of all animal protein. Two decades later Njiforti's (1996) study in north Cameroon resulted in a very similar conclusion: it estimated that respondents consumed annually almost 6 kg of bushmeat, representing 24 percent of the animal protein intake in the region (North African porcupine and guinea fowl were the most preferred species).

Besides equatorial Africa, high rates of wild meat consumption also prevail in some parts of Latin America, Southeast Asia, and Siberia. Although no reliable total can be offered, it would be rather conservative to assume that diets of several hundred million people are appreciably enriched by consumption of hunted and collected wild animal species.

In comparison with animal protein, collection of leaves, seeds, tubers, fruits, and nuts may be generally of much lesser importance but there are instances where such wild foods may, too, be the source of high-quality nutrients (Wilson 1985). An outstanding example of such a neglected contribution is the production of high-protein concentrate from the beans of *Parkia biglobosa*, a leguminous tree growing throughout Western Africa (Campbell-Platt 1980). After boiling, fermentation, and drying this concentrate (*dawadawa*) has about 20–50 percent of protein (with lysine content similar to that of egg), 30–40 percent of lipids (three-fifths unsaturated fatty acids) as well as plenty of iron and calcium.

As a frequent addition to the region's stew, *dawadawa* has far greater nutritional importance than its relatively small intake (usually no more than 20 g a day per capita) may suggest. Various African populations also use powdered leaves of baobab (*Adansonia digitata*) and *Vigna unguiculata* in their stews, extract "butter" from shea nuts (*Butyrospermum parkii*) and ferment *Sclerocarya birrea* fruits to prepare a beer. In Tanzania four-fifths of all leafy green vegetables (excellent sources of vitamins and minerals) eaten in rural areas are collected wild plants which accompany almost half of all meals.

According to Fleuret (1979) out of 114 species of trees and shrubs in the Sahelian zone of Africa, twenty-three are of great and another forty-six of a limited importance in human diets. In West Africa products of twenty-four out of 165 studied plant species are regularly eaten and in the continent as a whole more than five hundred different wild plants are consumed as food. In India a counterpart of African dawadawa is a very similar use of unripe pods of jant (*Prosopis cineraria*) in western semidesert areas (Gupta, Gandhi, and Tan 1974). Consumed after being pickled or dried, they can help to maintain adequate protein intake among the poorest peasants.

The rarely appreciated fact is that for most of the world's people we do not know either the food supply or actual individual consumption with satisfactory accuracy (say with an error of no more than +100 kcal a day per capita). These errors matter little inasmuch as the average supply of nearly all nutrients in virtually all rich countries is so high that even the food balance sheets with the lowest values will still indicate nothing but a country's nutritional adequacy.

As we will see in the next chapter, this is not so in the case of most low-income nations, especially all the poorest ones. If their real nutrient supplies were 10–15 percent lower than indicated by FAO's food balance sheets, then many of these countries would be in an even more critical situation than the widely quoted figures suggest. On the other hand, the cumulative uncertainties of balance sheets fashioned from estimates (or even guesses), may easily underestimate the real availability by similar, or, when one keeps in mind common omissions of some food categories, by even larger margins—and the situation of some population groups in many low-income countries may not actually be so desperate.

The existence of appreciable inaccuracies of average food supply figures has wide-ranging policy implications as it forces us to question the real nutritional situation in the poor world. What is the real supply of animal protein in Peru, iron in Zaire, or thiamine in Thailand? You can look up these values, the first one given to the nearest tenth of a gram, the second to the nearest tenth, and the third one to the hundredth of a milligram—but only by chance you would be looking at a correct mean. The real mark of our ignorance is not that one cannot offer a sensible estimate of a likely error—but that one cannot be even sure what is the most likely direction to correct many particular values!

In aggregate, it is highly probable that global food supplies, as well as average availabilities in most poor countries, are somewhat higher than indicated by food balance sheets. FAO's global mean of per capita food supply was about 2,750 kcal/day in the late 1990s; if the real value were just around 3 percent higher the mean would rise to over 2,800 kcal/capita. FAO's average for the poor world was about 2,500 kcal/capita; a 5 percent increase would bring it to just over 2,600 kcal. In order to judge the adequacy of this supply we must know what share of it is actually consumed, what is its nutritional quality—and how much food is really needed.

How Much Food Do We Eat?

Whatever their accuracy may be, food balance sheets do not tell us about actual average per capita food consumption. In order to represent reliably a nation's typical food intakes, such information must come from detailed household surveys that focus solely on food consumption of a representative sample of families during a period of a few days, preferably including also weekends. Less reliable data on average food intakes can be derived from information gathered as a part of regular income, expenditure, and budget surveys, during multisubject investigations of living standards, or during epidemiological studies.

The logical expectation is that actual food consumption averages should be lower than the rates derived from food balance sheet. The difference arises mainly because of the losses during wholesale and retail storage and transfers, kitchen waste (in preparation and cooking), and leftovers thrown away or fed to pets or domestic animals. Unfortunately, this obvious distinction is often unappreciated or ignored and readily available food supply means from FAO's global set of food balance sheets, or from national supply accounts, are presented as actual consumption figures.

Relevant examples abound. The French *Annuaire Statistique* carries food supply figures under a misleading title "Principales consommations des menages"; in *China Statistical Yearbook* they are a part of "Per capita consumption of major consumer goods"; and Statistics Canada publishes regular estimates of "apparent per capita consumption"—but all of these totals are supplies derived from food balance sheets. An uninformed reader taking such descriptions at face value would be making a substantial categorical error. And while FAO appropriately calls the figures in its *Production Yearbook* "supply," the wrong impression is left anyway. For example, when *The Economist*, a leading source of information to many decisionmakers, cited some of these food supply means, it introduced them under a title "Guzzling" and wrote about an average Irishman who "eats almost 4,100 calories of food a day, 40% more than the average Japanese."

In reality, neither the Irish—nor anybody else—could guzzle, on the average, 4,100 kcal a day: after considering the much lower food energy intakes of children and old people, this mean would translate to more than 5,000 kcal for every adult, resulting in a stunning extent of gross obesity in that hapless population (unless, of course, everybody were a hard-working lumberjack). Average Irish food supply is, indeed, among the world's highest—but in all rich nations the mean daily per capita availabilities are well over 3,000 kcal, implying means of over 4,000 kcal/day for every adult, clearly beyond any practically consumable rate (see figure 6.2).

Unfortunately, it is impossible to derive actual consumption means from supply figures by scaling down the latter by a fixed fraction accounting for the distribution and kitchen losses. Naturally, FBS data and information from surveys must correlate, but the strength of their relationship varies even within individual foodstuff or nutrient categories. An excellent illustration of this reality is a comparison of FBS-derived data for dietary fat with actual intake of lipids from more than fifty dietary surveys in nineteen countries (Sasaki and Kesteloot 1992). The two data sets had correlation of 0.82 for monounsaturated fatty acids, 0.74 for saturated ones, and mere 0.56 (explaining only about 30 percent of variance) for polyunsaturated fatty acids!

Consequently, even an adjustment based on a fairly high correlation for an overall nutrient intake may give misleading information about the consumption of particular, and in the cited case particularly desirable, nutritional component. Moreover, for many nations supply figures derived from balanced sheets are, as we have seen, too uncertain to serve as reliable bases for any such adjustment. Furthermore, differences in national habits, affluence, food distribution systems, and attitudes to waste will be obviously translated into highly idiosyncratic utilization patterns: compare, for example, the assiduous trimming of already very lean pork cuts by cholesterol-fearing Americans with the gustatory delight Chinese peasants take in consuming, literally, the whole pig.

Inevitably, there will be cases where the differences between the two rates will be insignificant and it is even possible that food balance sheets—because of underestimating commercial production or ignoring relatively large contributions from kitchen gardens, gathering and hunting—may give lower rates than those found by food consumption surveys. Indeed, a few such cases were uncovered by FAO's (1983a) comparative study of food consumption data from food balance sheets and from various national income, expenditure, or budget surveys.

Measuring Food Intakes

The need for reliable information on actual food intakes is obvious. How else can we say we have enough compared to basic nutritional requirements, how else can we guide sensibly agricultural development, how else can we intervene effectively with new public policies on behalf of those who do not have enough, how else can we draw valid conclusions about principal correlates of disease and health, and how else can we establish the links between environment and longevity—unless we know how much we actually eat?

But a moment's thought will show that getting this information is not at all easy. Forget about the necessity of having reliable nutrient content analyses of hundreds of basic foodstuffs and thousands of prepared meals—just think about what you ate yesterday. Try first to write down the items, and then try again to quantify them. How much meat was in the stew? What was the meat's average fat content? Was the sauce *roux*-based or light with wine? How much granola did you flood with how much milk? What was in that granola, nuts or raisins? How many drinks did you have? How sweet were they? How much and on what did you nibble between meals?

Many items will be forgotten, many quantities underestimated or exaggerated. And yet this twenty-four-hour recall is the basis of most food consumption surveys conducted irregularly around the world. And what will a single day tell you about a person's usual nutrient intake? Maybe all, providing he is a Trappist or a Buddhist monk living in serene monotony of prayerful days whose monotonous diet is dominated by bread and cheese or beancurd and rice. But maybe only something quite misleading when he is an Asian peasant during the days of plowing when he and his water buffalo walk in the deep wet soil, or during a harvest when she is bent over all day to cut handfuls of rice stalks near the ground with a sickle.

During these busy days peasants work hard, often for more than twelve hours a day, and their energy expenditures should be matched by appropriate intakes—while in slack periods they may sleep that much and eat only half as much as during the busy spells. For example, recent measurements of free-living energy expenditure in rural Gambian men showed that during twelve days of intensive farm work they averaged 3,880 kcal/day (Heini et al. 1996). This was about 2.4 times their basal metabolic rate, a level of activity substantially higher than the multiple of 2.1 seen as intensive labor by FAO/WHO/UNU (1985). And because most of the world's countries do not have periodical and representative food consumption surveys (many never had even a single limited inquiry) we do not know with satisfactory accuracy what are the actual food energy and nutrient intakes of their populations. But do we know the real food intakes in the countries that have done some nutritional surveys?

To begin with, not every kind of survey will do, and a closer look at two scores of recent, and hundreds of post–World War II, studies published worldwide (and reviewed in detail in FAO 1979, 1983b, 1986, and 1988) reveals that more than nine-tenths of them are not even true food consumption assessments. These studies are primarily income and expenditure surveys that are concerned with food consumption only as a part of total household budgets. They give almost always detailed breakdowns of spending for different income groups, but only rarely do they provide adequate breakdowns of kinds and quantities of consumed foods. Repeated attempts to use these data for estimates of nutrient intakes are questionable.

Such conversions would be most useful in poor countries lacking proper nutritional surveys, but even a cursory examination of household budget studies will show that rural populations in those predominantly agricultural nations are rarely sampled, and a closer look reveals that the urban poor are commonly undersampled. Moreover, income and expenditure underreporting in poor countries is common, as is subsistence food production and bartering whose values are rarely caught by monetary data.

True food consumption surveys measuring household or individual food intake fall into two principal categories: records of current food intake and recalls of past food consumption. Those surveys where the food is actually weighed before eating are certainly most accurate, but even they may result in underestimates of individual consumption owing to disruption of normal eating habits, difficulties in measuring portions taken directly from a single pot, and omissions of seasonal variations, extraordinary feast intakes, and food eaten away from home.

And, of course, these surveys are the most expensive and the least practical for large-scale, nationwide assessment involving samples of many thousands. One day of weighing may be sufficient to establish approximate group means, but several days (at least three, although even one week may be too short) are needed to represent adequately the current individual intakes. Japan's National Nutrition Survey is the best, as well as the oldest, continuous examination of this kind (Ministry of Health and Welfare 1995).

The annual study began in 1945, and for the first two decades it was conducted four times a year, switching to once a year frequency in 1964 (now in November). Some fifteen thousand households participate, and all food consumed by all household members is weighted and recorded by a member of each household during three consecutive days (excluding public holidays) and the records are checked every day by a dietician (Statistics Bureau 1999).

The only case where weighing may not give accurate individual rates is when such consumption averages are derived from household totals. Even if the latter values are very accurate, an observer coming to find out who eats how much in a family cannot do so, as Sen (1984) noted, without affecting the phenomenon itself. Inequalities of food consumption within households are common but poorly understood. Favoring the wage-earner with higher protein and lipid food was common in Europe and North America until about two generations ago, and from fragmentary information we know that in most poor countries children, and often also women, continue to subsist mainly on staple cereals and that meat, fish, vegetables, and fruit go preferentially to men.

Sen (1984) used studies from rural West Bengal and Calcutta to document the sex bias in food distribution resulting in greater undernourishment of girls in distress situations and the unusually high mortality of women during normal times. Kramer et al. (1997) documented recent intrahousehold inequalities in food allocation among more than a hundred rural families in Bangladesh: women received only 75–88 percent of recommended intakes, children 86–108 percent, but men 89–114 percent. Many people in poor countries still have no appreciation of the necessity of a balanced diet and protein requirements and unwittingly shortchange their children during their most vulnerable period of development. For example, van Steenbergen et al. (1984) found that in Kenyan households relying on *ugali* (a corn paste of solid consistency) children had considerable difficulties in masticating large volumes of the staple, and toddlers often missed the evening meal served when they were already asleep.

Dietary recalls, most often for the twenty-four hours preceding an interview, are favored in extensive surveys owing to their simplicity, and in poor countries also to their suitability of getting information from illiterate persons. Both of the major periodic food intake surveys in the United States—the Nationwide Food Consumption Survey (NFCS) conducted by the U.S. Department of Agriculture, and National Health and Nutrition Examination Surveys (NHANES) administered by the U.S. Department of Health and Human Services—have been based on recalls. Both surveys have been also recording only weekday consumption, a major fault considering the extent of clearly above-average weekend eating.

Accuracy of dietary recalls is clearly inferior to that of record surveys. They cannot be assumed to be reliable just because they may give reproducible results. Both under- or overestimation are the most obvious systematic drawbacks, as is the expertise of the interviewers (who must quantify the often fuzzy verbal information) and repeatability of the method (Block 1982; Kleges et al. 1987; Willett 1990). An increasing number of studies comparing recalls, or estimated dietary histories, with independent observations, weighed duplicate meals, or accurately measured energy expenditures found common, and often significant, underreporting of intakes, as well as substantial day-to-day variations, a reality arguing against single-day studies.

A fundamental check on the accuracy of self-reported intakes is to express them as multiples of basal metabolic rates (estimated by using body heights and weights that are usually accompany published studies)—and then compare this ratio with a specifically defined cutoff value representing normal energy expenditure of a sedentary lifestyle. Black et al. (1991) did this exercise by using published studies for sixty-eight groups of adults, and found a universal and strong bias tendency toward underestimation: the mean estimated ratio of 1.43 was significantly below the expected requirement of 1.55, with averages of just 1.37 for females and 1.5 for males. Not surprisingly, diet recall turned out to have the largest share (88 percent) of ratios below the expected level.

Although this revealing comparison unequivocally confirms the tendency toward underestimates, the uncovered disparities may not be actually as large as reported. The main reason for this lower disparity is the fact that Black et al. (1991) used equations by Schofield et al. (1985) to calculate male and female basal metabolism in each study group—and, as we will see in the next chapter, these equations tend to overpredict non-European rates. As soon as they began to be used, studies with doubly-labeled water (for a description of the technique see chapter 7) uncovered many cases of underreporting dietary intake among such diverse groups as obese women (whose reported intake was just 64 percent of measured energy expenditure) and cyclists during the Tour de France (Prentice et al. 1986; Westerterp et al. 1986). General applicability of these findings is uncertain, as the studied groups were very small (commonly less than a dozen of volunteers).

Most of the comparative studies have also found that underreporting correlates with obesity. For example, in a recent Swedish study actual food intakes in the nonobese males and females were insignificantly higher (a mere 2 percent) than the rates derived from questionnaires exploring diets over the past three-month period—but they were 35 percent higher in the obese (Lindroos et al. 1993).

Schoeller's (1990) meta-analysis of studies comparing self-reported energy intake and expenditure confirmed that obese people underreport significantly more than nonobese individuals, but some of the worst discrepancies applied to lean athletes of both sexes, and to hardworking, nonpregnant Gambian women whose reported intakes were only 40 percent of their actual expenditures. Lissner et al. (1989) found that lean American women were underestimating their intakes as much as their overweight counterparts. Schoeller concluded that individuals tend to report intakes that are closer to perceived norms for their sex, age, and activity level.

Not surprisingly, overestimates were noted in other cases where respondents want to be seen as doing a right thing, and confirmation of this attitude comes from an environment of extreme poverty as well as from America's affluent suburbs. Olinto et al. (1995) found that mothers of malnourished infants in two slum areas of Pelotas in southern Brazil had markedly overestimated food intake of their underweight children:

as they recognized the child's poor nutritional status they exaggerated its food intake in an apparent attempt at compensation. When asked to recall food eaten by their children during the previous day's lunch, middle- to upper-class fathers from Tennessee overestimated vegetable intake by 26 percent, milk drinking by 30 percent, and fruit consumption by 50 percent while underreporting bread by about 28 percent (Eck et al. 1989).

Studies of garbage from U.S. suburbia provided fascinating illustrations of how people misrepresent their food and beverage consumption (Rathje and Murphy 1992). American homemakers almost uniformly overreport the total amount of food consumed in a household, as well the amount of fresh produce eaten, while underreporting the quantity of prepared foods the family eats. For individual food items the relations between reported and actually consumed amounts mirror closely the perceived health or social merits of a foodstuff: beer, chocolates, and sausages are drastically underreported; asparagus, oranges, lean beef, and cottage cheese are hugely overreported (Rathje 1994). Similarly, Cote's (1984) examination of household refuse discovered underreporting of alcohol and frozen dinners and overreporting of fresh meat.

Alcohol consumption, often underreported in food surveys, can make a substantial difference in total energy intakes. For example, heavy male drinkers included in the Dutch National Food Consumption Survey consumed about 400 kcal of alcoholic drinks during midweek days and almost 500 kcal during weekends, the latter total amounting to nearly 15 percent of their overall food energy intake (Veenstra et al. 1993). Significantly, the energy derived from alcohol was not compensated by lower intake of other nutrients: heavy drinkers consumed 12–14 percent more food energy than abstinents.

As with the record surveys, one day recalls will be grossly inadequate to obtain a meaningful assessment of typical individual intakes. A few days of recall studies may give a fairly representative means in affluent societies with little seasonal variation of food intake—but similar results may be quite misleading in rural areas of poor countries where both supply and needs range widely, depending on fluctuating harvests and periods of heavy field labor. Higher food intake during festive periods will compound the errors, and the only way to correct these omissions would be repeated sampling during various seasons. An advantage of recall studies is the inclusion of meals eaten away from home—if they are well remembered, of course.

Other factors affecting the reliability and applicability of food surveys, no matter in what way they are conducted, range from substantial differences among ethnic or religious groups in multicultural societies (including various food taboos) to means available for food preparation (rural energy shortages, common in many deforested and arid regions of Asia and Africa, play a major role in this regard). Unless these concerns are addressed, it is unrealistic to claim that even otherwise well-designed studies provide accurate information about prevailing food intakes.

The only realistic conclusion is that we have no practical, affordable *and* reliable way to determine the true usual food consumption, that is intake representative over a long time; at best, we can get approximations of typical intakes limited to specific time periods. But because of the expense, most countries simply do not conduct any representative recurrent food intake surveys, and many of them never had even a single limited assessment. Consequently, we do not know with satisfactory accuracy the actual food intakes of most of the world population.

Expectations and Surprises

Keeping the weaknesses of food intake data in mind, we can still make a number of fairly unexceptional generalizations concerning national food energy consumption means. Average per capita values could only rarely surpass 2,500 kcal/day. Perhaps the highest published rate comes from New Zealand's dietary surveys showing an exceptionally high per capita intake of 2,750 kcal a day, compared to the supply of about 3,500 kcal (Shorland 1988). None of the reported national means comes anywhere close to this unusually high rate.

An overwhelming majority of average daily energy intakes in rich nations will fit between 2,000 and 2,300 kcal/capita. Differences in age structure will account for most of the variation. Interestingly, the rates just above 2,000 kcal/day do not differ appreciably from the best reconstructions of European food energy intakes during the nineteenth century industrialization (Toutain 1971; Lemnitzer 1977; Bekaert 1991). The Belgian mean during the 1840s was just over 2,200 kcal, the French average for the same decade was virtually identical (2,250 kcal), and the German mean for the 1850s was about 2,120 kcal/day. Closeness of the rates should not surprise: the combination of lower body masses and more physical labor during the nineteenth century produced about the same need as the lowered physical activity and larger bodies do today.

Nationwide means lower than 2,000 kcal/day will be common in the world's most impoverished countries. Because most of these nations never had even a single reliable food intake survey, it is impossible to know the lowest countrywide mean—but it is unlikely that it would be much below 1,700 kcal/capita, a level of Bangladeshi intakes. Two studies of nutrition intake by urban population of Bangladesh, based on actual weighing of consumed food, showed a decline from a daily average of 1,777 kcal/capita in 1962–1964 to just 1,682 kcal/capita in 1985–1986, a change accompanied by an even greater decrease (about 12 percent) in the average protein intake (Hassan and Ahmad 1992).

Lower per capita population means have been reported from many parts of sub-Saharan Africa during the periods of seasonal food shortages common throughout the 1960s and 1970s. These values were as low as 1,527 kcal/capita in August and September in Senegal, and between 1,520 and 1,610 kcal/day in parts of Cameroon and Kenya (Hulse and Pearson 1980). Expected sexual difference appears to be particularly large in affluent countries where female dieting has become very common. This is well illustrated by U.S. food energy intakes: surveys done during the past generation have found the difference between averages of daily adult male and female intakes to be no less than 800 and as much as 990 kcal/day. And North American female means appear to be quite low even in comparison with very conservatively calculated baseline energy need.

Contrasting with these very low rates are the relatively high intakes in many modernizing countries where large shares of population are still engaged in physically demanding farm labor, and where the level of mechanization in urban employment is also much lower than in affluent countries. For example, in spite of the average body mass being clearly well below the U.S. mean, China's first dietary survey undertaken during the period of economic reform found the average food energy intake at 2,485 kcal (Chen 1986). Physical demands of field labor keep the daily food energy intakes among China's huge agricultural labor force at about 2,600 kcal. Male-female differences in such traditional societies are strongly influenced by the division of heavy labor: males do a substantial share of it in China, but not in many parts of Africa.

Even a brief survey may give a fairly reliable information about the actual long-term intake in a rich country. But well-documented seasonal food shortages may result in very low average intakes in many traditional societies relying on subsistence farming in environments characterized by moderate to intense climatic seasonality. Impacts of these deficiencies—including cyclical loss of weight in adults, reduction of growth rates in children, and lower weight gains during pregnancy—is often aggravated by greatly increased workloads during the harvest period, or by seasonal outbreak of diseases (Branca et al. 1993).

Finally, there is a clear link between overall food energy intake and stages of economic modernization. As their purchasing power grows, countries in relatively early stages of modernization will see appreciable increases in average per capita intakes of both food energy and protein. Certainly the most notable recent example comes from China, where rural per capita rates have shown a steady rise since the beginning of farming privatization in the early 1980s: they rose from 2,300–2,400 kcal during the late 1970s to 2,600 kcal a decade later. But during the same time average urban intakes rose only marginally, from 2,100 to 2,200 kcal.

And, going a step beyond, it appears that high level of affluence and urbanization results in surprisingly low long-term variability as dietary changes affect the makeup of typical nutrition, but have little impact on the overall energy intake. Since the mid-1950s, once it recovered from its postwar lows, Japan's average daily rate fluctuated very narrowly, between 2,100 and 2,200 kcal/capita, while today's protein intakes are only about 10 percent higher than two generations ago (Statistics Bureau 1997).

U.S. intakes have been similarly stable. Between 1971 and 1974 the NHANES I found average daily male intakes at 2,561 kcal/capita, and female consumption at 1,571 kcal (National Center for Health Statistics 1979). Results of the latest NFCS differ little from those of a quarter century ago: in 1995 men twenty years and over averaged 2,479 kcal a day, the mean for adult women was mere 1,653 kcal, and the mean for all individuals of all ages was 2,017 kcal a day (U.S. Department of Agriculture 1997a). Similarly, average protein intakes have continued to vary narrowly around 100 g/day for males and around 60 g/day for females.

With uncertainties abounding within both the food supply and food intake accounts, it is inevitable that comparisons of the two rates—be they secular series for a single country or international cross-sections for an identical period—raise questions to which there are no clear answers. International comparisons turn out some interesting disparities as neighbors whose lifestyles appear very similar differ in their food intakes by relatively large margins. Temporal comparison may reveal irreconcilable trends between supply derived from FBS and average intakes estimated from income or intake studies.

When searching for the most fundamental concerns dominating these uncertainties, we see that the issues are different, as in so many other instances, for the world's affluent minority and for the countries that are either progressing rapidly on their modernization path, or as yet to embark on this complex journey. By far the most obvious result of comparing food supply and intake in affluent nations are enormous gaps between the two rates. Comparisons of food supply averages derived from food balance sheets with daily intake means derived from consumption surveys show that in rich nations the former values are no less than 20–25 percent higher than the latter rates, that the common difference is about 30 percent, and that the largest gaps are at, or slightly over, 50 percent of the supply total.

For the United States, the difference between FAO's FBS-derived supply for the years 1992–1994 (3,610 kcal/day) and intakes established by surveys was between 40 and 45 percent. The difference for Japan of the mid-1990s (approximately 2,900 vs. 2,050 kcal/day) was almost exactly 30 percent. Factors accounting for these gaps are easy to list: overestimations of supply, underreporting of intake, distribution and preparation losses, and deliberate discarding of some purchased food must provide most of the explanation. Ascribing reliable shares to these major variables is a different matter.

Even in the U.S. case the effort to explain fully the huge supply-intake gap generates more questions than answers. Is the increasing frequency of eating outside home responsible for a growing part of the gap, and what difference does the continent's dieting craze make? Why should the U.S. consumption surveys, based on twenty-four-hour recall which includes all food eaten anywhere, be so highly unreliable? Do Americans have very poor memories or are they just trying to appear less glutonous than they are? Or have Americans become so much more wasteful?

Rathje, whose Garbage Project has examined more than one million items from eight thousand household refuse samples between 1973 and 1985, concluded that in contrast to the year 1918, when the War Food Administration program estimated the losses at 25–30 percent of total solid food, current American household waste is only 10–15 percent of all purchased solid food (Rathje and Murphy 1992). And a new USDA study showed that in 1995 27 percent of the country's food was lost at retail, consumer, and food service levels, with fresh fruits, vegetables, fluid milk, grain products, and sweeteners accounting for two-thirds of the loss (USDA 1997).

But even after reducing the average U.S. per capita food supply by 27 percent, that is, from 3,600 kcal/day to about 2,600 kcal/day, the difference between latter rate and food consumption surveys is still more than 500 kcal/day. Only a part of this discrepancy can be explained by the fact that standard values of energy and fat content of red meat used in food balance sheets overestimate actual consumption of animal fats: garbage studies document clearly extensive trimming of fatty cuts. Similarly, French studies show that about 10 percent of all fat actually served on the plate is discarded (Dupin et al. 1984).

Disparities between apparent food availability and actual intakes diminish as one descends the developmental ladder: the gap is no larger than 15 percent in China, is well below 10 percent in India, and is apparently nonexistent in the poorest African countries. Food losses equivalent to 10–15 percent of total supply may be unavoidable, but there is no excuse for the enormous losses in affluent countries. If the rich world's food losses could be held to 20 percent of the overall supply, the annual savings (assuming that animal foods provide about 25 percent of all food energy) would be equivalent to at least 100 Mt of grain, or nearly half of all cereals on the world market.

7

How Much Food Do We Need?

What we need to know seems obvious: a fairly satisfactory understanding of how much food is required for a healthy life is the foremost concern. Without this knowledge we cannot assess the adequacy of our existing food supplies; with imperfect knowledge we run the risks of either exaggerating or underestimating the numbers of people suffering from malnutrition—and we offer dubious advice about nutritional optima. Without being confident about human nutritional needs we also cannot plan for necessary increases required by growing populations—and we cannot guide responsibly any long-term transitions toward rational diets.

If we were to judge the extent of our understanding solely by the amount of readily available information, we would seem to command a more than adequate foundation. Multidisciplinary literature on nutritional needs contains an enormous amount of impressively detailed quantitative information. But as a closer look will reveal, our understanding of food requirements is a complex mixture of solid comprehension, tentative conclusions, and continuing uncertainties.

Just ask around what should be a typical daily per capita food energy intake for adults—or, if you wish to make it more challenging, how much protein we should get every day. In the overwhelming number of cases your otherwise well-informed friends and acquaintances will give answers that are far too high. They should not be blamed for inexcusable ignorance, for it always takes some time for new scientific understanding to become a part of general knowledge: their answers will merely reflect the history of exaggerated dietary recommendations and curiously slow diffusion of new data showing much lower needs (and even lower actual intakes).

And if some knowledgeable individuals were to reply correctly, their answers would have to be carefully qualified: human food energy needs are not a simple function of one or two basic variables, but an amalgam of demands whose average rates vary fairly predictably for particular populations with sex, age, and body size, and can differ appreciably among certain groups and depart significantly from expected large-scale means for individuals, especially for adults. Inevitably, these complexities must be kept in mind as we evaluate the adequacy of existing food intakes and as we assess the outlook for desirable changes.

Researchers have identified at least fifty nutrients and food constituents to be of importance, and dietary recommendations have been established for a growing number of micronutrients. The latest consensus of the U.S. National Research Council (NRC) sets recommended daily allowances (RDA) for food energy and dietary protein—as well as for thirteen vitamins, three minerals, and nine trace elements (NRC 1989).

Food energy needs and dietary protein requirements dominate the nutritional concerns. Adequate food energy and protein supply provided in a reasonably varied diet rarely fails to carry sufficient amounts of miconutrients, and their major deficiencies are usually a mark of generally inadequate food provision. But not always, as average intakes of some vitamins and minerals may be insufficient even in societies with otherwise abundant food supply because of unbalanced diets eaten by many groups—or as the natural availability of these micronutrients is greatly restricted in some environments. Affluent North America offers excellent examples of the first category, as clinical and biochemical studies confirm that intakes of calcium, iron, and zinc are not adequate in some groups (Pennington 1996). Worldwide, the eradication of micronutrient deficiencies could exceed the impact of the global elimination of smallpox (Maberly et al. 1994).

Certainly the most important cases of environmentally induced deficiency are widespread shortages of iodine in mountainous or flood-prone regions of the poor world where the element has been washed out of soils. The World Health Organization (World Health Organization 1993) estimated that in the early 1990s about 1.6 billion people, or some 30 percent of the world's population, were at some risk of iodine deficiency. These people do not live only in poor mountainous regions of Latin America (Ecuador, Peru, Bolivia), Central Africa, and Asia (in a belt between Iran and Sichuan) but also throughout large parts of Europe where daily iodine excretions in urine are less than half, or even just a quarter, of adequate values.

Estimates of the world's population with goiter, the condition readily recognizable by a characteristic swelling of the thyroid gland and associated with some mental impairment, range from over two hundred million to more than six hundred million (WHO 1993; Lamberg 1993). Mild iodine deficiency causes intellectual impairment (reducing standard intelligence quotients by 10–15 percent), higher incidence of stillbirths, and higher infant mortality. WHO also estimated that iodine deficiencies during pregnancy have been responsible for at least twenty-five million seriously brain-damaged children and for nearly six millions cretins, whose severe mental retardation is combined with hearing loss or mutism and with abnormalities of body movements.

Other relatively common micronutrient deficiencies include shortages of vitamin A and iron. The worst consequence of vitamin A deficiency is the xerophthalmia syndrome, including night reversible blindness caused by lack of retinol in the eye's retina, corneal ulceration, and eventually irreversible loss of eyesight. In addition, low vitamin A levels are associated with higher mortality from respiratory and gastrointestinal diseases, and with their more severe course. Some forty million preschool children have vitamin A deficiency, and perhaps half a million of them go blind annually, and the total population at risk is well over half a billion.

Iron deficiency, with resulting anemia lowering resistance to infection and affecting learning and labor, is especially common among pregnant women, low-weight infants, and school-age children. Unlike other micronutrient deficiencies, its rates are relatively high even in affluent countries where anemia affects some 10–15 percent of women (Marx 1997). The highest estimates put the share of humanity affected by anemia at 50 percent, and iron deficiency is the most important reason for the condition.

But rectification of these shortcomings is a challenge very different from supplying adequate amounts of food energy and protein. Most micronutrients are needed at daily rates of 10^{-2} to 10^{-1} g and can be readily derived from abundant natural sources or can be affordably synthesized. Elimination of their deficiencies is thus either a matter of promoting appropriate food choices or making available necessary supplements, both being essentially public health, rather than food production, challenges.

Iodine deficiency disorder can be easily, and inexpensively, eliminated by adding the element to table salt, and the goal of universal iodization may be virtually complete by the year 2000 (UNICEF 1995). Shortages of vitamin A can be remedied by promoting consumption of foods rich in the vitamin and by providing supplements or fortifying some foods (Blum et al. 1997). Much like iodine, ferrous sulfate tablets (now also available in a convenient once-weekly dose) are cheap. In contrast, daily per capita protein needs range from 10^1 to 10^2 g, and consumption of carbohydrates, lipids, and proteins providing food energy adds up to 10^2 to 10^3 g every day—and only intensive crop cultivation and animal husbandry can supply these nutrients in sufficient quantities. That is why my focus in this chapter will be on food energy and protein.

Our understanding of human energy and protein requirements is based on short-term investigations of volunteers in laboratories, on animal models, on anthropological studies, and on epidemiological research. Each of these sources has its obvious limitations. Over the decades volunteers have been recruited overwhelmingly from healthy, young adults on the campuses of universities in rich countries. Animal responses to feeding experiments may be qualitatively different from human metabolism. Conclusions from anthropological studies cannot be usually applied with confidence beyond the specific time and space confines of such studies. And without deeper tracing of links and mechanisms, epidemiological evidence must be seen merely as suggestive, not causal.

I will look first at our evolving understanding of human energy and protein requirements. Then I will contrast the best recommendations of nutritional needs with actual intakes and discuss the implications of these revealing comparisons on levels ranging from individual to global.

Human Energetics

After more than a century of systematic studies of human energetics we have a confident understanding of basic rates and requirements, including their variations due to sex and age and their often surprisingly large inter-individual differences. And yet—in spite of the endemic preoccupation with weight and the presence of an enormous dieting industry—public misinformation abounds as even otherwise well-educated people are often unaware of some indisputable facts of human energetics. And a connection between this understanding and food production policy is almost entirely absent.

This section will review sequentially all types of human energy requirements. Most of the digested food energy is normally needed for basal metabolism. Thermogenesis induced by digestion of food and by cold or heat is only a marginal addition on a long-term basis. Similarly, energy for growth is only a very small fraction of lifetime food intake, but its adequacy, especially in infancy, is obviously critical for proper mental and physical development. For females additional energy costs are imposed by pregnancy and lactation. A highly variable category of energy needs for labor and leisure activities makes up the rest of normal food demand.

Basal Metabolic Rates

Basal metabolic rate (BMR) is the minimum amount of energy needed to maintain critical body functions. It varies with sex, body size, and age, and its measurements in adults show large individual departures from statistically expected means. The rate should be expressed in joules (J), the standard scientific units of energy, but kilocalories (kcal) have been so entrenched in nutritional studies that I will stay with them.

Determination of an individual's BMR requires a body at complete rest in a thermoneutral environment and at least twelve hours after the last meal. Neutral temperature (around 25°C) is required to preclude any shivering and sweating, and the wait eliminates any effects of thermogenesis caused by food digestion: thermic effect of food peaks about an hour after the meal when it can add up to 10 percent to the BMR, and becomes negligible three hours later.

Resting energy expenditure (REE) is higher than BMR because it includes the thermic effect of meals: that difference can be as little as 5 percent. The two rates are used interchangeably.

Thousands of BMR or REE determinations are available for both sexes. They have been acquired mostly by the direct measurement of oxygen uptake: on the average, one liter of consumed oxygen equals about 83 kcal/g of oxidized nutrients. Unfortunately, this global data base is biased: although it includes statistically significant coverage for all ages as well as for adults of different body masses and statures, it is made up largely of individuals from affluent nations. People from poor countries are clearly underrepresented.

The relationship between basal or resting energy needs and body size can be expressed by simple linear equations (figure 7.1). Classic equations of Harris and Benedict (1919) widely used in the United States include both body weight and height to express REE. Regression equations derived by Schofield et al. (1985) from almost eleven thousand measurements done during sixty preceding years were favored by the

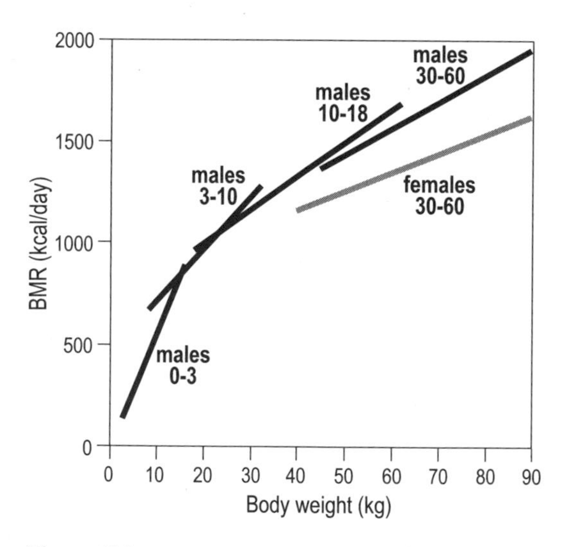

Figure 7.1
Linear predictions of basal metabolic rates (in kcal/day) from average body weights (in kg); calculated from equations in FAO/WHO/UNU (1985).

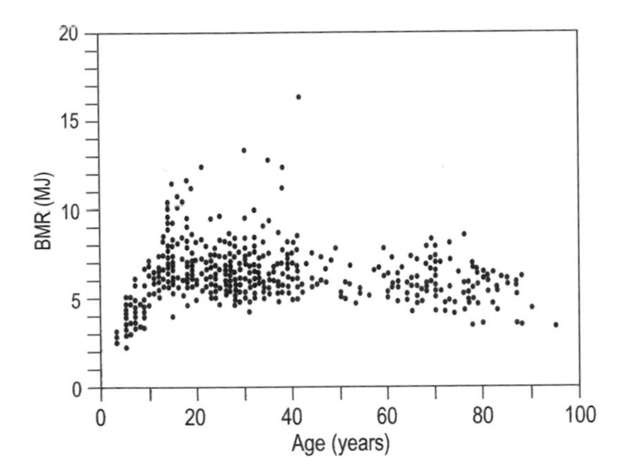

Figure 7.2
Distribution of energy expenditures among 574 free-living individuals aged 2–95 years (Black et al. 1996).

FAO/WHO/UNU (1985) expert group; they give similar results for BMRs without including body height. In order to avoid excessively high recommendations of average food energy intakes, desirable weights for given heights (rather than actual mean or median body weights) should be used in these regressions in any country where obesity is widespread, notably in North America (Pellett 1990a).

Regression equations recommended by FAO are excellent predictors of individual BMRs only among children (correlation coefficient of 0.97 for children less than three years old) and adolescents (coefficient still as high as 0.9 for teenage boys); in contrast, for men between ages of 30 to 60 years the correlation explains only about a third of the variance (figure 7.2). More importantly, using these equations may lead to exaggerated estimates of energy requirements of populations outside the affluent world in general, and outside Europe in particular. In their original compilation of worldwide BMRs Schofield et al. noted that the rates for Indians included in the database were significantly lower (by nearly 13 percent) in comparison with their matched European and North American counterparts. Henry and Rees (1991) found that Schofield's equations overpredicted BMRs of people living in tropical regions, with the difference being largest for adults over thirty years of age.

The average rate of overprediction for twelve sampled populations was 9 percent for males (ranging from just 1.5 percent for Maya to 22.4 percent for Ceylonese) and 5.4 percent for females (and up to 12.9 percent for Indian women). Explanations for these differences have ranged from varying abilities in producing muscle relaxation to BMR-lowering effects of warm climate. The latter conclusion is supported by BMR changes in individuals moving from temperate to tropical climates. Piers and Shetty (1993) revisited these disparities by measuring BMRs of Indian women. They found that the equations both of Schofield's group and Henry and Rees overpredicted the basal metabolism by, respectively, 9.2 and 4.2 percent. Schofield's equations, favored by FAO, also overpredict BMRs of American women, young Australians, and nighttime energy expenditures for European males, and they appear to be accurate only for predicting European BMRs (de Boer et al. 1988; Piers et al. 1997).

Undoubtedly, adult BMRs vary appreciably not only among individuals of the same population but also among different population groups. To the question Widdowson (1947) first asked two generations ago—"Why can one person live on half the calories of another, and yet remain a perfectly efficient physical machine?"—can be added a no less intriguing query: "Why some populations need 5, 10 or even 15 percent less energy for their basal metabolism than others?" As yet we have no indisputable, comprehensive explanations of these disparities. A major reason for disparities among different populations may be simply due to different shares of body mass made up of metabolizing tissues, that is, muscles and, above all, internal organs. In newborns the brain's needs dominate, claiming more than 40 percent of their BMR (Durnin and Passmore 1967).

Metabolically the most active organs in adults are, in descending order of relative magnitudes, the heart, kidneys, liver, and brain. In absolute terms, these four organs account for just over half of adult BMR, with the liver claiming the largest share of the total, almost one-fifth (figure 7.3) (Aiello and Wheeler 1995). Where the shares of active tissues may be very similar, the difference must be attributed to a lowering of metabolic activity.

Although individual basal metabolic rates vary widely, their lifetime course follows a uniform trend. After the peak between three to six

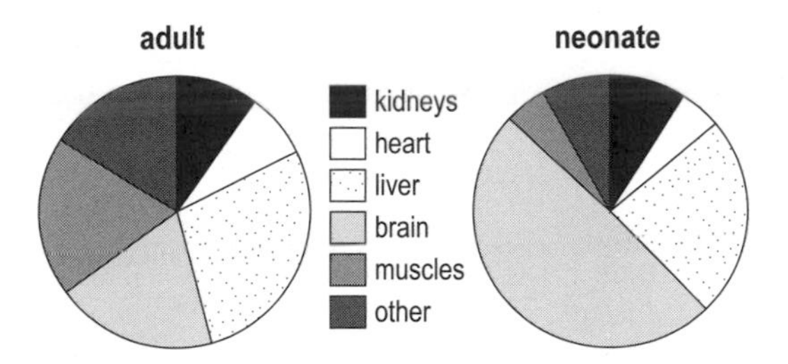

Figure 7.3
Metabolic shares of internal organs in newborns and adults (Smil 1998).

months of age comes a steep decline to about half of the maximum value by late adolescence followed by an extended plateau. Metabolic decline resumes during the sixth decade, and the rate diminishes for the rest of life. This decline is due to a steady loss of metabolizing lean tissue, a process strongly affected by pronounced sexual dimorphism expressed primarily in different rates of storing fat.

After adolescence, the body's fat content increases steadily in both sexes, but the difference widens with age. While fat makes up around one-seventh of body weight in well-nourished young adult males, it accounts for about one-fourth in females; by the seventh decade of life the respective shares are almost one-fourth and a bit over one-third. This means that, on the average, females are adding 0.3–0.4 kg of fat a year, compared to 0.15–0.25 kg for men. Loss of muscle mass is the countertrend of fat gain. After the third decade of life males lose their muscle mass more rapidly (2–3 kg per decade) than females (about 1.5 kg per decade). After the age of seventy both sexes have about 40 percent less muscle than they had as young adults.

Aging of Africa's and Asia's population will tend to increase average food requirement of the two continents during the next two generations, perhaps by as much as 5–7 percent in the first case, by some 2 percent in the latter one. Increased heights of healthier populations may add another 1–2 percent to food demand in most modernizing countries. In contrast, continuing urbanization, resulting in lowered physical activity, will reduce mean intakes by 2–4 percent. This means that even if the global population instantly stabilized itself there would be a slight increase in global food demand.

Energy for Growth and Activities

Overall energy conversion efficiency of human growth depends on the ratio of newly stored proteins and lipids: proteins are always more costly to synthesize. The average food energy need is about 21 kJ/g of new body mass in infants, a rate implying conversion efficiency of nearly 50 percent. Adult growth efficiencies, based on overfeeding experiments, are up to one-third lower. Body growth is fastest during the first nine months of life when its rate is almost perfectly linear. By the end of their first year healthy babies almost triple their birth weight. Afterward, the growth lines begin curving gently for both girls and boys. Body growth claims nearly a third of all food energy intake during the first few month of life, but the share drops sharply to just between 5 and 6 percent by the end of the first year, and to a mere 2 percent by the end of the first decade (figure 7.4).

Adequate energy intake in early childhood is essential not only for transforming us from the state of immobile dependency to one of active exploration, but also to make us human: neonate brain (at about 350 g

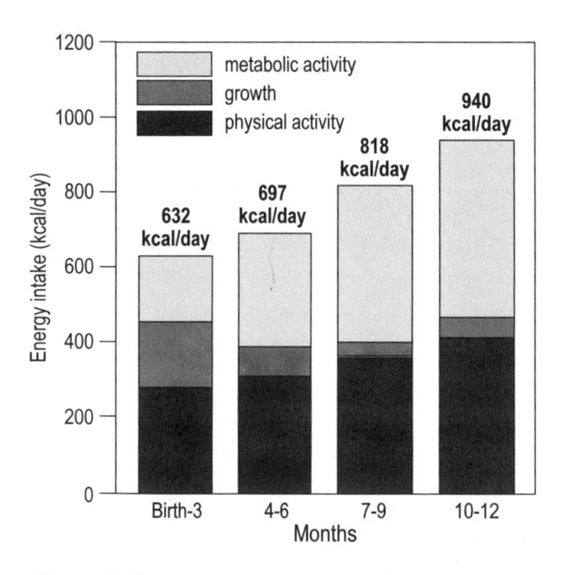

Figure 7.4
Allocation of food energy spent on metabolism, growth, and activity during the first year of life (Glinsmann et al. 1996).

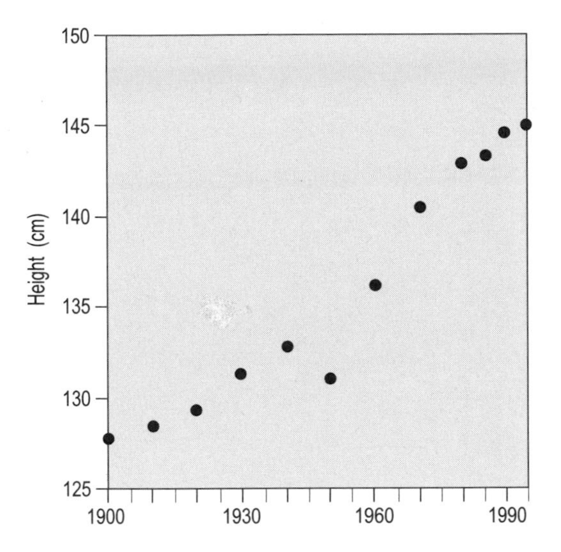

Figure 7.5
Average heights of 11-year-old Japanese schoolboys (Smil 1998).

twice as large as that of a newborn chimpanzee) enlarges 3.5 times by the age of five to become more than three times as massive as the brain of our closest primate species (Foley and Lee 1992).

Average body weights at particular ages differ appreciably among countries, and better nutrition has been pushing them higher in most affluent nations over time. Long-term Japanese trends are an excellent illustration of this universal trend: between 1900 and 1995 eleven-year-old Japanese boys gained about 20 cm of height, with the hungry years of World War II representing the only interruption of a steady rise (figure 7.5). After a temporary doubling during a growth spike of the early teens the share of food energy claimed by growth drops back to about 2 percent before the age of eighteen and in a couple of years it falls to a small fraction of 1 percent. This minuscule rate is sufficient to repair and regrow adult tissues.

Determination of energy costs of numerous labor and leisure activities used to be a much greater challenge than finding BMR or REE: rigging up and wearing the apparatus necessary for accurate respirometry—masks, hoods or mouthpieces, and hoses needed to measure gas exchange—is no problem when testing a cyclist on a stationary bicycle, but it is a challenge for measuring the costs of coal mining or mountain climbing. Three portable systems that can be used for continuous measurements of oxygen uptake for long periods are in widespread use (Patton 1997), and a large number of diverse and interesting activity values was gathered by decades of dedicated research (Durnin and Passmore 1967).

Two convenient methods now allow nonintrusive measurement of actual energy expenditures. The first one, developed in the mid-1950s (Lifson et al. 1955) and used subsequently to measure energy expenditure in small animals, has been applied to the monitoring of human energy needs since the mid-1980s. This technique relies on monitoring exponential losses of the doubly labeled water (DLW, water marked with stable heavy isotopes of ^{2}H and ^{18}O) drunk by subjects over a period of one to two weeks. When the two heavy isotopes are washed out of the body and replaced with dominant ^{1}H and ^{16}O, loss of deuterium measures water flux and the elimination of ^{18}O traces not only the water loss but also the flux of CO_2 in the expired air (DeLany 1997).

As the gas is the final product of oxidizing food substrates, the difference between the washout rates of the two isotopes (obviously, ^{18}O will be eliminated faster) makes it possible to find integrated value for energy expenditure over a period of time by using standard indirect calorimetric calculations. There is no interference in activities of free-living subjects monitored by this technique except for periodic urine or saliva sampling. The other method relies on continuous monitoring of heart rate and its conversion to energy expenditure by using calibrations previously determined by a laboratory respirometry.

In any case, given the substantial differences among individuals, it is preferable to express activity costs—frequently called physical activity levels (PAL)—as typical ratios of total energy expenditures (TEE) and BMR or REE rather than in absolute values. Minimal survival requirements including metabolic response to food, and energy needed to maintain basic personal hygiene would give PALs between 1.15 and 1.2.

As far as specific energy expenditures are concerned, thinking is at the bottom of activity scale. The brain's high basal metabolism, claiming a fifth of the adult food energy intake, goes up only marginally even when a person is engaged in the most challenging mental tasks: typical cost would be merely 5 percent above the BMR. Sitting and standing rank next: they require the deployment of large leg muscles resulting, respectively, in rates of 1.15–1.20 times and 1.3–1.5 times BMR. They are followed by light exertions typical of numerous service jobs that now dominate in all modern economies. Tasks as diverse as secretarial work and truck driving, or food retail and car repair, belong to the light exertion category and do not typically require more than 2.5 times REE per unit time of activity. A great deal of modern manufacturing and construction, as well as most tasks in highly mechanized farming, belongs to the same category.

Activities requiring moderate (up to five times REE), and sometimes even heavy (up to seven times REE), exertion are still most common in traditional farming (ranging from plowing with animals and hand weeding, to manual transplanting of rice and cleaning of irrigation canals), forestry, and fishing. For some individuals the highest multiples may be 20–40 percent higher. But even for the physically most demanding occupations longer-term (daily to annual) energy expenditures may belong just to a medium-demand category, as the spells of taxing exertions are interspersed with periods of less demanding activity or rest. The FAO/WHO/UNU (1985) consensus chose PAL of 1.55 for males and 1.56 for females as the average daily energy requirement of adults whose occupations require light exertions, PALs of 1.64 for females and 1.78 for males engaged in moderately demanding work, and rates of 1.82 and 2.1 for individuals or groups in jobs demanding heavy exertions.

Intensity of exercise is a critical variable in assessing energy cost of sports and recreational activities. Recreational swimming or tennis may require only moderate exertions—but in competitions these, and many other sports, are clearly in the heavy-exertion category. Cycling is a perfect example of this difference: its energy demands can range from very moderate in sedate pedaling to extremely heavy in record-breaking races. Because it deploys very efficiently the body's largest leg muscles, cycling is also the fastest mode of human locomotion: the fastest human-powered machines on land, on water, and in the air have all been propelled by accomplished cyclists (Whitt and Wilson 1993).

Individual sports normally requiring high investment of energy range from long-distance running to rowing, while basketball and soccer rank high in group sports category. Untrained individuals engaged in these activities come close to their maximum metabolic scopes, which are about 10 times REE. Trained endurance athletes can do much better: their metabolic scopes of up to twenty-five times their BMR are surpassed among mammals only by canids.

Calculations of average food energy needs usually assume that labor and leisure exertions over a longer period of time (not necessarily every day) are an equivalent of light to moderate activity, and rates for inactive adult males and females should be reduced by 600 and 500 kcal a day, while those for exceptionally active individuals should be boosted by 800–1,000 kcal. The increasing number of total energy expenditures measured with DLW technique provides the most accurate check of expected requirements. An analysis of all such measurements done in affluent countries and published during the first decade of the technique's use (Black et al. 1996) included only those values where total energy expenditures were also accompanied by measurements of BMRs, enabling to isolate the activity component. Generally, the measured TEEs were similar to FAO/WHO/UNU recommendations.

The highest activity factors, 4.7 and 4.5, were found, respectively, in cyclists taking part in the Tour de France and in men pulling a sled across the Arctic. Such maximal sustained metabolic scopes in humans, as well as in animals, are limited by energy cost of the most metabolically active tissues, and range between 4 and 7 (Hammond and Diamond 1997), a fraction of maximal short-term rates. Activity markups found by DLW studies for individuals leading normal lifestyles corresponded very well with expected multiples, ranging between 1.2 for sedentary subjects to 2.5 for physically active individuals.

Reports of food intakes equivalent to PALs lower than 1.45 BMR should be considered inaccurate, insofar as they could not support normal healthy life and would be accompanied by weight loss and reduction of physical activity (Livingstone et al. 1990). PALs between 1.6 and 1.8, means for all adult subjects in the analyzed studies, would tend to encompass activity markups for most healthy, active individuals. Most, but certainly not all: a UK study found that healthy housewives had PAL of 1.6, but the mean was a suspiciously low 1.3 in another group (Prentice et al. 1985).

Finally, because people in affluent societies spend most of their time indoors, in temperatures kept within the comfortable range of 18–25°C, and are usually well insulated by clothes against cold winter temperatures, climate-induced food energy markups for these populations are generally marginal. Wearing extra clothes in low temperatures will boost the usual energy requirement of individuals working in cold climates by 5–10 percent. Naturally, energy needs would increase further with cold-induced shivering in inadequately clad people.

Heavy exertions in hot environments, especially those above the normal body temperature of 37°C, which preclude radiative loss of body heat, also need an additional expenditure of energy in order to safeguard the organism from overheating. This is done primarily through sweating. Without it an average-sized adult would be able to lose, through respiration and skin diffusion, body heat at a rate just over 20 W; with normal profuse perspiration (at a rate five times higher than in the horse!), a hard-working man can lose more than 500 W and keep himself from overheating, and peak sweating rates can be more than twice as high (Smil 1991).

The first international recommendations of desirable food energy intakes used the concept of reference individuals, a twenty-five-year-old male weighing 65 kg and an equally old women weighing 50 kg, both consuming well balanced diet, engaged in moderate activities and living in the temperate zone (FAO 1950). Their average food energy needs were put at, respectively, 3,200 and 2,300 kcal/day, and they were reaffirmed by FAO's second committee on food intakes (FAO 1957a). Various formulas were given to adjust the intakes for body weight, age and ambient temperature. In 1971 the third meeting, conducted jointly with WHO experts, slightly redefined the activity levels of the two reference adults and lowered their average daily intakes to, respectively, 3,000 and 2,200 kcal/day (FAO/WHO 1973).

FAO's 1985 recommendations on human energy requirements were framed in terms of weight- and age-specific BMRs adjusted for different activity levels (FAO/WHO/UNU 1985). Using multipliers of 1.78 and 1.64 times BMR for, respectively, average male and female daily energy expenditures involving moderate level of activity, this new approach yielded rates of 2,978 kcal for the previously used reference male (virtually unchanged compared to previous recommendations), and 2,018 kcal (reduction of about 9 percent) for his female counterpart.

Additional allowances should be made for energy costs of pregnancy and lactation. The history of these recommendations epitomizes early exaggerations of dietary needs and their subsequent secular decline. In 1943 the first NRC report recommended that food energy intake be increased by 19 percent during pregnancy, and in 1950 the first British report advocated a 31 percent rise during the second half of pregnancy; in contrast, the latest versions of these recommendations ask, respectively, for a 14 and 10 percent increment (National Research Council 1943, 1989; British Medical Association 1950; Committee on Medical Aspects of Food Policy 1991). In absolute terms this translates to as little as 250 kcal a day during the second and third trimester—or even less as many women in advanced pregnancies sharply cut, or virtually eliminate, any physical activity beyond basic survival chores. Such a need can be satisfied by increasing food intake by an equivalent of two bananas a day.

As for lactation, when its energy costs were considered for the first time, the FAO expert committee assumed that the efficiency of human milk production is about 60 percent, and it recommended that breastfeeding women should increase daily intake by 1,000 kcal (FAO 1957a). Not surprisingly, the report concluded that some women find that such an increase is not easily achieved, and that an additional 800 kcal may be a more reasonable estimate. This recommendation ignored the fact that a significant portion of energy needs for lactation comes from depleting fat reserves deposited during pregnancy. Recognition of this fact, and the assumption of 80 percent efficiency of milk formation, led FAO/WHO/UNU (1985) to recommend additional 500 kcal a day during six months of lactation, a rate echoed by the latest US guidelines. Later in this chapter I will show that the real need may be lower still.

Protein Needs

Food proteins are required to supply a group of amino acids (ten in children, nine in adults) that cannot be synthesized in adequate amounts by mammals but are indispensable for humans in order to construct body proteins, as well as nitrogen for production of such compounds as hormones and neurotransmitters.

Inefficiencies and oxidations entailed by relatively rapid breakdown and reutilization of proteins in human body, excretions of nitrogen (mainly in urine, feces, and sweat) and shedding and cutting of skin, hair, and nails add up to small, but continuous, losses of the element and of essential amino acids that must be replaced by eating (FAO/WHO/UNU 1985; Pellett 1990b). Obligatory nitrogen losses in urine and feces remain fairly constant throughout adult life, averaging about 53 mg per kg of body weight (even the observed range, at 41–69 mg/kg, is not too wide). Other sources of nitrogen loss add another 8 mg/kg to that total.

Protein requirements are relatively highest during infancy when the rapid growth requires plenty of essential amino acids to synthesize new tissues; at that time breakdown rates of proteins will be also high. Amino acid requirements of infants can be best established by noting the consumption of breast, or cow's, milk that supports satisfactory growth. For older age groups most of the evidence for setting the recommended levels has come from direct studies of nitrogen balance. Unlike fat, protein cannot be stored in body tissues, and its excessive intake results in higher excretion of nitrogen-containing metabolites (moderately excessive intakes have no harmful effects, but extremely high-protein diets have been associated with a number of potentially serious disorders).

Requirements can be set by feeding protein below and above predicted adequate intake and then interpolating to the zero balance level. Perhaps the greatest shortcoming of this approach is that all but a few balance studies have been fairly brief. But a handful of long-term studies, running for one to three months, generally confirm that healthy adults can easily maintain nitrogen balance with daily intakes no higher than 0.6–0.7 g per kg of body weight as long as proteins come from animal foods.

Changing Standards

Although the approaches to setting protein requirements may appear to be fairly straightforward, finding lasting consensus has been an elusive task as protein controversies continue even after more than a century of relevant research. Although the long tug of war between the advocates of very high and relatively low recommended protein intakes has been basically settled in favor of more modest needs, uncertainties remain, particularly as far as the needs for specific amounts of individual amino acids and overall protein digestibilities are concerned.

The earliest studies of protein needs pointed out in opposite directions: some argued for very high, others for surprisingly low dietary requirements. Max Rubner, Germany's leading physiologist, argued that 165 g a day should be the norm for German soldiers (Rubner 1909). The requirement translated to no less than 340 g of beef a day per capita, and trying to fulfil this ruinously high rate greatly aggravated the country's food shortages during World War I as the requisite feeding of cereals and tubers to produce meat diverted staples from direct consumption, military and civilian.

Germany's General Staff might have done a better job, at least in securing adequate nutrition for the army, if it noted the work of Russell Henry Chittenden (1904), who concluded that 35–50 g of protein a day are adequate. Altogether some one hundred values of desirable protein intakes have been published since Chittenden's time, showing a great deal of variability, but no particular secular trend. When expressed in relative terms, these rates range from 3 to 4 g of ideal protein per kg of body weight for infants, to 1.5–3 grams for teenagers, and from 0.3 to 1 g/kg for adults.

FAO's first expert committee set the rate at just 2 g/kg for infants up to six months, and a steady decline of value for higher ages was interrupted by a bump to 0.8 g/kg during puberty (FAO 1957b) (figure 7.6). Less than a decade later the second assessment eliminated this short-term increase, lowered slightly the rate for children two to four years old—and raised by 60 percent the adult requirement (FAO 1965). At that time FAO and other United Nations agencies became engaged in a highly publicized campaign to avert what they saw as the impending protein crisis:

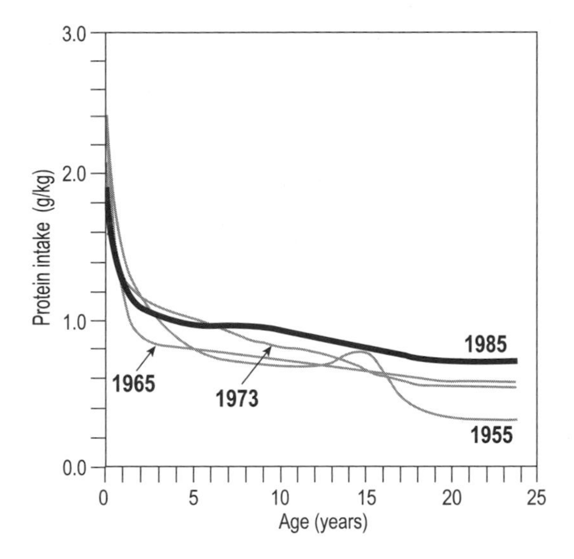

Figure 7.6
FAO protein recommendations.

according to that appraisal protein shortage was the most widespread disorder known to medical and nutritional science (UN Advisory Committee 1968).

Almost immediately, this concern was shown to be ill-founded. Cicely Williams—who first described protein-energy malnutrition syndrome, also known as kwashiorkor, among children in Ghana—pointed out that crippling protein deficiency accounts for only a small part of the hunger problem, mostly in regions where protein-poor cassava is a staple. Adding to these arguments, Sukhatme (1970) focused on the interdependence of energy and protein intakes: increasing the latter without satisfying the former offers no resolution of the problem.

In fact, India's average protein intake of the late 1960s was about 10 percent above the rate recommended by FAO's second assessment—but there were widespread food energy shortages. Eventually, the matter was found to be even more complex. As Bhattacharyya (1986) explained, because protein-energy malnutrition is a multifactorial disease it is impossible to isolate dietary factors from other causes, including infections. Protein and food energy shortages are among the causal dietary factors (micronutrients are usually also deficient), but energy intakes in some areas may be adequate, or even excessive, in relation to body weight or age.

The third FAO consultation on protein needs (FAO/WHO 1973) recommended a marginal lowering of adult rates while raising the requirements for children less than twelve years old (figure 7.6). Finally, in FAO/WHO/UNU (1985) set a standard of 0.6 g of high-quality protein per kg of adult weight, and then raised the rate to 0.75 g/kg in order to cover demand variations within a population.

Similarly, half a century of U.S. RDAs for adults shows a wavy pattern, while the rate for young children has been greatly reduced since 1958; this decline appears even steeper when the protein requirement is expressed as a percentage of total energy intake (Webb 1994) (figure 7.7). Even after it was reduced from 2.2 to 2.0 g/kg a day, the latest U.S. RDA for young children remains nearly 20 percent above the level that was found to support satisfactory rate of growth in healthy youngsters. The latest U.S. RDA also set a much lower recommendation for additional protein intake during pregnancy, just 2, 10 and 17 g/day during the respective trimesters, compared to the previous 30 g/day for the entire pregnancy.

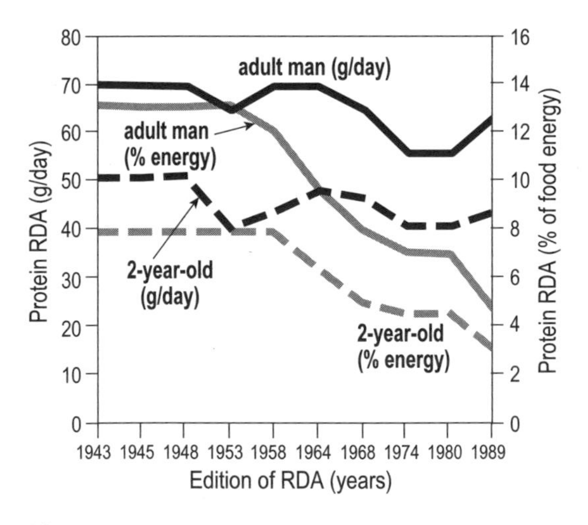

Figure 7.7
Changes in the U.S. recommended daily allowances for protein for 2-year-old children and 30-year-old men, 1943–1989 (Webb 1994).

Protein Quality

Adjusting RDAs to appropriate equivalents of actual everyday intakes of food protein is the main reason why uncertainties concerning protein needs would not end even if changing recommendations were replaced by immutable consensus. All recommended rates are given in terms of ideal, or reference, protein, or, more realistically, in terms of an actual protein possessing the dual advantage of having adequate amounts of all essential amino acids and being easily digestible; eggs and cow milk have been the two most common choices, but meat and fish proteins would also qualify. Naturally, proteins containing a suboptimal amount of one or more essential amino acids, or those that are not easy to digest, will not be able to support growth equally well.

The essential amino acids that are likely to limit the protein quality in typical mixed diets include lysine (present in relatively low amounts in cereal grains), sulfur-containing methionine and cysteine (relatively low in leguminous grains), threonine, and tryptophan. This means the value of actually consumed proteins must be adjusted according to the least abundant amino acid. This is particularly critical for children with their relatively high demand for essential amino acids needed for growth.

Mixed diets with plenty of animal foods, where all amino acids are always present in more than adequate amounts, will get fairly high scores; vegetarian diets including rice, beans, and corn may score almost as high—but those based on wheat, sorghum and cassava will, mostly because of relatively low amounts of lysine in those staples, score only around 70, or even as low as 60. In order to assure adequate amounts of all essential amino acids, infants have to consume 40–70 percent more of the actual dietary protein than if they would consume dairy products, meat, and fish.

As for the digestibilities, proteins in eggs, milk, and cheeses rate 95 percent and higher, and rates for meat and fish are nearly as high. Most staple grains have means in high 80s, but fine wheat flour averages 96 percent, whereas proteins in beans have digestibilities below 80 percent. Rates for mixed diets rise with higher shares of animal foods and highly milled grain: the U.S. mean has been around 95, but the average for vegetarian Indian diet may be just short of 80.

Net protein utilization, a product of amino acid score and digestibility, is well above 80 for typical U.S. diets, but in rural Asia, with diets dominated by staple grains and vegetables, it may dip below 50 (Huang and Lin 1981). Vegetarian wheat-based diets of poor Mediterranean countries would score similarly low. As a result, children growing up on such diets should consume daily 2.0–2.3 g of protein, rather than 1.0–1.1 g of milk or meat protein that would cover the needs of youngsters in more affluent settings.

Until a few years ago it was generally agreed that adult requirements for specific amino acids are much lower than those of infants, and hence they can be met by all normal mixed diets, including the largely vegetarian ones in poor countries. In practice it meant that as long as total protein and food energy needs were adequate there should be no concern about securing needed amounts of all essential amino acids for adult maintenance.

But three years after FAO published its latest recommendations, Millward and Rivers (1988) concluded that the estimates of adult amino acid requirements are too low, allowing either no more than minimum oxidative losses, or, in some cases, being below their obligatory level. However, they were unable to quantify that deficit. This task was eventually done by a group of Massachusetts researchers proposes, based on theoretical expectations and on ^{13}C-labeled amino acid tracer studies.

They suggested that the minimum values for amino acid requirements in adults should be set two to three times higher than the current, internationally accepted, standards (Young et al. 1989; McLarney et al. 1996). To buttress their argument they also noted that the currently recommended total protein requirements for adults seem to be anomalous when compared to other species: they are much lower than for any other mammal whose protein needs have been studied, while in infancy they are among the highest (figure 7.8).

A Massachusetts team concluded that the amino acid requirements recommended by FAO/WHO/UNU (1985) for adults are of questionable validity for practical use in nutritional studies and proposed an upgrade (Young and Pellett 1990). This upgrade roughly triples total daily amino acid requirements for adults, and the new amino acid pattern closely resembles the one recommended by the Joint FAO/WHO Expert

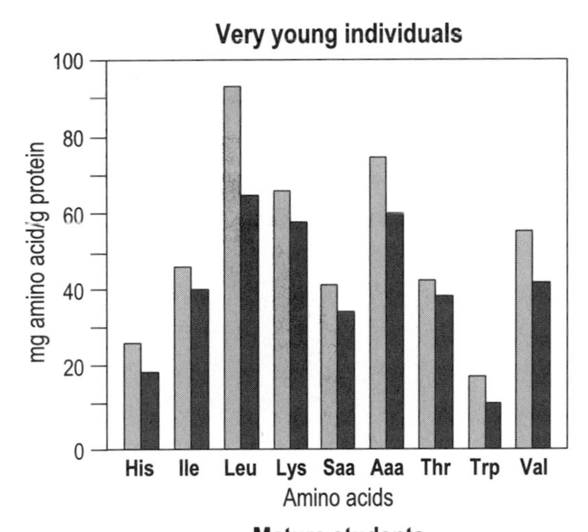

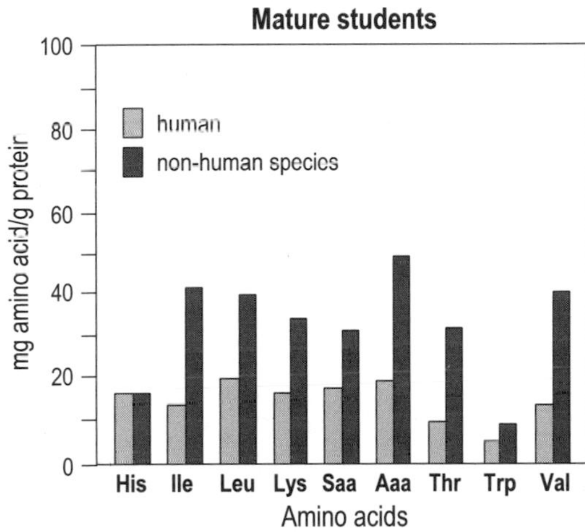

Figure 7.8
Comparison of specific essential amino acid requirements for humans and non-human mammalian species (McLarney et al. 1996).

Consultation in its 1991 report on protein quality evaluation for preschool (two- to five-year-old) children (FAO/WHO 1991). Echoing the arguments of the Massachusetts group, the consultation recommended that this preschool pattern should be used for assessing protein needs for all groups beyond infancy (for that age group the pattern based on human milk should prevail).

Needless to say, the new amino acid pattern would obviously tend to increase the demand for dietary protein among those adult populations who are largely vegetarian. But where mixed diets contain a large share of animal proteins such adjustments would hardly matter. In those cases there is actually no need for accurate calculations of amino acid scores and net protein utilization ratios, procedures requiring a great deal of requisite biochemical and nutritional information: a simple protein/energy ratio—the share of total food energy derived from proteins—serves as a handy approximation of the nutrient's adequacy.

Because food energy needs change a great deal depending on individual's activities, while protein needs remain much more constant, the ratio cannot be an accurate indicator of protein adequacy in individuals, but it works well for populations. If, as in the U.S. case, protein RDAs translate to protein/energy ratios between 7 and 11 percent, while actual intakes range between 14 and 18 percent, there can be little doubt about the overall adequacy of the nutrient.

Yet another set of new findings would tend to increase average protein needs of some groups. For most of the twentieth century nutritionists maintained that protein needs are not changed by exercise—and yet high-performance athletes have been consuming diets high in protein. Lemon (1996) investigated this contradiction and found that those adults engaged in regular endurance exercise should consume between 1.2 and 1.4 g of reference protein per kg of body weight, that is up to 25 percent above the normal RDA. Those practicing heavy resistance exercise that stimulates muscle growth should increase their protein intake to as much as 1.8 g/kg. But not all new research on protein requirements points to higher specific needs. Dewey et al. (1996) proposed that the amino acid requirements of infants between three and six months of age should be much lower than set by the FAO/WHO/UNU (1985). Arguments about protein intakes and amino acid patterns are far from settled.

Comparisons and Implications

Evaluation of basic nutritional status on a national level—findings of approximate balance, excess, or shortfall—must begin with representative averages. On the supply side, the ideal situation would be to have reliable, repeated food consumption surveys which would also give a fairly detailed information about average intakes not only according to basic demographic categories but also for various socioeconomic groups. In the absence of this information the FAO generates national distribution curves of access to food by combining the per capita dietary energy supply (DES) calculated from food balance sheets with information derived mostly from general income and expenditure surveys.

On the requirement side, acceptable calculations of human food energy needs on the national level cannot be done without knowing the population's age and sex structure: everything else being equal, populations with large shares of children or old people will have lower average per capita requirements than those dominated by adults in their productive years. Although age and sex structure may be readily available from the latest census data, average BMRs for various groups must be based on regression equations of, as we have seen, often questionable value, and prevailing levels of daily activity are just approximations of actual typical expenditures.

Even greater uncertainties complicate our attempts at quantifying the total number of people experiencing malnutrition and hunger. Looking at the history of these estimates, it is clear that their derivation became easier only in one respect: today's calculations usually do not consider protein intakes and are based solely on measures and assumptions of average food energy availability. This is quite appropriate because we have learned that adequate intake of food energy is only rarely associated with inadequate provision of dietary protein.

National Food Energy Needs

Calculations of average national food energy requirements thus begin with the latest available census data on the population's age and sex structure and with average body weights taken from anthropometric surveys or estimated from other sources (James and Schofield 1990).

Expected BMRs for specified groups are then adjusted for prevailing activity levels, and for additional needs due to pregnancy and diseases. For countries lacking good demographic and anthropometric information these calculations must rely on many simplifying assumptions.

FAO has been calculating regularly such national approximations. When assuming only light activity, relatively young populations and low body weights typical of low-income countries translate to daily per capita food energy needs ranging from just over 1,700 kcal to about 2,000 kcal during the early 1990s. Requirements of moderate physical activity raise these average per capita rates to between 2,000 and 2,400 kcal/day. I have done these calculations for six of the world's most populous nations and came up with the weighted per capita mean of about 2,100 kcal/day in 1995. Bender's (1994) estimate for more than 130 countries with 92 percent of global population averaged almost exactly 2,000 kcal/day.

Taking FAO's FBS-derived global food supply mean of 2,700 kcal/capita, adjusting it slightly in order to correct for the supply of unaccounted foods (to 2,800 kcal/capita), and comparing it to a generous estimate of average needs (2,200 kcal/capita), means that even groups and individuals whose intake would be 20 percent below the supply mean would be receiving enough food. This means that the current global supply is sufficient to provide adequate food for the world's population with a moderately uneven access to food. Most of the difference between harvest and availability rates is due to feeding of some 700 Mt of grain to domestic animals, an input which produces most of the 400 kcal/day available in animal products; post-harvest wastes are equivalent to about 600 kcal/day, and losses during distribution and retail and at the institutional and household level add up to about 800 kcal/day (table 7.1).

Moreover, Bender (1993) concluded that the current global consumption would have to increase by less than 2 percent if all people were to get enough food energy intake for their height. Obviously, more food would be needed to eliminate stunting which has reduced national food energy requirements by more than 15 percent in some countries. But even this change would call for relatively small intake adjustment: if everybody's height and weight were increased to desirable levels, global

Table 7.1
Estimated global per capita averages (kcal/day) of food harvests, supply, losses, consumption, and requirements in 1990

Edible crop harvests	*4,600*
cereals	3,500
tubers	300
pulses	100
vegetables, fruits, nuts	150
oils	350
sugar	200
Animal feed	*1,700*
grain	1,300
grain milling residues	300
other crops and processing residues	100
Apparent post-harvest losses	*600*
Food availability	
FAO total	2,700
plant foods	2,300
animal foods	400
FAO's food balance sheets +3%	2,800
Apparent distribution, retail, institutional, and household losses	*800*
Food consumption	*2,000*
Food requirements	
no stunting	2,050
no stunting and high activity levels	2,200

Source: Calculated and estimated by the author from data in FAO (1999) and Bender et al. (1993).

food energy consumption would rise by less than 8 percent. In aggregate, adjustments for desired weight for actual height (an increase which could be put into effect immediately) and the worldwide elimination of stunting (necessarily a change that would take a generation) would require less than a 10 percent increase of global food energy consumption.

Even if all of the poor world's populations would raise their health status and activity level to those prevailing in rich nations, the average 1990 global food energy consumption would have to increase by only

10 percent, and it would still remain well below the available mean supply. This conclusion "starkly confirms that the current pattern of malnutrition and hunger is unrelated to food availability, but is instead a function of global entitlements to food" (Bender 1994).

Even when assuming that China's average per capita food energy supply has already reached the global mean of 2,800 kcal/day (if not, it is very close), more than two billion people live in countries whose average DES is below that level. Even if none of these countries would have the average food requirement above the global mean, most would obviously experience different degrees of nutritional shortfall as average intakes of disadvantaged groups would be less than 80 percent of mean per capita food supply.

National, local and individual food deficits, ranging from marginal to crippling, are only rarely caused by absolute physical shortages that have not been effectively managed through transportation and distribution from surplus areas. This phenomenon arises most often amid wars (Afghanistan, Angola, Ethiopia, Mozambique, Somalia, and Sudan have been the worst recent entries) or in aftermath of major natural catastrophes (repeated washing away of Bangladeshi harvests). Much more commonly, insufficient individual or group access to food is strongly related to social status and income. This is common both in the richest as well as in the poorest countries. The extent of hunger and malnutrition in affluent nations has been a relatively understudied subject—even though the problem appears to be far from negligible (Riches 1997).

Estimates of the total number of Americans who cannot afford to buy enough food to maintain good health are particularly high, ranging from twenty-two to thirty million people (Poppendieck 1997). These findings have been questioned, but the link between poverty and inadequate nutrition is undeniable and even the most conservative estimates acknowledge that anywhere between ten and twenty million poor Americans could not feed themselves adequately without assistance, and that far from all of them are actually getting it. Surprisingly little attention has been also paid to very low levels of average food intakes coexisting with rising shares of obesity not only among adults but also among children are perhaps the most interesting outcome of comparing food consumption and nutritional requirements in rich countries.

Particularly low levels of average food energy intakes have been recorded by every dietary survey conducted in North America since the 1970s. Although FAO/WHO/UNU (1985) would recommend at least 1,840 kcal/day as a survival requirement with minimal physical activity for a 65 kg adult female, no U.S. food consumption survey has returned an adult female mean higher than 1,700 kcal/day during the past generation. The mean for the NHANES II was only about 1,550 kcal/day for women between twenty-five and fifty years of age, nearly 25 percent below the NRC's recommendations. Canada's one and only national nutritional survey, Nutrition Canada, put average food energy intakes of adult females at 1,752 kcal/day, while a survey of women older than thirty-five years, done as a part the Nova Scotia Heart Health Program (1993), found a much lower average rate of 1,540 kcal/day.

And yet these exceedingly low intakes coexist with an unprecedented frequency of adult female obesity, a situation strikingly obvious to anybody returning to North America from a visit to Asia or Europe. Where then does the bulk of the error lie: in exaggerated food supply, in underreported food intake, in unrealistically appraised metabolic requirements? Implications of this reality for making the whole food chain more efficient and for adopting more rational diets are obvious—but the finding raises no concerns about the adequacy of average supply.

The situation is quite different in low-income countries. There, too, average intakes are often surprisingly low—but given the combination of commonly narrow, or even nonexistent, gap between supply and intakes and higher nutritional needs of populations whose average physical exertion is still so much more common than throughout the rich world, this reality brings unavoidable concerns about the extent and degree of malnutrition.

Quantifying Malnutrition and Hunger

Not surprisingly, a great deal of effort has gone into estimates and calculations of the total number of people in low-income countries suffering from hunger and malnutrition. The resulting figures have been cited prominently, and usually without any enlightening qualifications, in debates about the world's food prospects. Their comprehensive reviews, including changing concepts and methodologies, can be found in Pellett (1983), Foster (1992), and Poleman (1996).

FAO's estimates of the global share of undernourished people have ranged from the high of two-thirds in the late 1940s to less than one-seventh in the early 1990s. The latest estimate amounts to about 840 million people, a reduction of 65 million in a decade, and a relative decline from 20 to 15 percent of the global total. Meadows et al. (1993), in their repeat display of meaninglessly aggregate modeling, concluded than one billion people had less food than required in the early 1990s.

On national basis, FAO estimated the highest shares of undernourished population in the early 1990s for Afghanistan and Somalia (73 and 72 percent, respectively), and the rates were put at 21 percent for India and 16 percent for China (FAO 1996). For the same period the FAO (1996) put the number of stunted children (with low height-for-age) at 215 million, underweight children (low weight-for-age) at 180 million, and wasted children (low weight-for-height) at 50 million.

In order to appreciate the reliability of these figures comparing food supply with recommended requirements, it is necessary to understand the quality of basic supply and consumption figures and the method of deriving the distribution of access to food. Previous discussions should have made it clear that any of the published food flow figures—be it for supply or intakes, in rich or poor nations, originating from international or national calculations—can be correct only by accident. Only when numerous errors and biases cancel each other we may have an accurate figure, but we do not know when that infrequent instance occurs.

The multitude of errors and biases inherent in the generation of these figures makes it a matter of an overwhelming probability that the final per capita values are neither the correct expressions of average food supply nor the representative levels of usual nutrient intake. The critical fact is that for most of the poor world's people we do not know with satisfactory accuracy (say with an error of no more than ±100 kcal a day per capita) either the food supply of actual individual consumption. This fact has implications for accepting any supply and intake figures, as well as for judging the reliability of estimates of access to food.

Even highly accurate data on actual average per capita food consumption would inform us about the extent of national nutritional inadequacy only if strict food rationing would ensure egalitarian access. Situations closely resembling this condition occur only rarely: World War II–era Britain and most of pre-1979 China came fairly close. Otherwise, even when total food supply may easily exceed the aggregated individual needs, disadvantaged segments of the population will be consuming inadequate amounts of food.

The two key assumptions behind this process of comparing supplies and needs include log-normal distribution of per capita food energy intakes (this makes it possible to calculate the levels of food consumption simply from the mean and the standard deviation) and quantification of the minimum per capita food energy requirements below which the average person's intake is seen to be inadequate. Shares of the population below the minima calculated on this basis are multiplied by a nation's population total to get an estimate of people with inadequate access to food.

Uncertainties inherent in this method abound. Those on the supply side have been discussed already, and they cut in both directions: real DES figures for many low-income countries may be anywhere between 5 and 15 percent higher than the rates derived from FAO's FBS, but actual intakes may be also appreciably lower due to waste not accounted for in food balance sheets. As for the typical food energy requirement, assumption of uniformly low activity may lead to some underestimates of actual energy needs. But by far the most important consideration acting in the opposite direction is the existence of metabolic needs significantly lower than those calculated by using BMR equations by Schofield et al. (1985).

As we already discussed, there is a body of quantitative evidence questioning the universal use of standard regression equations. I agree with Piers and Shetty (1993), who concluded that this practice leads to significant overestimates of food energy requirements of well-nourished populations outside Europe. Offering a single corrective value would be misleading, but available research indicates that reductions on the order of 5–10 percent would seem to be appropriate. For a well-nourished Asian woman of smaller stature engaged in light activity this would cut as much as 200 kcal a day, and a moderately active male would need up to 300 kcal a day less than predicted by standard calculations. These reduced rates would bring large numbers of people closer to, or above, the average DES.

No less importantly, many people who are not, according to standard assessments, well nourished have been able to make some remarkable adaptations to lower food supply. Nothing illustrates this reality more strikingly than different energy costs of pregnancy and lactation. The additional basal energy expenditure estimated by Hytten and Leitch (1964) at some 80 Mcal over the entire pregnancy has become entrenched in the literature of human energetics, but the total certainly does not have any universal validity. The first careful study questioning the value came from periodic measurements of BMRs in pregnant and lactating Gambian women (Prentice 1984). Reconstructed estimates for a twelve-month period show averages of just 1,485 kcal/day during pregnancy and 1,815 kcal/day during lactation. The BMR of these women would have to be no more than 90 percent of the value predicted by standard equations.

Later research found that those women who received supplemental nutrition increased their BMRs to the level suggested by Hytten and Leitch only when they were near delivery (Lawrence et al. 1988). Even more surprisingly, women who did not get any additional food actually lowered their resting metabolism during the early weeks of their pregnancy, and their pregnancy came almost energy-free, increasing their basal metabolism by a mere 1 Mcal!

Then a comparison of resting energy expenditures in a group of lactating Gambian women with a carefully matched control group of nonpregnant, nonlactating females led to the conclusion that the mean efficiency of their milk production was 94.2 percent (Frigerio et al. 1991). And, as at least three recent European studies cited by the authors show, this high rate is not an adaptive African singularity evolved in response to marginal food energy intakes. In fact, mean efficiency found in those three studies was at least 97 percent, leading the authors of the Gambian project to recommend an average value of 95 percent for the calculation of energy cost of human lactation.

As already noted, standard Western dietary recommendations call for increasing food intake by between 15 and 20 percent during pregnancy, and 25 percent during lactation. But reliable food intake measurements have repeatedly shown than in various poor countries women give birth to healthy babies while consuming much less food energy. They maintain genuine energy balance on what seem to be, by Western standards, incredibly low food intakes, not merely 5–10 percent, but 20–40 percent and even close to 50 percent lower than the expected requirement. Major weight loss, very low birth weights, or inadequate milk production would be undesirable consequences of reduced food energy intakes—but clear evidence of harmless adaptations means that there is a range of long-term averages of energy intakes compatible with regulation of expenditures without exceeding the homeostatic limits. And the same appears to be true for whole populations.

General appreciation of this reality still leaves one surprised at some actually measured rates (Benefice et al. 1984). Among Kenya's pastoral Turkana, average intakes of adult males were measured at only about 1,900 kcal, and among females at only about 1,400 kcal/day during the late dry season. Senegalese peasants had a seasonal food deficit of nearly 300 kcal/day compared to standard recommendations. Yet this deficit was not accompanied by any significant increase in malnutrition or by clinical signs of food deficiency. Children appeared to be relatively protected, and adults adapted by losing some subcutaneous fat and some weight. Similarly, in rural Ethiopia Branca et al. (1993) found insignificant seasonal changes of children's weight-for-height.

There can be no doubt that drawing the lines for undernourishment overwhelmingly on the basis of Western nutritional research, and setting single reference values for population energy needs ignores the existence of considerable human variation and of effective adaptation to lower energy intakes. This adaptation can take both physical forms (slower rate of growth and reduction in adult body mass) and behavioral adjustments (reduction of activity). But do not these changes carry serious economic and, more importantly, regrettable human costs?

Of course, these adaptations are not without costs: although they help to avoid both endemic starvation and restrictions of economic productivity, they limit body's fat reserves and they may weaken the chances of coping successfully with additional external stress. But, as shown by incontrovertible evidence, their overall impact is surprisingly modest. People from low-income countries are commonly shorter and lighter than an average person in an affluent country, but their fat-free mass may be only marginally reduced, or it may be actually higher than in individuals leading sedentary lives. Lower energy needs must be then explained by reductions in specific BMR and diet-induced thermogenesis, by curtailed free-time physical activity, and by increases in efficiency of muscular activity and absorption of dietary energy (Prentice 1984).

Many observations confirm that workers with low-energy intakes are often as productive as those with high food consumption, that a long-term adaptation to lower food availability can maintain good health (albeit in smaller bodies—but they will be energetically more efficient), and that there is hardly any correlation between food energy intake and time spent actually working (Edmundson and Sukhatme 1990). As Poleman (1993) concluded, "poor people long ago discovered how to allocate their resources so as to get by on what by the standards of the industrialized world is very little. It serves no purpose to deny this ingenuity."

But the matter of adjustments may become instantly emotional when one talks not about smaller bodies but about stunting. That very word carries a connotation of permanent and regrettable damage—but in reality, stunting has been a common human condition: indeed, even throughout today's affluent world most people born before 1950 grew up without expressing their full growth potential and have been noticeably outgrown by their children. The conclusion is clear: evidence of successful adaptations to reduced food availability confirms that not only exceptional individuals or particular groups, but whole populations can use their food energy much more efficiently than the standard expectations would have it; they can maintain energy balance with surprisingly low rates of food energy intake, amounting to as much as 20–30 percent less than Western-set metabolic expectations (Pollitt and Amante 1984; de Garine and Harrison 1988; Garby 1990).

Recommendations of energy intakes are then a pursuit of a moving target whose position changes not only due to differences in individual or populationwide metabolic efficiencies but also because of culturally conditioned work habits and attitudes, seasonal fluctuations in staple diets, and a host of genetic and environmental factors controlling the adaptive process. Clearly, there is no single preset minimum of food energy supply applicable to all populations (Borrini and Margen 1985). And in the short term there may be no obvious pattern relating intakes and expenditures of individuals; in the long run the two values necessarily coincide but large individual differences in homeostatic levels make any a priori applications of standard calculations even more error-prone than in the case of populations.

These basic uncertainties make it highly questionable to offer any confident estimates of national, and even more so of global, prevalence of malnutrition based on contrasting the known means of food energy supplies with expected requirements calculated by using standard BMR equations and activity markups. All we know is that *Homo sapiens* is a flexible convertor of food energy responding with altered metabolic efficiencies to different diets, environmental conditions, specific tasks, and health states. The question about food requirements is not simply "how much," but rather "for what" and "in what context." These questions remove the search for food requirements from the realm of quantifiable considerations to the much larger and largely unquantifiable setting of cultural preferences and social expectations.

Human energetics is so contextual and so value-laden precisely because it concerns humans. Borrini and Margen (1985) summed up its challenge well: before defining specific food requirements it is imperative to appreciate the perceived needs and wants of the people and their customs, the structure and dynamics of their society, and the ecology of their environment; in synthesizing these determinants we may discover that there may not be an answer at all, let alone a single answer!

Finally, all assessments of food inadequacy are critically influenced by the manner in which the distribution of energy intake is specified. We know that the distribution is unequal, but we have no direct measurements of this inequality in countries rich or poor. Coefficients of variation, measures capturing this inequality, must be derived from information gathered by household consumption surveys, preferably those that include explicit information on total income and food expenditures.

Such surveys contain many other sources of variation irrelevant to the distribution of per capita energy intakes. Not all of the variability caused by these influences can be eliminated.

Recent Chinese figures illustrate the extent to which unequal access to food influences nutritional status even in a country that has been extraordinarily successful in boosting its food production. In spite of the countrywide improvements, China's satisfactorily high nationwide food supply mean hides great urban-rural differences and major regional and socioeconomic disparities. Fairly reliable disaggregations show that in the early 1990s per capita food energy consumption in at least 5 percent of rural households, or of about forty million people, is less than 1,700 kcal, or not even 70 percent of the recommended intake (Ge et al. 1991).

These shortages show up in the extent of stunting: a growth survey conducted in 1987 showed that while 17 percent of children younger than six years were stunted in cities, the average rate was 40 percent in the countryside, and up to 60–80 percent among children in families with incomes below the poverty level (Ge et al. 1991). Moreover, derivation of coefficients of variation for those countries whose household surveys contain only data on total incomes and expenditures will be heavily influenced by assumptions concerning the elasticity of food demand. And for many countries there is not even any proxy information on food expenditures.

The linkage between poverty and food shortages is obvious, but complex: it is both very strong and surprisingly weak. Problems of definitions complicate the understanding of the link. Both poverty (from below-average incomes to abject destitution) and food shortages (from undernutrition to famine) are distributed along nondiscrete continua. Single-value cutoff points for minimum energy availability and minimum income do not respect this reality.

The degree of linkage between poverty and food shortages owes more to the capacity of a government able to set priorities, to bureaucratic competence, to the extent of cooperative farming practices, to cultural cohesion, and to relatively high educational level than to absolute per capita incomes and average daily food intakes. More than one of these desirable attributes must be present in order to safeguard food security, but a wide range of combinations is possible: just compare authoritarian and fairly cohesive China with democratic and culturally diverse India—or land-reforming Taiwan of the 1950s with Cuba of the 1990s. In general, it undoubtedly helps when a poor nation is democratic and has a free press: it may not be getting rich very fast, but effects of this combination on food security are clearly positive. Malnutrition may be fairly common, but famines do not happen.

There is also no universal normative solution for the food security challenge. Government intervention has worked repeatedly to increase food security: food rationing may be an imperfect, and inelegant, tool, but it can be highly effective for years and even decades, offering an opportunity for a gradual transition to better arrangements. Of course, it cannot work without basic bureaucratic competence. The same is true for public food-for-work projects (India could do them, Somalia cannot).

Taking care of food security by following surely the most touted economic model of the last generation—rapid export-oriented growth—has worked very well for its pioneers in Asia. But the high level of food security in Japan (few people realize that its per capita food consumption was below China's mean as late as 1953), South Korea, and Taiwan is predicated on massive food imports, a strategy not open to China or to sub-Saharan Africa.

Conclusions drawn from these realities do not make it easier to offer better estimates of undernourished people, but they should both provide a keener appreciation of widespread, and not easily reducible, uncertainties that preclude reliable quantifications and support a prudent consideration of qualified ranges. Even that task is highly challenging—and any global estimate of population receiving inadequate amount of food energy will have a considerable margin of uncertainty.

Assuming that FAO's recent estimates of undernourished people represent realistic maxima, plausible minima may be not just 5–10 percent, but even 20–25 percent lower. The most convincing argument for such lower values comes from long-term trends of infant and adult mortalities and average life expectancies. Although there is a broad correlation between these values and average food supply, the latter is often only a weak determinant of the former. Except for temporary breaks caused by wars, even the poorest nations have experienced steady improvements in all of these vital statistics. Some Asian gains have been very impressive: average life expectancies in China and Indonesia rose respectively, from around fifty years in 1950 to sixty to seventy years in the year 2000.

FAO estimated that among Sri Lankan children under the age of five years 38 percent were undernourished, 24 percent stunted, and 16 percent wasted, and that every fourth adult was undernourished. Yet the country's infant mortality is comparable to that of poorer European nations, and its life expectancy (just over seventy-two years) is close to the mean of industrialized countries. And it is also well ahead of life expectancy in Brazil (sixty-six years) or Egypt (sixty-four years) whose average per capita food supplies (2,800 and 3,300 kcal, respectively) are much higher than Sri Lanka's 2,230 kcal, and whose shares of undernourished and stunted children are estimated by the FAO to be much lower. Too many people around the world still experience food shortages. But, as Edmundson and Sukhatme (1990) conclude, "it is unwise and perhaps even immoral to examine the complex problems of human health only in terms of human energy needs."

How much food do they want?
What happens when poor Indians
or Vietnamese move to the U.S.?

8

Searching for Optimum Diets

There is no need to wait for the resolution of various uncertainties and controversies surrounding human nutritional requirements in order to begin promoting diets that minimize the risk of food-induced disease and maximize the chances of long and active lives. But the quest for optimized diets must recognize two key sets of realities that considerably complicate the task: nutritional transitions, the long-term shifts in typical eating patterns, have taken virtually all modern societies further away from what the broad consensus now sees as the most desirable diet; and this consensus itself is an evolving construct, made up of some basically undisputed cornerstones but also containing some questionable and controversial ingredients liable to be modified or even abandoned in the future.

The first reality means that attempts aimed at a greater acceptance of rational diet will be running against a powerful current of choices promoted by food processing industries and by providers of (overwhelmingly nutritionally undesirable) fast food services. The second fact means that we must eschew detailed quantitative recommendations in favor of general guidelines. In order to appreciate the impact of these two factors I will first outline major trends of often undesirable nutritional shifts accompanying industrialization and postindustrial modernization. Then I will look at some links between food, health, and disease, ranging from solid evidence of negative effects resulting from excessive, or inadequate, intake of particular nutrients or foodstuffs to speculative suggestions of great rewards of specific diets or micronutrients. Only then I will present recommendations for desirable diets.

Nutritional Transitions

As already noted (see chapter 1), gradual shifts in total food availability, in actual intakes and in the makeup of typical diets can profoundly change the basic patterns of food production, resulting in widespread diffusion of previously marginal crops, in marginalizations of old staples, in new preferences for different animal foodstuffs, and in wholly unanticipated concerns about long-term effects of nutrition.

Popkin (1993) identified several major factors determining the form and the pace of these nutritional transitions: demographic variables, above all shifts in population growth, age structure, and place of residence (urbanization); food industry (above all extensive grain milling) and state intervention (common promotion of animal husbandry); socioeconomic transformations bringing changes in women's roles (different time allocation promoting processed foods); changes in public understanding of diet's role in health (a factor with potential for very positive developments, but also one producing dubious results in populations searching for alternative healing through nutrition). I would add the effects of growing international trade and globalization of tastes.

The pace of these changes has been obviously both nutrient- and country-specific, but a wealth of historical data reveals a number of fundamental underlying patterns and several nearly universal trends marking the nutrition transition. On the most general level, total per capita food energy supply has increased with progressing affluence from preindustrial means of less than 2,500 kcal/day to satiation levels between 2,800 and 3,000 kcal/day: virtually all of the production above that level appears to be wasted. These satiation levels are achieved during early stages of modernization, at per capita incomes less than one-fifth of today's rates for the rich world (that is, generally below US$4,000/capita).

European rates illustrate well this reality: there is a striking similarity of average per capita food availability in spite of still very substantial income differences between the northern and southern parts of the continent. The top rates, all over 3,600 kcal a day per capita, come, not surprisingly, from Denmark and Belgium, but Ireland and Greece are also in that group, and Italy and Germany are not far behind. The lowest rates, besides Portugal's 3,100 kcal, are in the United Kingdom, the Netherlands, Spain, and France, all about 3,300 kcal a day. Total food energy supply among poorer Mediterranean countries is thus hardly distinguishable from the availability in affluent Atlantic Europe.

This overall rise of food energy intakes includes substantial increases in the supply of animal foods and hence also in intakes of animal proteins and lipids. In most preindustrial societies meat, fish, eggs and dairy products supplied no more than 10 and often less than 5 percent of all food energy; in North America and Europe they now provide around 30 percent of all food energy, while in those East Asian countries that have reached satiation levels (Japan, Taiwan) they supply around 20 percent of the total (figure 8.1). Changes in the level and composition of food intakes have been reflected in changes of body size and makeup. Diets rich in animal protein assured that the genetic potential for growth had become fully expressed in majority of young adults in affluent nations, and continuing gains in height and weight have marked the development of children in all rapidly modernizing Asian and Latin American countries.

Decline of Carbohydrate Staples

The rise of food supply to satiation levels also hides an even more impressive decline in typical consumption of starchy staples. Indeed, none of these nearly universal dietary changes has been as far reaching as this large decline in direct consumption of carbohydrate staples (Poleman and Thomas 1995). While staple carbohydrate foods—cereals, tubers and, to a lesser extent, pulses—provided at least two-thirds (and often over four-fifths) of all food energy in preindustrial societies, their share in modern economies is mostly below one-third, with minima just around fifth (figure 8.2). And this quantitative change has been accompanied by an important qualitative shift, particularly as far as cereals are concerned.

Dominance of cereal grains in all extratropical diets has been due to a combination of fairly high yields, relatively high food energy density, and good nutritional quality. In traditional agricultures cereal yields were commonly at least twice as high as for legumes; in modern farming the

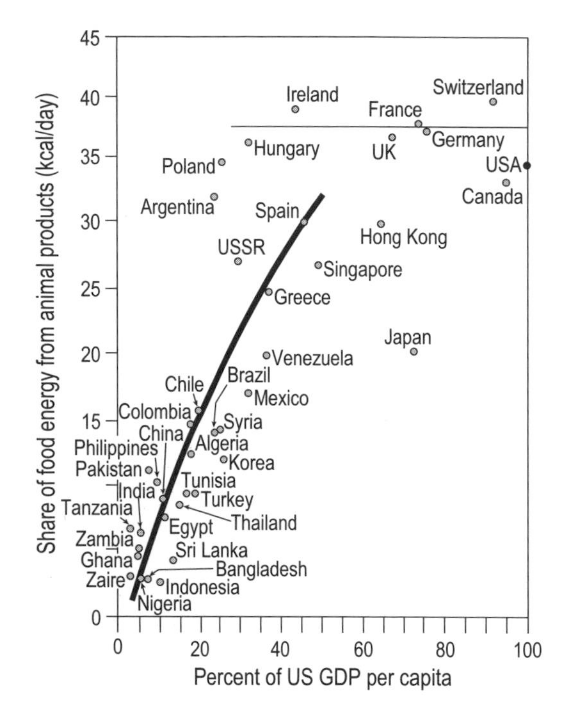

Figure 8.1
Shares of total available food energy derived from animal foods in relation to national GDP measured in terms of the purchasing power parity (Poleman and Thomas 1995).

difference is even larger, usually threefold (8 t/ha for Iowa corn vs. 2.5 t/ha for soybeans, 8 t/ha for English wheat vs. 2.5 t/ha for field peas). Energy density of cereal grains—averaging around 15 MJ/kg and differing by less than 10 percent among major cultivated species—is roughly five times higher than for tubers, and equal to moderately fatty cuts of meat. Milling does not change overall energy density of cereals and it increases palatability, but it also decreases nutritional value by removing vitamins and minerals present in grains' outer layers, and it reduces the beneficial intake of indigestible fiber.

Bulk of cereal food energy comes from carbohydrates, mostly in the form of highly digestible polysaccharides (starches). And cereals also contain appreciable amounts of protein: the typical mean of around 10 percent is about five times higher than that of tubers, and more than

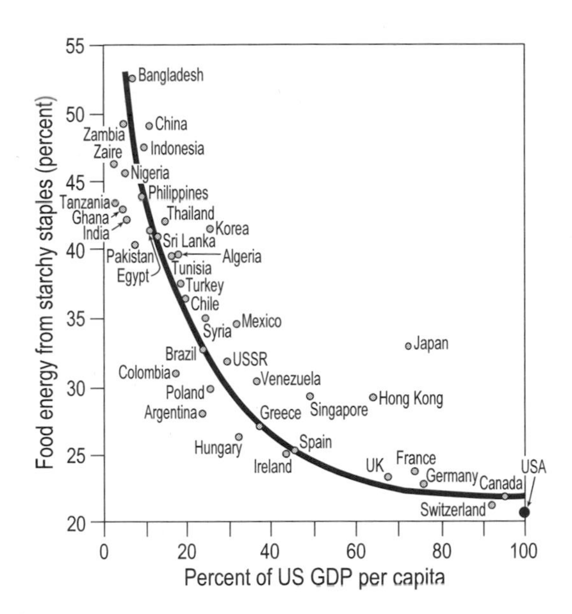

Figure 8.2
Shares of total available food energy derived from starchy staples in relation to national GDP measured in terms of the purchasing power parity (Poleman and Thomas 1995).

order of magnitude above the values for vegetables. Cereals are also extraordinarily versatile foodstuffs. Their grains can be consumed whole, crushed, or milled, and can be turned into an enormous variety of foods ranging from breakfast cereals to odd-shaped pastas by steaming, simmering, boiling, roasting, and baking.

Staple food cereals—wheat, rice, or corn—still supply a major part of average per capita energy intakes in affluent countries, but the recent shares of between 20 and 30 percent are invariably less than a half, and in some cases just a quarter, of levels prevailing in traditional settings. For example, the temporary rise during World War II aside, British consumption of grain fell from 49 percent of all food energy in 1880 to 22 percent by the early 1970s (Hollingsworth 1983).

The retreat of cereal consumption does not affect grain staples equally. Direct consumption of coarse grains—corn, millets, oats, and rye—is displaced by eating more rice and wheat. Cultivation of smaller coarse grains has been falling, but rising use of feed corn has been greatly surpassing any declines in harvests of the food grains. In a worldwide shift, consumption of wheat and rice is being displaced by higher indirect intakes of corn and soybeans, providing efficient feed for higher production of animal foods. Average rice intakes have declined with particular rapidity in the modernizing countries of East Asia. During the past generation Japan's average rice consumption has decreased by more than 25 percent, to 75 kg of milled grain in 1995 (Statistics Bureau 1995). During the same time consumption in South Korea fell by more than a third (to 120 kg in 1995), as did the Taiwanese intake–and these trends still continue, albeit now at a slower pace.

Wheat competes directly with rice as a more flexible foodstuff convertible to noodles, bread, and a multitude of leavened or unleavened baked products (Faridi and Faubion 1995). Since the 1950s leavened bread, the dominant staple of European diets for millennia, has made major inroads in every part of the world where its baking was traditionally absent. Diffusion of bread, together with expanding sales of a large variety of other baked goods, has been a key ingredient of adopting Western diet. Urbanization, and higher female participation in the work force, are major factors behind that diffusion: accordingly, populous and rapidly urbanizing nations will demand much more wheat.

The peculiar physical properties of two major proteins in wheat flour—of glutenin and gliadin—explain bread's unique qualities. White flour makes no nutritional sense (its production entails higher losses of total protein as well as the lower quality), but it makes baking a lot easier: the endosperm proteins are nutritionally inferior, but when combined with water they form a gluten complex that is sufficiently elastic to permit stretching and shaping of an unleavened and rising of a leavened dough, and yet strong enough to retain carbon dioxide bubbles formed during the yeast fermentation.

Yet although bread consumption is rising steadily in many non-European countries, it is generally declining throughout the continent where the food was by far the most important staple for millennia. French per capita bread consumption dropped from 600 g a day in 1880 to 170 g by the early 1980s, a 70 percent drop (Dupin et al. 1984), and Dutch bread eating fell by 40 percent during the first post–World War II generation (den Hartog 1992). Germany has shown a mixed trend, with per capita annual consumption falling from about 125 kg before World War II to a low of about 70 kg in the mid-1970s but rising since then to over 85 kg (Faridi and Faubion 1995).

Drastic declines of tuber consumption have had no exceptions throughout Europe or in newly affluent parts of Asia. French potato consumption was more than halved since the end of World War I (Dupin et al. 1984), and by the early 1990s Japanese potato consumption fell by about two-thirds compared to the year 1940 (Statistics Bureau 1995). Sub-Saharan Africa has been a notable exception to this universal trend: there the root and tuber crops—mainly yams and cassava—remain as major staples, and even higher income groups have not abandoned them in favor of grains. But these crops have received too little attention from modern breeders and their yields are far below realistically achievable means. This is particularly unfortunate because of the contribution of traditional tuber and root crops to food security: they can bridge seasonal food gaps, provide food reserve against cereal crop failures, and ease the decline of per capita food supply in deteriorating environments.

In this respect the decline of cassava cultivation is particularly unfortunate. Cassava (*Manihot esculenta*) has been for millennia an important source of food energy in South America and the Caribbean, and its post-1500 introduction to Africa and Asia has made it a leading pantropical crop (Cock 1982). Cassava toxicity (due to the presence of cyanide) rarely causes acute poisoning. Detoxification of the cyanide by appropriate processing (mashing, squeezing, roasting) eliminates the risk.

Although yields can be as high as 80 t/ha (about 30 t/ha of dry roots), traditional harvests are often only a fraction of that level; improvements should be possible simply by planting more productive clones. Even low-yielding cassava is a highly efficient producer of carbohydrate mainly because of its exceptionally high harvest index: dry roots may account for as much as 80 percent of the total plant mass. And because cassava has an indeterminate harvest period, it could become an important food reserve even in drier parts of Africa (Romanoff and Lynam 1992).

Not surprisingly, Prudencio and Al-Hassan (1994) found that cultivation of cassava in Africa has been positively related to population density and to the risk of climate-induced crop failure and negatively related to land quality, use of modern farming inputs, and access to markets. Given these facts, and given Africa's rapid population increase, the question of how sub-Saharan Africa's dietary pattern will change during the next two generations is of particular importance.

While the contributions of cereals and tubers to modern diets have been weakened, those of leguminous grains (pulses) have been reduced to marginal roles. Nutritional appeal of legumes has always rested on their unusually high protein content, and on the fact that their amino acid makeup complements that of staple cereals. Whereas protein levels in wheat, rice, and corn range between 7 and 14 percent, most legumes contain between 20 and 25 percent, and soybeans around 40 percent. These proteins are deficient in sulfur-containing amino acids (methionine and cysteine) but have abundant lysine, the amino acid pattern exactly opposite to the amino acid makeup of staple cereals (figure 8.3). Overwhelmingly vegetarian preindustrial societies relied on this protein complementarity—millennia before we understood its biochemical basis—by combining staple cereals with soybeans and beans in East Asia; with lentils, beans, and peas in India, the Middle East, and Europe; with peanuts and cowpeas in West Africa; and with beans in the Americas.

The nutritional quality of legumes is further enhanced by their relatively high levels of riboflavin, thiamine, calcium, and iron, and by provision of excellent dietary roughage by mature seeds. And, of course, several leguminous seeds, above all soybeans and peanuts, are rich sources of plant oils high in polyunsaturated fatty acids. Legumes are also flexible foods. Besides their long history as common staples, they can be eaten as vegetables (either as immature pods or seeds, or after sprouting), as uncooked dry or roasted seeds make good snacks (peanuts are a global favorite), or as various fermented products (Wang et al. 1979; Rose 1982).

And yet legume consumption has been falling virtually everywhere. Food balance sheets available for the pre–World War II period for a small number of non-Western countries reflect traditional patterns of legume

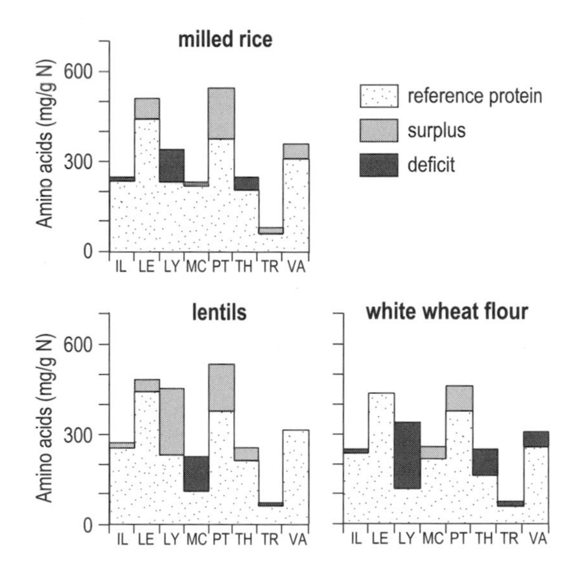

Figure 8.3
Amino acid composition of rice, wheat, and lentil proteins (Smil 1987).

consumption with annual per capita intakes as high as about 25 kilograms per year in India and well over 10 kilograms in most of Latin America. Such intakes provided between 5 and 15 percent of all food energy supply—and 15 to almost 30 percent of all protein. In contrast, in rich countries the pre-1939 supply of legumes already slipped to a marginal position, with legumes supplying no more than 0.5–2 percent of available food energy and 5 percent of all protein (den Hartog 1992; Dupin et al. 1984; U.S. Department of Agriculture 1920–1995).

After 1945 legume intakes fell on every continent. Given the much higher increases of other staple crop yields and rising availability of animal foods, relative importance of leguminous foodstuffs has fallen even more rapidly. Current average per capita supplies are just above one kilogram a year in the European Union, below 3 kilograms in North America and Japan, and below 10 kilograms throughout Latin America. And although India is the only major country whose per capita demand for pulses is increasing (official statistics show availability of about 14 kg during the latter half of the 1980s and about 19 kg during the first half of the 1990s), even there legumes now provide only about 5 percent of all food energy and 15 percent of all protein.

In many Latin American and African countries the importance of legumes has been reduced to levels prevailing in pre–World War II Europe, and throughout the industrialized world the intake of pulses has sunk to a marginal category that includes the disparate worlds of underclass poverty (canned beans) and voguish vegetarian experimentations (bacon-flavored beancurd). Outside India, lentil has been a major exception: its production increased by more than 70 percent during the 1980s, mainly because of expanded cultivation in Turkey and Canada (Muehlbauer et al. 1995). However, average yields remain low, mere 800–1,000 kg/ha in North America, and less than 700 kg/ha in India.

In sum, in terms of large-scale averages, legumes hold their own as staple foods only on the Indian subcontinent; they are becoming marginal in their traditional Latin American stronghold; and their nutritional contribution is inconsequential throughout the Western world. The principal reason for this demise is the unatractiveness of legumes in societies commanding plentiful supplies of animal protein. Nutritional detractions of pulses, above all complications in cooking and digesting, have been known since antiquity. Minimum cooking times for many mature seeds are much longer than those for cereal seeds and flours, on the order of two hours if the seeds are not presoaked. Although the food energy content of legumes is nearly identical to that of cereal grains, their lack of gluten complex makes it difficult to process them into such staples as bread and noodles.

Increased flatulence following higher intakes of legumes is due to the presence of indigestible oligosaccharides. Because the human digestive tract lacks the requisite enzyme (α-galactosidase), these sugars can become subject to anaerobic microbial fermentation resulting in an excessive gas production (Augustin and Klein 1989). All food legumes contain at least one, and most of them several, antinutritive factors (Ferrando 1981; Matthews 1989; Jones 1992). Enzyme inhibitors, mostly those interfering with the functioning of proteases (above all with trypsin and chymotrypsin), are especially common. Ironically, protease inhibitors make it more difficult for our bodies to use the proteins present in food legumes, which may be eaten specifically for their high amino acid content!

Concentrations of lectins (seed toxins agglutinating red blood cells) are relatively high in virtually all legumes, while goitrogens, cyanogens (glycosides releasing HCN), estrogens, antivitamin factors (most notably those causing rachitis and vitamin E antagonists) and toxic amino acids are less widespread, but prominent in some species. Of course, traditional cultures learned to live with the poorer digestibility of legumes, and they also discovered how to destroy, or at least substantially eliminate, many antinutritive factors by prolonged cooking and by specific processing. But the resulting inconvenience and the remaining risks accompanying insufficiently modified foodstuffs do not add to the mass appeal of legumes. Foods not requiring such elaborate preparation yet providing equal, or better, nutritional value, will be obviously preferable.

As soon as increased purchasing power enables people to buy more convenient and less risky foodstuffs, they follow one of the most notable universal nutritional shifts and begin reducing their consumption of legumes. Although the initial stages of incipient affluence may be accompanied by a slight increase of demand for pulses (a similar upturn is also often noticed for staple cereals), later developmental stages bring a progressing decline in the eating of traditional pulses.

The virtually universal retreat of complex carbohydrates eaten as cereals, tubers, and legumes has been accompanied by a nearly as universal growth of consumption of simple sugars, above all sucrose, a disaccharide commercially derived from sugar cane and sugar beets. Before the widespread cultivation of sugar beets, sugar was a marginal source of food energy in Europe, with annual per capita intakes uniformly well below 5 kg. By the beginning of the twentieth century the combination of spreading cultivation of sugar beets and rising imports of cane brought the average per capita rates in most of Europe above 10 kg—and by its end the typical intakes are ranging between 40 and 60 kg. Sucrose now supplies up to 20 percent of all food energy in Western countries, compared to less than 5 percent in East Asia's rice-eating nations.

In addition to the rising consumption of sucrose, there has been also a major increase in the use of dextrose (or fructose) and corn syrup produced from wet milled corn starch. In many Western countries these sweeteners have largely replaced sucrose in a wide variety of processed foods, ranging from pastry to ice cream, and in soft drinks whose consumption is yet to show any signs of saturation. The latest U.S. food intake surveys indicate a continuing increase of drinking carbonated sugary beverages—the rise was an astonishing 72 percent during the 1990s, to about 250 g a day per capita. Syrupy colas, beverages designed to provoke rather than to quench thirst, are becoming increasingly popular throughout the poor world, and particularly in modernizing Asia.

But, in energy terms, overall share of carbohydrates in affluent diets has still declined: higher consumption of sucrose and fructose has only partially compensated for declining intakes of starches in cereals, legumes, and tubers. Higher lipid and animal protein intakes have filled most of the resulting gap. The decline in average per capita food energy requirements, a result of less strenuous physical work in increasingly urbanized societies, obviously contributes to lower consumption of staple carbohydrates. Higher incomes pushing the eating pattern up the food chain—resulting in higher consumption of animal foods—are the most important reason. This development represents a major nutritional departure for an overwhelming majority of affected populations because diets high in both animal protein and in animal fat have not been the norm throughout human evolution.

Rising Consumption of Animal Foods

Given the great diversity of foraging societies, it is obvious that there was no typical gatherer-hunter diet. Intakes of both animal protein and fat were high among maritime hunters consuming plenty of fatty fish and marine mammals. The same was true about collectors of seashore mollusca, and some land hunters also enjoyed, at least seasonally, a fairly abundant supply of animal protein (Smil 1994b).

For example, Ulijaszek's (1991) models of pre-Neolithic and Neolithic nutrient intake in the Near East and Mediterranean region assume that in those populations where more than a third of food energy came from or originated in animal sources, proteins supplied at least 30 and up to nearly 50 percent of total all food. But average fat intakes among land hunters were modest as nearly all small, and most medium-sized animals are very lean (Smil 1994b). And diets of many foragers, both in arid grasslands and in rain forests, were low in both animal protein and fat.

The extremes of daily animal protein intakes reliably assessed in twentieth-century foraging populations ranged from exceptionally high rates of more than 300 g/capita among for Inuit feeding on whales, seals, fish and cariboos to less than 20 g a day for foragers in arid African environments subsisting mainly on nuts and tubers. Eaton et al. (1997) used nutrient analyses of wild plant and animal foods eaten by recent gatherers and hunters in order to estimate the dominant composition of prevailing preagricultural diets, which were very different from modern intakes (figure 8.4).

The representativeness of dietary patterns derived from these data is arguable, inasmuch as those populations of foragers that survived into the twentieth century and could be studied by modern anthropological methods were found mostly in environments where foraging existence is

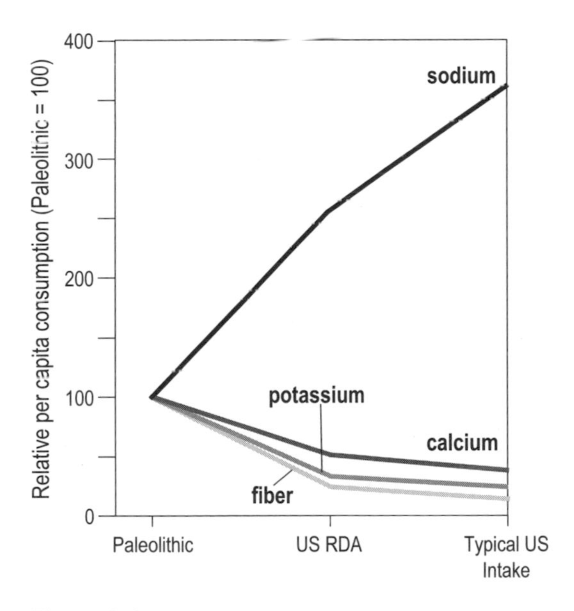

Figure 8.4
The greatest differences between modern and Paleolithic diets are in the average intakes of minerals and fiber; plotted from data in Eaton et al. (1997) and NRC (1989).

often precarious. Counterintuitively, this is also true of tropical rainforests. Because most of the animals in these forests are arboreal, and hence relatively small and inaccessible in high canopies, their hunting yields only low energy returns. Bailey et al. (1989) found no unambiguous ethnographic accounts of foragers who lived in tropical rainforest without some reliance on domesticated plants and animals.

With these limitations in mind, and taking into account the just outlined range of foraging diets, it is possible to make the following generalizations. High-fat maritime diets aside, preagricultural societies consumed no more than 20–25 percent of food energy as fat, with saturated fatty acids supplying well below the currently recommended maximum of 10 percent of daily food energy. As we have already seen, per capita means of meat consumption remained low during the whole preindustrial era, averaging usually no more than 10 kg/year. This means that animal foods provided rarely more than 15 percent of all protein intakes, and that saturated animal fats supplied just around 10 percent of all food energy. This pattern of low meat consumption persevered during the early phases of European modernization during the nineteenth century.

We do not have data that would enable us to reconstruct reliably the inequalities of the continent's meat consumption at that time, but given the low means of absolute intakes even highly egalitarian distributions would mean that the bottom half of the French population ate only about 20 kg of meat a year during the 1860s, and that the rate for the same group would have been barely above 10 kg in England (Fogel 1991). Remarkably, average British meat supply during the mid-1930s was less than 5 percent above the mean of the 1890s (Perren 1985)—and it was only after World War II when even Europe's richest countries achieved per capita meat consumption levels reached by the United States more than a century earlier.

Consequently, annual average consumption of meat equal to, or greatly surpassing, adult body weight is—few inevitable exceptions aside—a fairly new experience. Because of it all affluent countries now derive more than half of their protein from animal foods (shares between 55 and 65 percent are most common), and just over one-third of their food energy from lipids. The latter share is almost twice the rate for low-income nations whose fat intakes in average diets were equivalent to about 18 percent of all food energy in the early 1990s (FAO 1999).

The income-dependent nature of the transition to diets higher in fat is illustrated by recent Brazilian figures: shares calculated for the rural area of the Northeast (Brazil's most impoverished region), for suburban Rio de Janeiro, and for urban São Paulo (the country's most industrialized region) were, respectively 12, 23, and 28 percent (Chevassus-Agnes 1994). Increasing consumption of eggs and dairy products has almost always accompanied a country's rising meat consumption. It, too, is highly income-dependent.

Countertrends to eating more animal foods—driven mainly by increased health concerns among aging populations—are now clearly discernible in a number of countries. In the United States, eating of high-fat red meat declined by nearly 20 percent during the 1980s. Per capita consumption of animal fat has been going down in the United Kingdom and Canada, and low-fat foods are now widely eaten in the Netherlands and Scandinavia. In Norway the share of energy derived from animal fats fell by 20 percent during the last generation (Milio 1991). During the 1980s Americans reduced their share of food energy originating in lipids from about 40 to 34 percent, but that leaves the level still above the recommended maximum of 30 percent (U.S. Department of Agriculture 1996).

The critical determinant of future levels of animal food consumption will be the degree of westernization of diets in modernizing countries, particularly in populous Asian nations. To begin with, in spite of widely shared trends and broadly similar shares claimed by major foodstuffs, there are substantial differences in meat and fat intakes among Western countries. While per capita GDPs and disposable incomes among the richest European countries are very close, per capita meat supplies range from less than 60 kg/year in Norway to almost 100 kg/year in France (FAO 1999). National peculiarities accompanying westernization of diets are no less surprising. Japan's per capita consumption of dairy foods, nonexistent in 1945, now equals about 40 percent of the high U.S. level—but the country's mean annual per capita intake of all sweeteners is still less than half of the U.S. rate (Statistics Bureau 1998).

Seckler and Rock (1995) suggest that there are two distinct patterns of food consumption representing two alternative attractors in forecasting future composition of per capita food supplies. The Western pattern has the daily mean of more than 3,200 kcal/capita, and more than 30 percent of food energy comes from animal foodstuffs; the Asian-Mediterranean pattern has overall inputs below 3,200 kcal/capita and animal products supply no more than 20–25 percent of food energy.

Historical perspectives show that the consumption of animal foods in the countries with Asian-Mediterranean diet is not moving rapidly toward the Western pattern. In Egypt and Turkey the proportion of meat in average diets has hardly changed in thirty years. Taiwan's average per capita meat consumption had more than tripled in three decades, from 23 kg in the early 1960s to 72 kg in the early 1990s; but the Japanese one has stabilized at around 40 kg, similar to the Malaysian mean. Trend lines of the overall per capita consumption of animal foods in North America and in the European Union have been mostly negative, moving in the direction of the alternative attractor.

Asian populations will grow in coming generations, and as the desirability of Mediterranean diets becomes even more appreciated throughout the Western world, there is a considerable probability that the Asian-Mediterranean attractor will prevail. If these trends continue then the widespread assumption of high income elasticity of demand for meat will not be realized. Naturally, this reality would greatly lower the future demand for feed grains and bring somewhat higher need for food cereals.

Nutrition, Health, and Disease

A variety of old and new data shows that the heritability of life-span accounts for no more than 35 percent of its variance in mammals, including humans (Finch and Tanzi 1997). This leaves a great deal of opportunity for influencing nonheritable environmental factors of which nutrition must be a key one. Indeed, the modern Western diet—made up of highly processed foodstuffs containing a large share of fat in general and of its saturated variety in particular; too much refined sucrose, cholesterol, and salt; and too low levels of several vitamins, minerals, and indigestible roughage—has been singled out as a key reason for the epidemic of obesity, a disease that induces and aggravates a variety of serious ailments. And even among people who are not obese this diet has been blamed for the rising incidence of several civilizational diseases.

Diets containing more than 25 percent of all food energy in animal fat—combined, naturally, with much reduced physical activity—are thought to be one of the main causes of widespread obesity, be it in North America or China. Because of pronounced human sexual dimorphism, fat tissues make up normally some 15 percent of body mass in young adult males but almost twice as much, about 27 percent, in young females (Bailey 1982). Group variability is much larger: total fat content may be less than 10 percent in elite athletes, but more than 40 percent in overweight North American females (their maxima go up to 55 percent).

Obesity is usually defined as having at least a 35 percent excess over ideal body weight. U.S. national health and nutrition surveys show that the prevalence of obesity, which was basically stable between 1960 and 1980 (at about 25 percent of the adult population), increased by 8 percent during the 1980s (Kuczmarski et al. 1994). By the early 1990s the mean weight gain of 3.6 kg had made every third U.S. adult obese, with the highest weight increases among men over fifty and women between thirty and thirty-nine and fifty and fifty-nine years of age (Flegal 1996). The situation is only slightly better in Canada (van Itallie 1997): half of the country's population is overweight, one in four people is obese.

An even more alarming picture emerges when looking at actual weights associated with the lowest mortality: for U.S. adults these values from life insurance statistics are close to the 25th percentile weights for height, putting three-quarters of adults above the optimal body mass. While the actual median body weight of adult males is 79 kg, the weight associated with the lowest mortality for the median height is only around 70 kg; for females the difference is smaller, 62 vs. 59 kg.

Higher incidence of obesity has been also recorded in many European countries, and it is not limited to affluent nations: its prevalence in some lower-income countries is as high or higher than in the United States and is increasing rapidly (Popkin and Doak 1998). Germany, the United Kingdom, Mexico, Egypt, and South Africa are some of the populous countries with high shares of adult obesity. In China the recent shift to a relatively high-fat and high-sugar diet is already being blamed for rising obesity. A survey in Beijing discovered that more than 60 percent of women aged forty-five to sixty-nine were overweight, and the nationwide proportion of overweight city adults rose from 9.7 percent in 1982 to 14.9 percent in the early 1990s, with the capital's rate at over 30 percent, similar to North American incidence (Ge 1991; Cui 1995). I hasten to add that in China a growing degree of urban obesity coexists with a surprisingly high incidence of undernutrition and pronounced infant stunting (Popkin et al. 1993).

Of course, obesity is not just a matter of carrying excess fat: considerable body of research confirms that it is an insidious diseases whose rising incidence reduces longevity and is strongly associated with a variety of health problems (Garrow 1988; Belfiore et al. 1991; Cassell 1994). Epidemiological studies link obesity with such major metabolic consequences as insulin resistance, hyperglycemia, hypercholesterolemia, hypertension, and gallbladder disease; obesity's most common structural impacts are orthopedic impairment, pulmonary difficulties, and surgical risk.

Increased mortality is thus due mostly to type II (non-insulin dependent) diabetes, coronary heart disease (CHD), stroke, and certain malignancies (colon, rectum, prostate, breast, ovary). To illustrate just a few well-established links, figure 8.5 shows a curvilinear relationship between body mass index, blood cholesterol level and diastolic blood pressure (Bray 1997). Up to 90 percent of non-insulin dependent diabetes, between a quarter and a third of CHD and cancers, and nearly a quarter of total premature mortality can be attributed to obesity among nonsmokers. For the United States Bray and Gray (1988) estimated that CHD incidence could be cut by 25 percent and congestive heart failure and brain infarction by 35 percent if the country's entire population were at optimal body weight. Birmingham et al. (1999) estimated that the total direct cost of obesity in Canada corresponds to 2.4 percent (and perhaps up to 4.6 percent) of the country's health care expenditure for all diseases.

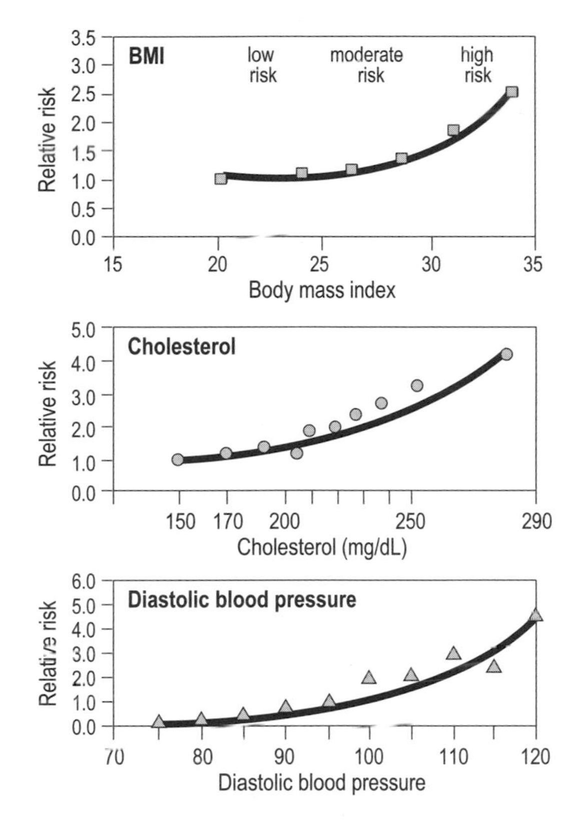

Figure 8.5
Relationships between relative risk of the CHD, body mass index, blood cholesterol, and diastolic blood pressure (Bray 1997).

Obesity cures appear to be exceedingly difficult in an environment saturated with a heavily promoted supply of palatable food and fostering an extremely sedentary life style. Percentage of body fat, its distribution (upper- vs. lower-body obesity), family history, past dieting experience, and underlying medical conditions must be all considered in formulating weight-reduction strategies that could make a real difference: a "magic bullet" strategy for the treatment of obesity is unrealistic (Dausch 1992; Woods et al. 1998). Return to excessive weight after a period of successful dieting is a norm among people trying repeatedly to reduce their mass. The most obvious prevention is not to adhere to any of scores of dubious diets but to balance total food energy intake (in the form of normal, varied meals) with basal metabolic and activity expenditures (not necessarily on a daily basis).

Besides increasing the incidence of cardiovascular diseases and cancer indirectly, through their contribution to more widespread obesity, modern diets in general, and some of their ingredients (or lack of them) in particular, are also considered to be either a major cause of these two leading causes of mortality in modern societies, or at least a very important contributing factor to their development.

Coronary Heart Disease and Diets

Although in most Western nations the total rates of cardiovascular mortality have been declining, coronary heart disease (CHD) still claims every year more lives than all cancers combined. For middle-aged and old men in the rich countries heart attack (myocardial infarction [MI], the death of heart-muscle tissue) remains the leading cause of dying—or of becoming seriously incapacitated while awaiting the return of the dreaded episode.

Some past attributions of higher incidence of CHD to inappropriate diet were unequivocal: "There is no doubt that diet and diet alone is responsible for the vast majority of all coronary heart disease in Western society" (Rees 1983). According to what might be called a standard explanation of atherosclerosis, the dangerous multifocal process starts with the deposition of fat, particularly cholesterol ester, in the inner layer of arteries and it culminates in extracellular cholesterol ester crystals reducing and eventually blocking the blood flow; fibrous and calcification of the arterial deposits are considered to be inflammatory reactions to cholesterol's presence.

In the late 1940s and the early 1950s Ancel Keys of the University of Minnesota emerged as the most influential proponent of linking this degenerative arterial process with environmental factors, prominently including nutrition (Keys 1948, 1952), and the outgrowth of his early work was the Seven Countries Study initiated in 1958 and completed twenty-two years later (Keys 1980). Certainly the other most influential of several long-term prospective projects linking CHD and diet has been the Framingham Study, whose initial examinations began in the spring of 1950 and whose twenty-four-year results were published in 1980 by Dawber (1980). The study followed just over five thousand men and women aged thirty to sixty-two years and provided "overwhelming evidence" that the blood cholesterol level is a "powerful factor" in development of CHD.

Although the study found that differences of dietary intake did not account either for the intrapopulation differences or for the effect of cholesterol level on the relative risk within the population, Dawber (1980) argued that "in order to ascertain the effect of diet on cholesterol level . . . we must look outside this population" because "the usual dietary intake has a powerful effect in determining the average cholesterol level." This has been, of course, also one of the principal conclusions of the Seven Countries Study.

Although Keys (1980) could not demonstrate correlation between individual serum cholesterol levels and estimated nutritional cholesterol and fat intakes within the cohorts consuming relatively homogeneous diets, he argued that such studies can reliably characterize "the central tendency of the group" so that for the cohorts with markedly different diets the relationship between blood cholesterol and nutrition becomes demonstrable. Keys concluded that CHD rates show "a strong direct relationship with saturated fat in the diet."

Perhaps the most impressive of these differences and implicit relationships came out in Keys's comparisons of Americans and Finns with their high CHD mortalities and high saturated fat intake on one side and the low incidence of myocardial infarction in Japan, a country of minimal animal fat consumption. Not surprisingly, both the Framingham Study and the Seven Countries Study have received considerable media attention during their decades of progress and their fundamental message—desirability of lowering cholesterol intakes—was translated into some significant dietary shifts.

In the United States consumption of butter fell to half of the 1950 value by 1970 while that of plant oils, mostly eaten as margarine, rose by 70 percent; demand for eggs weakened by about 20 percent while poultry sales rose nearly twofold; these trends have continued, albeit at slower rates, after 1970 (figure 8.6). Dietary intervention as a key

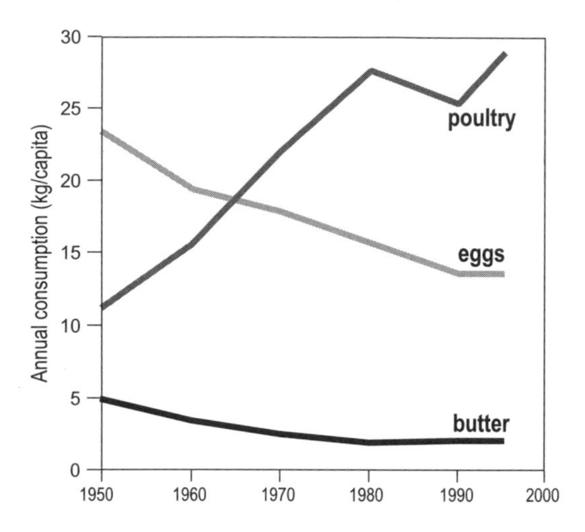

Figure 8.6
Trends in U.S. butter, egg, and poultry consumption; plotted from average supply data in U.S. Department of Agriculture (1950–1996).

prevention effort to cut the incidence of CHD had been a constant advice for a generation as many epidemiologists, heart foundations, and media kept spreading the message.

There has been no shortage of subsequent research proving the diet-cholesterol-CHD link. Perhaps most interestingly, a unique set of one hundred years of Norwegian dietary data shows a doubling in fat's contribution to total food energy (from just 20 percent in 1890 to 40 percent in 1975), a change paralleled by an increase in serum cholesterol corresponding to a 60 percent rise in risk for coronary heart disease (Johansson et al. 1996). A fall in lipids' contribution to 34 percent of all food energy by 1992 then brought a 30 percent reduction in CHD risk.

There has been also no shortage of research questioning the degree and the consequences of the CHD-cholesterol link. For example, Renaud et al. (1995) demonstrated that the cardioprotective effect of Cretan diet was not related to serum concentrations of cholesterol (total, LDL, or HDL) but to changes in plasma fatty acid resulting from higher intakes of linolenic and oleic acids, and also to higher plasma levels of antioxidant vitamins C and E. A WHO-sponsored study identified low blood level of vitamin E as the most important factor in MI, more important than high cholesterol, elevated blood pressure, or smoking (Gey et al. 1991). Inhibition of lipoprotein oxidation is the most likely mode of the vitamin's action (Tangney 1997).

And a survey done by the American Dietetic Association (1991) revealed that because cholesterol was seen as the single most important nutrition concern, one-fifth of all respondents believed that they should eliminate fat from their diets. As Harper (1996) noted, this has the potential of recreating a problem that has been solved in a modern society, namely an adequate supply of fat and food necessary to prevent retarded growth of children.

This clash of conclusions is hardly surprising given the complex etiology of CHD. At the beginning of the 1980s Hopkins and Williams (1981) compiled a list of no fewer than 246 CHD risk factors, to which new items are being steadily added. The best known, and certainly the most important items on the list, are family medical history, being a male over the age of forty, having high blood pressure, elevated blood sugar, smoking heavily and drinking immoderately; the most intriguing ones are not drinking plenty of Bordeaux or not eating large quantities of mackerel.

Wine drinking became the best explanation for the French paradox, the fact that consumers of diet as high as or higher in saturated fat than the one eaten by Americans have a much lower CHD mortality (Perdue et al. 1992; Bloomgarden 1994). Cardioprotective value of regular consumption of moderate amounts of wine has become a widely accepted explanation, but debates continue regarding the kind of wine, efficacy of other alcoholic beverages, and their optimum daily intakes (Smil 1989; Klatsky et al. 1997). Still, there is no significant difference in overall rates of mortality and life expectancy between the two countries, or between any other two affluent nations with different drinking patterns.

I will mention just one more epidemiological example showing that diet may not be as critical a factor as is often portrayed: a recent Spanish paradox. Spain's CHD mortality has been declining even as meat and dairy intakes (and hence saturated fats) have been increasing and consumption of olive oil and foods rich in complex carbohydrates has been dropping (Serra-Majem et al. 1995). The best explanation appears to be expanded access to health care, improved control of hypertension, and reduced smoking; increased consumption of fruit and fish may be a contributing factor.

The evidence is also far from clear-cut as far as the individual foodstuffs are concerned. Oat bran mania swept a number of affluent countries during the 1980s, but the lipid-lowering effect of oat products, found to be substantial in metabolic ward studies, appears to be only negligible (lowering the total cholesterol by 2–5 percent) or absent in free-living subjects (Ripsin and Keenan 1992). A meta-analysis of studies of the relationship between ischemic heart disease and markers of vegetable and fruit consumption showed that the risk of the disease is about 15 percent lower at the 90th than the 10th centile of intakes (Law and Morris 1998).

Critique of the mass dietary changes does not imply any endorsement of high-fat diets. As already stressed, there are other major Western health risks whose links to a high-fat diet have been demonstrated much more convincingly than in the case of CHD and whose early onset (often much earlier than in the case of CHD) could be delayed and whose incidence could be reduced by diets lower in total and saturated fats (U.S. Department of Health and Human Services, 1988; Lichtenstein et al. 1998). Lowering of fat intakes and replacing a large share of simple sugars with grains, vegetables, and fruits should remain the most desirable shift in Western eating patterns. Although this transformation may not add any years to most lives, it can enhance the quality of life in later years by considerably reducing the burdens of chronic ailments and improving personal mobility.

Cancer and Nutrition

The earliest descriptive epidemiological studies looking at changing rates of cancer mortality among migrants provided a number of relatively strong correlations between incidence of various cancers and per capita consumption of specific foods. Doll and Peto (1981) estimated that diet might contribute to anywhere between 10 and 70 percent of all U.S. cancers. Soon afterward the National Research Council Committee on

Diet, Nutrition and Cancer published its first dietary guidelines to reduce the probability of cancer (National Research Council 1982).

If every claim made about anticarcinogenic benefits of particular foods, drinks, and herbs made since that time would be taken at its face value then this section could be just a list of items. Such claims come not only from the expected sources—from traditional food lore, ancient pharmacopoeias, and the burgeoning alternative medicine and wellness movement—but increasingly also from refereed scientific publications and state-of-the-art summaries of the latest medical understanding. I will note, with a single exception, only those cancer-nutrition links that are generally accepted, or at least seen as highly probable candidates, on the basis of solid epidemiological and physiological evidence.

The exception concerns what was initially portrayed as perhaps the most persuasive connection between a malignancy and diet, involving a food ingredient which we cannot actually digest. Burkitt et al. (1972) noted that cancer of the large intestine is almost unknown in tropical Africa—and attributed its high incidence in affluent countries to decreased consumption of indigestible dietary fiber, which prevents the disease by causing a more rapid transit of the contents of the large intestine. Their findings were widely accepted and recommendation of regular intake of high-fiber foods has been among a few constant items on the list of desirable dietary practices.

Nutritional justification for this advise appears to be solid. Reconstructions of preagricultural diets indicate that fiber intakes surpassed 100 g/day, and diets of preindustrial societies contained at least 40–50 g/day (Eaton et al. 1997). Diets rich in fiber also have low energy density and low fat and refined sucrose content but are rich in starch (Birch and Parker 1983). In contrast, most people in rich countries consume diets with less than 20 g of fiber a day. Higher intakes of dietary fiber are not so difficult to achieve—but their anticarcinogenic effect are a different matter. Wasan and Goodland (1996) concluded that claims that up to two-thirds of the variation in incidence of colon cancer are ascribable to differences in total fiber and fat intake are wrong—and they actually warned against using fiber-supplemented foods.

The existence of different kinds of fiber—ranging from insoluble cellulosic compounds in whole grains to soluble pectins and gums in legumes, oats, fruits, and vegetables—complicates the appraisal of its efficacy in diseases prevention. Epidemiological evidence is unclear. In South Africa colon cancer remains rare even in urban areas that have experienced increased intakes of food energy and fat and falls in dietary fiber, changes paralleled by substantial spread of dental caries, obesity, hypertension, and diabetes (Walker and Walker 1995), and European rates of colon cancer differ widely in spite of similarly low intakes of fiber.

Two basic recommendations for dietary cancer prevention are identical to those for lowering CHD risks: avoid obesity and reduce total fat intake. Cancers of uterus, gallbladder, kidney, stomach, colon, and breast have statistically significant correlation with obesity, while cancers of colon, prostate, and breast have been linked (not always strongly in every published study) with excessive fat intake. On the other hand, case-control and prospective studies in humans have provided little evidence of a link between dietary fat and cancer (Carroll 1991). In contrast, by the mid-1990s nineteen studies found that a higher consumption of at least one category of vegetables or fruits is associated with lower risk of various cancers (Kushi et al. 1995; Steinmetz 1996).

More specific recommendations, going back to the first NRC report on diet and cancer adequate intake of vitamins A and C, and regular eating of cruciferous vegetables, that is mostly many varieties of genus *Brassica*, which includes, among others, cabbages, broccoli, cauliflower, and kohlrabi (National Research Council 1982). Sulforaphane—and other related isothyocyanates present in cruciferous vegetables, carrots, and green onions—are potent inducers of enzymes involved in inactivation of carcinogens (Prochaska et al. 1992).

Of course, all fruits and vegetables contain complex mixtures of pro- and antioxidants, carcinogens, and anticarcinogens: the best evidence is that when eaten as part of sensible, that is balanced and varied diet, their anticarcinogenic effect dominates. Although the American vitamania has been particularly exuberant (Apple 1996), eliminating the intake of unwanted compounds in fresh fruits and vegetables and turning instead to megadoses of synthetic vitamins is not advisable. Vitamin C is a perfect case in point: as a redox agent it acts usually as an antioxidant in moderate quantities but often it is an oxidant in megadoses (Herbert 1993).

Enstrom's (1993) analysis of standardized mortality ratios and vitamin C intakes concluded that both deaths due to all causes and due to all cardiovascular diseases were significantly lower among males taking more than 50 mg of vitamin C supplements; the effect was much weaker among women. As a result, he argued for regular supplements up to a few hundred mg a day, but not for megadoses on the order of several grams. Similar arguments have been made in the case of vitamin E, an antioxidant protecting cells from free radicals whose damaging actions have been implicated in a large number of clinical conditions. High blood levels of vitamin E have been associated with statistically significant lowering of risks of several cancers (lung, stomach, breast, cervix). Typical intakes of foods high in vitamin E (plant oils, nuts, and whole grains) could be supplemented by daily doses of 100–400 international units.

Since the 1970s the spreading habit of taking vitamin supplements developed a distinct subculture of vitamin megadoses, a shift spearheaded and made credible to many by Linus Pauling, a double Nobel laureate who consumed daily 18 g of vitamin C, that is, three hundred times the recommended dose of 60 mg for adults (Pauling 1986). In assessing the effect of megavitamin therapies it is extremely difficult to distinguish a truly independent contribution of vitamin supplements from their role as surrogate markers for a number of variables associated with a generally healthy lifestyle and better access to medical care. In sum, CEOs take supplements, homeless drifters do not, and blanket claims that megadoses of vitamins are harmless are both wrong and irresponsible.

The dangers of rushing into specific recommendations are illustrated by the recent controversy concerning the effect of green tea. Jankun et al. (1997) concluded that green tea prevents various cancers by inhibiting urokinase, an enzyme that is commonly overexpressed in human cancers. But Yang (1997) found that hypothesis a misleading and unlikely explanation, and cautioned against any sweeping conclusions concerning the inhibition of tumorigenesis by tea, black or green.

Optimized Diets

The need for good dietary guidelines, for their effective promotion and widespread acceptance is obvious in both rich and poor countries. In the United States, amid the excessive food energy supply available in the widest imaginable choice of foodstuffs, tens of millions of people are malnourished. The resulting obesity and micronutrient deficiencies contribute to a higher frequency of health problems. And in 1997 a third of the China's primary and middle school students were reported to be undernourished—not because of lack of food, but due to unbalanced diets in school canteens (*China News Digest*, May 22, 1997).

Formulating sensible advise for healthy eating is not as simple as it may seem. How age- and sex-specific such advice should be? Going into quantitative details for individual macronutrients and for several major food groups may be counterproductive—but can an effective advise be limited to strictly qualitative recommendations? And what do such admonitions as adopting the Mediterranean diet really mean when that diet itself has been changing rather rapidly? And then there are the matters of ingrained foodways, rates at which innovations can diffuse, and time needed for populationwide health impact. On the first two accounts, evidence cuts both ways. Many traditional food and drink preferences—butter in France, raw fish in Japan, beer in Bohemia—are highly inertial, and higher incomes have only increased what have been traditionally high rates of consumption. But, as we have already seen, some dietary changes can happen quite fast.

An interesting illustration of both the stability of eating preferences and the opportunities for relatively rapid dietary change can be seen by ranking prestige value of foods. A survey of U.S. university students revealed that in just fifteen years the prestige of skim milk more than doubled to nearly rival that of whole milk (Crockett and Stuber 1992). Other fast-rising foodstuffs have included cauliflower, grapes, strawberries, and chicken, while tomato juice, brussels sprouts, whole milk, and sherbets suffered the largest prestige declines. French bread, apples, and shrimp were among the items that retained their high prestige without virtually any change—while pork, herring, and lima beans (all excellent nutritional choices!) remained distinctly unappreciated.

As for the time required to see some populationwide effects of better eating, the evidence also allows room for caution and optimism. Decades may be needed for some shifts to become highly effective—but many prospective studies comparing prudent diets with typical ways of eating have shown significant differences in a matter of years. In any case, although we should not pretend that we can offer detailed quantitative recommendations for every kind of nutrient, we are now reasonably sure of the key qualitative needs and we can give sensible advice regarding healthy, no-regret diets.

Key Ingredients

We have no doubt about how to provide the perfect nutritional beginning of our lives: mother's milk is the best diet for newborns. Breastfeeding, being also a truly sustainable source of food with no environmental costs and with clear social benefits, thus deserves the widest possible support (Draper 1996). The decline in breastfeeding, so noticeable in every affluent country after 1950, has been thus an unwelcome development. As expected, national means differ, but some show a very rapid retreat; for example, the share of Dutch infants breastfed at three months of age fell from 70 percent during the 1930s to 35 percent in the 1950s and to mere 11 percent by 1975 (den Hartog 1992).

Benefits of breastfeeding for the protection of infants against various infections are well documented (Goldman 1993). Molecules of antibodies, the most abundant of which is secretory immunoglobulin A, are passed in mother's milk to a breastfed child and by binding to microorganisms they keep them away from body tissues. Antibodies transmitted to a child are targeted to disable the infectious agents that are most likely to threaten its health while ignoring useful bacteria whose growth can displace the undesirable species—and they do so without causing any inflammation. Other helpful molecules in human milk—oligosaccharides, proteins, and fatty acids—intercept pathogens, limit their spread, display strong antiviral capacities, and promote growth of beneficial bacteria.

Attempts have been made to estimated the economic value of breastfeeding (Almroth and Greiner 1979; Meershoek 1993). This depends heavily on assumptions concerning the valuation of diseases prevented in the absence of formula feeding and on monetization of the opportunity cost of breastfeeding. In any case, the economic argument is not needed in order to promote six to twelve months of breastfeeding as the optimum nutrition (supplemented by solid foods after six months of age, but not by formula feeds) for babies.

Milk should remain an important component of diets beyond infancy. As we have seen, cow milk is a foodstuff whose production uses feed energy very efficiently. Nutritionally, cow milk is an excellent foodstuff: while human and cow milk have a very similar energy content (750 vs. 690 kcal/L), cow milk has three times as much protein (3.3 vs 1.1 g/100 mL), and it is, of course, an excellent source of vitamin D and calcium.

But are not the two great advantages that could make milk a much more important ingredient of global nutrition—high efficiency of its production and high nutritional quality—substantially negated because of a high global prevalence of lactose malabsorption? To a different degree, this condition—also known as lactose intolerance or lactase deficiency—affects most of the humanity (Buller et al. 1990; Suarez and Savaiano 1997). Lactose, a key nutrient in mammalian milks made up of glucose and galactose, is readily digested by all infants except for some that are born prematurely or have a rare clinical syndrome. But in about 70 percent of the world's populations the intestinal presence of lactase, the enzyme hydrolyzing lactose, declines sharply around the age of five years.

When those individuals drink milk, the sugar is not absorbed in their small intestines and its bacterial fermentation in colon produces gases and short-chain fatty acids; the gases cause symptoms ranging from mild abdominal discomfort to nausea, bloating, and cramping, and the acids can produce variable degrees of diarrhea; in children and adolescents vomiting is not uncommon. Complex global pattern of lactose malabsorption is almost certainly the result of evolutionary selection that began some ten thousand years ago and maintained lactase activity in adults among pastoral populations consuming large quantities of milk (Simoons 1978; Friedl 1981).

Consequently, malabsorption peaks in populations of nonmilking cultures throughout East Asia and sub-Saharan Africa, where its rates are between 85 and 100 percent. The rates are above 70 percent among

African Americans and native Indians in both Americas, between 50 and 60 percent in Mediterranean populations; they are below 25 percent throughout non-Mediterranean Europe and in Africa's pastoral regions, and reach their minima (1–2 percent) in Northwestern Europe, and among migrants from those countries to the Americas and Australia (Simoons 1978). Persistence of lactose digestion beyond early childhood is a mutation of dominant genetic trait that has evolved during more than ten thousand years in at least four regions of the world—in Europe, the Indian subcontinent, parts of Middle East, and some regions of Africa—where populations have been consuming cow, sheep, goat, or mare milk for thousands of years since the time they ceased to be foragers and became settled farmers or pastoralists.

And yet this reality does not mean that the majority of the world's people can not consume appreciable amounts of dairy foodstuffs, or that they can do so only by buying low-lactose milk (where the sugar had been prehydrolyzed) or relying on lactase capsules now readily available only in affluent countries. To begin with, lactase synthesis is not completely turned off in most individuals, and most people in populations with lactose malabsorption can tolerate small amounts of the sugar at a time, and hence all they have to do is to divide their drinking of fresh milk into a number of smaller doses. Consumption of fermented dairy products causes even fewer—or absolutely no—problems.

Such fermented foodstuffs as buttermilk, sour cream, and yogurt have only a slightly lower lactose content than fresh milk, but because they also contain bacterial enzymes, which can largely substitute for the lack of the endogenous lactase, their digestion causes fewer upsets (Suarez and Savaiano 1997). And eating cheeses is even less problematic. Cheesemaking—a microbially aided dehydration of milk resulting in six- to twelvefold concentration of caseine, lipids, and minerals—takes away nearly 90 percent of initial water and with it nearly all lactose. Unripe cheeses (cottage, ricotta, cream) have less than 30 percent of the original lactose, and the sugar is hardly detectable in ripened varieties, be they soft blue-veined or Camembert cheeses or hard types such as Cheddar or Edam (Hui 1993).

Nor is a traditional absence of dairy foodstuffs in a particular society a real obstacle. Studies of foodways and cultural preferences tend to overemphasize the importance of such cultural prejudices, many of which do not seem to pose any insurmountable barriers to major, but necessarily gradual, dietary changes. Japan is a perfect example of an impressively rapid shift. When the U.S. army occupied Japan in 1945 the country's consumption of dairy foods was practically zero. Forty years later their average annual per capita intake was well over 50 kg. Rapidly modernizing China should be the next country to experience this dairy invasion. With China's mean intake of dairy foods at mere 2 kg, it is easy to see the potential for major gains in the future.

Nearly all of the future increase of meat production should consist of poultry, particularly chicken, and pork. As one of a few truly universal foods, chicken faces no cultural prejudices and a combination of its lean meat and high feed conversion efficiency makes it a most desirable choice both from the nutritional and from the production perspective. But chicken, a granivorous bird, cannot consume (without complicated processing) a great variety of phytomass, which can be fed to omnivorous pigs instead of being discarded as an objectionable waste. And besides fitting into agroecosystems as efficient scavenges, pigs also convert concentrate feeds with high efficiency (see chapter 5).

Unlike chicken, wider adoption of pork faces some deep cultural taboos. Pig prohibition practiced by Islam, Judaism, Hinduism, Ethiopian Christianity, Lamaism, and other faiths denies pork to roughly a quarter of humanity inhabiting a huge swath of land from the Atlantic coast of Africa to the Southeast Asia. By far the most persuasive argument shows it to be a matter of cultural, religious, and social choices, and not one of a rational response to agroecosystemic realities of the region (Diener and Robkin 1978).

Where culturally acceptable, modern pigs now produce meat that fits the demand of health-conscious consumers. The traditional perception of pork as a meat high in saturated fat, sodium, and cholesterol—the image combatted by the "other white meat" advertising campaign launched by the U.S. National Pork Producers Council in 1987—is simply false. Modern lean pig breeds produce meat with a mere trace (0.5 percent) of fat in muscles—but even fat pigs have only about 3 percent of intramuscular fat, most of their fat being stored in subcutaneous layers. As the eating quality of pork (tenderness and juiciness) is directly related to the amount of intramuscular fat, even leaner pigs might be acceptable only to those consumers willing to accept some deterioration of palatability in return for extremely lean meat.

Pork is higher in polyunsaturated fatty acids (they make between 10 and 15 percent of pork fat) than beef, lamb, or milk, and the ratio of polyunsaturated/saturated fatty acids in lean pork is 0.8, very close to the recommended value. Moreover, pork's fatty acid pattern resembles that of the dietary lipids, and hence the choice of feed rich in polyunsaturated oils can raise the share of desirable acids in the animal's fat (Whittemore 1993). And pork has 10 percent less cholesterol than beef, and much less than aged cheese or shrimp.

The meat also has no more sodium and at least as much iron as beef, and it is richer in most micronutrients (the differences being 40 percent for phosphorus and 110 percent for thiamine). Pork consumption has been going up on every continent, with shares in total meat consumption ranging from 60 percent in Scandinavia to a third in North America, to mere 5 percent in Argentina and to none in the most orthodox Muslim countries. The risk of infection with *Trichinella spiralis*, perhaps the most important nonreligious barrier to pork's even wider acceptance, has been greatly reduced throughout the affluent world (Miller et al. 1991). But trichinosis, resulting from eating infected pork that has been insufficiently cooked, remains a problem in poor countries. Thiabendazole provides effective treatment for both infected people and animals.

Perhaps the only serious nutritional argument against higher consumption of fish is the possibility of excessive intake of heavy metals, particularly mercury, by populations fishing in polluted waters. Egeland and Middaugh (1997) showed that this danger is exaggerated. Environmental arguments against more frequent eating of fish are a different matter, both because of overfishing and because of the inevitably higher share of fish coming from often destructive and polluting aquaculture (see chapter 5).

Finfish from sustainable netting and from carefully managed aquaculture is a perfect source of protein that is either almost totally unburdened by fat or whose lipids contain fatty acids that appear to be beneficial for preventing CHD (Dattilo 1992; Morris et al. 1995). A major challenge is to expand the acceptability of different species in societies traditionally limited to consuming only a handful of finfish and crustaceans.

Yet another desirable option to supply high-quality protein is a more widespread cultivation of mushrooms: they contain 30–50 percent protein on a dry-weight basis, and because they are heterotrophs their proteins have amino acid balance almost as good as milk or meat. The most important production advantage is that mushrooms can be grown on often otherwise useless agricultural and industrial wastes (above all cereal straws, but substrates range from banana leaves to saw dust), they require little land, and the used substrate makes a good soil fertilizer (Wuest et al. 1987; Chang and Miles 1989; Mahler 1991). Today's commercial cultivation is still dominated by *Agaricus bisporus* (the button mushroom, first domesticated in France around 1650) and *Lentinus edodes* (the shiitake mushroom, whose cultivation began in China eight centuries ago), but many other edible mushrooms already are, or can be, commercially cultivated.

Little has to be said about daily eating of fruits and vegetables as is stressed by all modern dietary guidelines. Recent trends indicate almost universal increases in both categories, and particularly for fruit consumption (since the mid-1960s it has quintupled in China and went up by nearly two-thirds in the United States). Common vegetables, much cheaper than fruits, should also play a much greater role in combating micronutrient deficiencies, especially those of vitamins A and C, and of iron and calcium. Ali and Tsou (1997) showed that only three of fifteen Asian countries had average daily intakes above the recommended level of 200 g/capita.

Although most long-term trends still indicate that we should expect further decline of their consumption, intakes of complex carbohydrates in affluent countries should be at least maintained at their current levels. Two recent trends evident in nearly all Western countries should be helpful in this respect: availability of a greater variety, and better quality, of freshly baked products, and a growing popularity of cuisines based on high carbohydrate intakes, ranging from varieties of Mediterranean offerings to Asian meals. Converting this qualitative understanding to quantitative recommendations must be done with a great deal of caution: this is necessary not only in order to avoid any harmful repercussions but also to avoid promoting diets whose efficacy will be shortly disproved by new research. This admonition applies not just to much-publicized cases of contentious megavitamin regimens.

No-Regret Diets

Dietary recommendations must make the best possible effort to distinguish between solid understanding and appealing, but unproven, conjectures. Determination of food energy intakes provides an excellent example of this necessity. Designs of optimal diets for adults in rich countries must start with the recognition that the recommended intakes should be pegged to body masses associated with the lowest mortality for a given age (Kushner 1993). For adults this means weights corresponding to body mass indices between 19 and 27. This choice will eliminate excessive recommendations that would result from using actual body weights in populations where obesity is endemic, and it would make a rather large difference to the total food energy needed by American and European adults. Although higher mortalities are also associated with relative underweight, smoking and alcohol intake, rather than lower weight, are the main explanations.

At the same time, it appears that pegging recommended food energy intakes to the body weight corresponding to the lowest mortality could still lead to excessive energy intakes if the goal of proper eating were to be the extension of maximum life span. The fascinating link between energy intake and longevity, discovered in rats at Cornell University during the 1930s, requires the combination of adequate levels of all nutrients with significantly reduced energy intakes; restrictions of specific nutrients unaccompanied by overall energy reduction do not produce the effect. Besides extending the average life-span of tested groups and maximum life span of individuals, low-energy diets in rodents postpone the onset of many major diseases, retard development of some cancers, lower their blood pressure, help to clear blood glucose faster, and protect immune responses.

Significantly, most of these effects persist even if the reduced-energy diets begin later in life. But this intriguing evidence does not mean that we should start indiscriminately promoting low-energy diets. Confirmation of the full range of these effect in primates must await the completion of ongoing trials with monkeys. Preliminary results show, much like in rodents, lower blood pressure and glucose levels compared to control animals. Moreover, epidemiological evidence in populations consuming low-energy diets adequate in basic nutrients tends to support both the expectations of longer life-spans and reduced frequencies of high blood pressure and glucose, as well as lower mortality due to some cancers (Weindruch 1996; Sohal and Weindruch 1996). Even if further evidence provides an unassailable case for reduced-energy diets, clear guidelines would be tricky to formulate. Children should be excluded, for reduced energy diets retard normal growth.

In fact, nearly half of the world's children would benefit from eating more. FAO estimates that in the early 1990s about 40 percent of the world's children (or about 190 million) were underweight, the condition defined by the WHO as weight-for-the-age lower than two standard deviations. The peak rates were put at 62 percent in South Asia, 38 percent in Southeastern Asia, and 32 percent in West Africa. Restrictions would be impractical for most people whose jobs demand hard physical work and for any adults engaged in regular prolonged aerobic exercise. Experience with low-energy diets in females also makes it quite clear that there could be negative effects on fertility, later increase of osteoporosis, and higher loss of bone and muscle mass. Lowered immune response and greater difficulties in coping with stress may be other common effects.

Even without these complications, adherence to reduced energy diets would no be an easy matter in populations where a rising share of individuals finds it impossible merely to maintain normal weight. As already stressed, I have not called the attention to possible beneficial effects of caloric restriction on aging in order to argue for eventual promotion of reduced energy intake—but rather in order to demonstrate that in societies where basic nutrient needs are covered, energy intakes somewhat lower than those recommended on the basis of ideal body weights are not going to imperil the health of many, if not most, adults and may even improve it. Recommending an occasional fast may be the best way to deal with the intriguing link between low-energy diets and longevity.

In contrast to low-energy diets, recommendations for desirable shares of macronutrients are not controversial. Bringing the share derived from lipids closer to our preindustrial norm is highly desirable. For most people in affluent countries this would mean lowering it from the current 35–45 percent to no more than 30 percent, although 25 percent, or even 20 percent, might be even more desirable. Using U.S. and Dutch data, Glanz et al. (1997) demonstrated that a large share of adults lacks accurate awareness about how much fat they actually consume.

Although perfectly sufficient to supply essential fatty acids, low-fat diets may be very hard to follow for many adults. More frequent eating outside of the home and the rise of fat-based fast food make the compliance with low-fat diets more difficult. Fats provide desirable mouth feel (smoothness, creaminess, richness) and produce satisfactory satiety, the two food qualities eschewed only by ascetics. Fast-food empires were built to take advantage of these universal cravings and their global presence is a formidable obstacle to reducing fat in modern diets. Although fast food has existed in virtually all traditional societies, homogenization and globalization of the practice is new.

Although the choice of meals has been increasing, and many chains now offer relatively low-fat meals, the most popular items—hamburgers, pizza, fried chicken, doughnuts, quasi-Mexican dishes—have much more than 30 percent of their food energy in fats. How much this can be changed by the use of fat-reduction ingredients is still unclear. These compounds can be carbohydrate-, protein, or fat-based. Carbohydrate-based modified starches, gums, and cellulose gels have been widely used to mimic the mouth feel of fat. Olestra, introduced by Procter & Gamble in a variety of savory snacks, is a sucrose polyester whose molecule is not absorbed by the body and hence it has zero energy content. In contrast, Benefat, produced by Nabisco, is a real fat (a triglyceride with stearic acid as its major long-chain fatty acid) that is easily digestible and gives full-fat taste with almost half the calories of normal fat.

Although the current recommendations are very much in accord with the diet most of our ancestors consumed during thousands of generations of preagricultural evolution as far as the share of saturated fatty acids is concerned, they differ a great deal in terms of protein, micronutrient, and fiber intake. In foraging societies proteins provided around one-third of all food energy, prorating to between 2.5–3.5 g/kg a day.

These intakes were both much above the current consumption rates (typically almost 20 percent of all food energy) and above the RDAs (see chapter 7), but similar to the rates observed in wild primates (Eaton et al. 1997). This comparison does not call for higher protein intakes: eating less fat and more cereals (bread flours are up to 14 percent protein) would increase the protein shares of the diet.

Intake of vitamins and minerals was commonly well above the currently recommended levels. Sodium intakes were generally very low but potassium consumption was much above our current RDAs, so the proportion of these two minerals was reversed compared to our diets. In addition, noncereal carbohydrates—from fruits, roots, legumes, and nuts—supplied commonly 40–45 percent of all food energy, and they were consumed shortly after gathering, and often uncooked. Preferably, regular eating of cereal products, legumes, and tubers should be kept at levels supplying no less than 35–45 percent of daily food energy, the rates of current Italian and Japanese intakes (for comparison, the U.S. mean is now just a bit above 25 percent).

Another sensible conclusion to be derived from these comparisons is to eat a greater variety of plant foods, including nuts. For example, cabbage has nearly as much vitamin C as oranges, and peppers have two to four times as much (Jaffe 1984). Insufficient diversity of diet is also the main reason for widespread vitamin A deficiency, which makes at least 250,000 children blind every year (and then kills some two-thirds of them in a matter of months). Severe subclinical deficiency exists in such populous countries as Indonesia, Brazil, and Mexico, and clinical shortage of vitamin A exists in India and most of sub-Saharan Africa (FAO 1996). With proper understanding, natural sources can be used effectively in areas deficient in vitamin A, obviating the use of supplements (Kuhnlein and Pelto 1997).

Instead of dealing with shares and rates, most of the desirable features of a no-regret diet could be summed simply by saying: eat as the Mediterranean cultures do. This has been a popular advice ever since Ancel Keys established the largely plant-based Mediterranean diet as the prototype of healthy eating for prevention of CHD (Keys and Keys 1975). His recommendations have been reflected in current dietary guidelines of many affluent countries (Nestle 1995), and the Mediterranean diet has been evoked as the main explanatory factor for low rates of chronic diseases and long life expectancies observed in southern Europe and North Africa, often in spite of relatively modest economic means. But the concept of a Mediterranean diet is an artifact distilled from a complex pattern of different foods and foodways (Helsing 1993).

Even more importantly, average food intakes of Mediterranean populations have been moving in the direction of less healthy northern diets due to increased intake of meat, fish, and cheese, and decreased consumption of bread, fruit, potatoes, and olive oil (Nestle 1995). Some of these shifts have been both rapid and substantial. Italian eating, still widely seen as the very embodiment of the diet, has been moving away from its traditional pattern and acquiring some unmistakeable features of typical Western diet. A detailed comparison of the Italian food supply shows that between 1961 and 1992 the country's consumption of cereals fell, the overall food energy supply rose by more than 15 percent, mainly because fat availability increased by almost 60 percent, which means that the share of energy derived from lipids was at 39 percent in 1992, compared to just 25 percent in 1961 (Zizza 1997).

Moreover, composition of this higher fat intake also changed: olive oil dominated a generation ago, but now other oils supply more than half of all liquid lipids, and the supply of solid lipids had nearly doubled (figure 8.7)! Trichopoulou (1993, 1995) offers the Greek diet as a better choice for the Mediterranean paradigm. Until recently this diet was resistant to change, so much so that in rural areas a generation ago it was still remarkably similar to the ancient past. Basic elements of the traditional Greek diet—its overall frugality dominated by cereals, pulses, vegetables, fruits, olive oil, cheese, and milk, with occasional fish and meat, and with moderate drinking of wine—explain why a population with a modest health service has one of the world's longest life-spans, second only to Japan.

Trichopoulou (1995) argues that no single factor in the traditional Greek diet determines the high survival rates, that it works only as a whole: plenty of grains, bread, beans, yogurt, feta cheese, and vegetables cooked in olive oil together with moderate drinking of wine. Those people who are not enthusiastic about beans and yogurt, or feta and grape leaves, would inevitably ask: how much could be left out of the

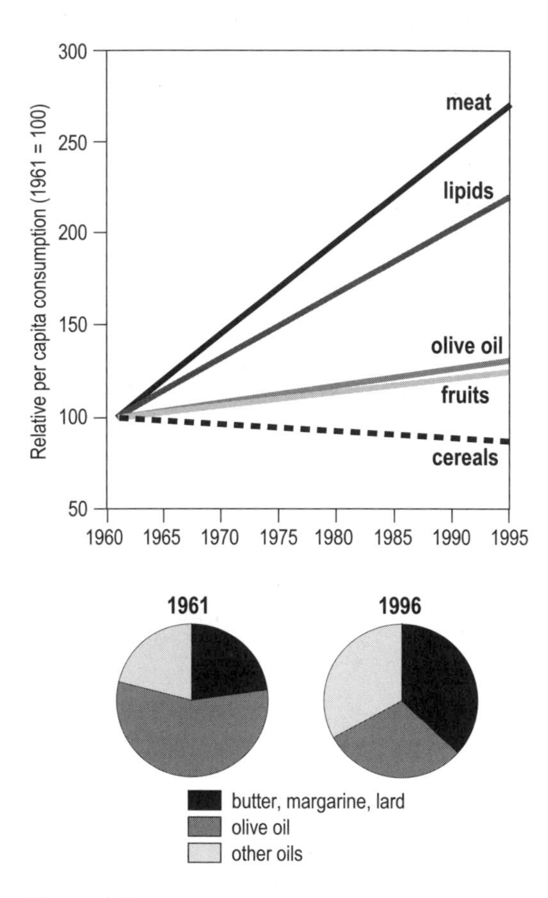

Figure 8.7
Changing Italian diet, 1960–1998; plotted from FAOSTAT data (FAO 1999).

package, or to what extent it could be modified, without substantially reducing its beneficial effect?

Whatever the recommended pattern may be, it can make a difference on a population scale only with widespread adoption. How effective a widespread adherence to a more desirable diet can be is well illustrated by the British experience during the two world wars. During World War I nutrition was not a subject for government intervention and average food intakes declined across the board, with meat consumption down by 27 percent, milk by 26 percent, and vegetables by 6 percent. In contrast, rationalization of production and rationing during World War II brought a 28 percent increase in milk consumption compared to the 1934–1938 mean, and a 36 percent increase in vegetable intakes; meat was down by 21 percent, but eggs only by 6 percent (Hollingsworth 1983).

With grain and potato eating up and oil, sugar, and fruit supply down, the overall trend was toward reduced variety and lower palatability—but total energy intake remained within a few percent of the prewar mean, protein consumption rose, and there were no micronutrient deficiencies! But the British experience also shows the powerful craving for refined sugar. Once the rationing of sweets and chocolate was removed in 1949, the demand was so high that rationing had to be reimposed for another four years!

Some peacetime, and hence voluntary, shifts have been also appreciable. In every affluent country there are certain segments of population clearly exhibiting health-conscious choices in selecting food. For example, a Danish study found that during a ten-year period both men and women shifted to a more healthy diet with less animal fat, meat, and white bread and more low-fat margarine, fruit, vegetables, coarse bread, and oatmeal (Osler et al. 1997). But they also drank less milk and ate more cakes and candy!

On the other hand, a survey conducted in 1995 and 1996 in the European Union found that 71 percent of respondents thought they do not need to make changes to the food they eat, "as it is already healthy enough" (Kearney et al. 1997). And Walker and Segal (1996) concluded that compliance with the dietary requirements for lessening the risk for colon and breast cancers seems a forlorn hope. Lack of time and lack of willpower are the most important reasons people give for not shifting to healthier diets (Lappalainen et al. 1997).

9

If China Could Do It . . .

This book concentrates deliberately on widely, if not universally, valid realities whose understanding provides a basis for setting up more rational food systems and for managing them with higher efficiencies. Covering these matters of general import in some detail is preferable, I believe, to a lengthy series of inevitably cursory regional or national portraits. At the same time, I would like to convey a specific complexity of the challenge by a closer look at a nation's challenge to feed itself. There could be hardly any better example than that of China.

The country is a near perfect embodiment of the concatenation of worrisome changes that complicate and undermine the quest for higher food production. The combination of realities that weaken its food production capacity—its, in absolute terms, still very high population growth, limited (and declining) availability of farmland, widespread and intensifying shortages of water, serious air and water pollution and extensive ecosystemic degradation, reduced growth rates of staple grain yields, and rapid dietary changes—is behind the recent questioning of the country's ability to feed itself (Crook 1994; Brown 1995a; Smil 1995).

At the same time, China's entire food system offers some of the world's most convincing examples of widespread inefficiency and waste. Even a relatively modest effort to eliminate these failures would go a long way toward securing adequate food for coming generations. If a conservative assessment can show that China should be able to meet the challenge, then we may feel much more confident about most of the rest of the world.

China's Predicament

The same factors that have made the task of feeding China an uncommon challenge in the past will remain influential during the coming fifty years. First is the necessity to feed the world's largest population (its 1.25 billion people in 2000 equaled just over one-fifth of the global total). Second is the limited availability of farmland exacerbated by losses of cultivated land to urban and industrial expansion. The third decisive factor is the deteriorating state of China's agroecosystems, including growing shortages of water and spreading environmental pollution. Two new concerns also matter: dietary transition driven by much higher disposable incomes, a trend particularly pronounced in China's richest coastal provinces; and declining productivity of farming inputs, above all the falling response of staple cereal yields to intensifying applications of synthetic fertilizers.

Rising Demand for Food

The country's relative population growth, averaging just 1.1 percent during the first half of the 1990s, is quite low in comparison with the mean for all modernizing nations (1.77 percent during the same period), and considerably below the rates for either Southeast Asia or South America, which stood, respectively, at about 1.7 and 1.6 percent (United Nations 1998). But the huge absolute increase of China's population during the past generation means that this relatively low growth still translates into historically high level of nearly twenty-five million births a year and to net additions of more than thirteen million people. These totals will not decline appreciably at least for another decade, and some two hundred million people will be added during the next two generations, about as many as Indonesia's total population of the late 1990s.

Obviously, merely maintaining the existing food consumption rates would call for at least 1.1 percent increase in annual grain harvests; during the last years of the 1990s this would mean adding about 5 Mt every year. But if the food supply is to keep up with rising expectations, the actual rates would have to be much higher. Since the beginning of economic modernization in the early 1980s China's average intakes have moved very rapidly up the food chain as major per capita consumption increases brought the typical dietary pattern much closer to those of Japan and Taiwan.

By 1985—just five years after the beginning of farming privatization—China's average per capita food availability rose to 2,700 kcal/day, the rate less than 5 percent behind the Japanese mean food supply, and it has remained just slightly above that clearly adequate rate ever since (Smil 1985, 1995). This quantitative rise has been accompanied by major qualitative gains. Consumption of coarse grains and tubers declined as intakes of more finely milled wheat and rice increased. Milling rates for rice returned 70 percent or less, compared to 85 percent typical of pre-1980 years.

Per capita intakes of traditional nonstaple favorites have multiplied severalfold. Between 1980 and 1995 average annual consumption of plant oils nearly tripled, as did the consumption of poultry, eggs, and freshwater fish (figure 9.1). Pork purchases doubled nationwide and tripled in the cities—per capita consumption of pork among the highest income groups in coastal cities surpassed the Japanese national mean—

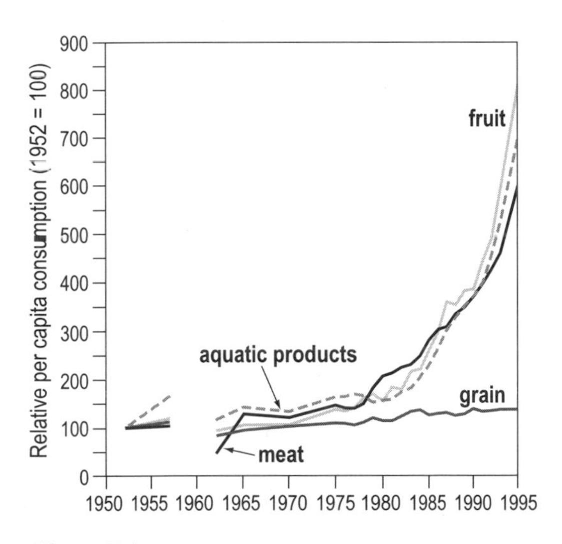

Figure 9.1
Indices of China's average per capita food supply, 1952–1995 (1952 = 100); calculated from data in State Statistical Bureau (1980–1998).

and drinking of alcoholic beverages rose fourfold (State Statistical Bureau 1980–1998). Future rates of consumption increases will slow down, but given the still low rate of China's urbanization (about one-third of all population was urban in 1999) and great urban-rural disparities of food intakes, the pattern will not stabilize soon. Consumption surveys show that the expenditure elasticities for rice and coarse grains are declining, but those for wheat, meat, alcohol, and vegetables are increasing.

And then there is the matter of large regional disparities: national means are impressively high because the majority of people in coastal provinces in general, and in their large cities and periurban areas in particular, may be now eating as well as their compatriots in Taiwan—but average diets are still barely adequate in inland provinces. Even official Chinese admissions class more than fifty million people as living with lower than acceptable levels of nutrition (Chen et al. 1990; Ge et al. 1991; China News Digest 1997). Rising demand for meat and alcohols allows us to foresee easily a doubling of grain demand during the next generation. Advancing westernization of urban diets will push up the demand for wheat, sugar, and oils. At the same time, increased grain harvests will have to come from the shrinking amount of farmland and cope with precarious availability of irrigation water as well as with a more widespread environmental pollution.

Environmental Constraints

According to official claims the country lost about 15 percent of its farmland between 1957 and 1990 (Smil 1993). Given the country's intervening population increase, the average farmland availability was thus more than halved, from about 0.18 to just 0.08 ha/person. Rapid post-1980 modernization brought a spate of new rural and urban housing construction and unprecedented expansion of export-oriented manufactures and transportation links. New peasant houses are rarely built on sites of old structures, new factories usually take over highly productive alluvial land, and government policies promote multilane freeways instead of rapid trains. Not surprisingly, annual farmland losses have been averaging at least 0.5 Mha since 1980, and they have been mostly

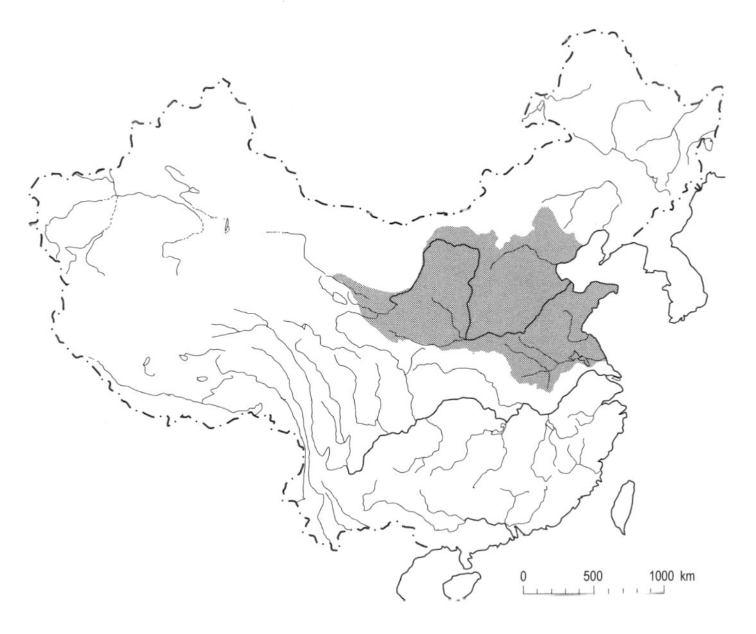

Figure 9.2
Area of North China affected by serious water scarcity.

concentrated in rapidly developing coastal provinces where the intensity of farmland use is the highest.

During the past decade water shortages have become seasonally acute throughout most of the North China Plain (figure 9.2) Large-scale irrigation of the Plain began only in the 1960s with the introduction of the first shallow tube wells, and by the late 1980s the plain had more than two million tube wells irrigating over 11 Mha of farmland (O'Mara 1988). Initially, pumping helped to lower the formerly high water table and hence to reduce the extent of soil salinization, but soon it began causing excessive exploitation of aquifers accompanied by spreading ground subsidence (Smil 1993).

During the early 1980s roughly a third of irrigation water on the Plain came from the Huang He River, but the combination of recurrent droughts and higher agricultural, urban, and industrial demand began exhausting the stream long before it reached the sea. The river's total runoff has recently fallen to as low as two-fifths of the long-term mean as the normally very low summer flow had repeatedly ceased altogether for hundreds of kilometers from its mouth for a period lasting between one to four months. Diversion of Huang He water, amounting to more than a quarter of its total flow during dry years, also reduces the silt transport to the ocean; a heavy sediment load is thus deposited on the river's bed, particularly in Henan and Shandong provinces. Irrigation thus aggravates the elevation of the river's bed above the surrounding countryside.

Water shortages in the North now affect an area extending over some 600,000 km^2, a total about 10 percent larger than France. But water shortages are not limited to the Plain: they have become a near-chronic reality in every northern and northwestern province. At a basin level the Hai-Luan basin has the highest water stress, followed by the Huai River basin (Nickum 1998). On a provincial basis Shanxi (particularly its southern part) and peninsular Shandong face the greatest water shortage; in Shanxi even drinking water is often scarce, and about 10 percent of the province's peasants suffer chronic shortages of its supply. Planned expansion of surface coal mining and construction of large coal-fired power plants will further strain the inadequate supply. And although expanding cities are now claiming substantial volumes of water used previously in agriculture, urban water shortages have become the norm in the capital and two hundred other municipalities in the region.

Environmental pollution accompanying China's industrial and urban expansion is both widespread and severe (Smil 1993). With about 1.2 billion tonnes extracted annually, China is now the world's largest consumer of coal, producing more SO_2 and particulate matter than all of Europe outside Russia. More than 80 percent of its waste water is discharged without treatment, and irrigation waters in the most intensively cultivated periurban areas have been polluted by industrial wastes. An even greater water pollution threat comes from hundreds of thousands of new village and township enterprises that have been absorbing rural surplus labor. China's Environmental Protection Agency can only guess at the total amount of untreated waste leaving those factories.

Degradation of ecosystems has an even greater impact, with worsening shortages of water, extensive soil erosion (causing silting of reservoirs and irrigation canals, and aggravating annual flooding), salinization and waterlogging of farmland, overgrazing and pest infestation of grasslands, and disappearance of the remaining mature forests. Huang and Rozelle (1995) estimated that erosion, salinization, and losses of farmland may have cost China recently 6 Mt of grain a year, more than the additional output needed to keep up with the country's population growth. My more comprehensive, but still incomplete, survey of economic costs of China's environmental pollution and ecosystemic degradation shows that they are equivalent to at least 10 percent of the country's annual GDP, and that roughly one-fifth of that cost is attributable to losses of agricultural production (Smil 1996a).

While some problems have been eased by higher investment in environmental protection—above all through waste water treatment in large cities, installation of effective particulate matter controls at large stationary combustion sources, and private afforestation of slopelands—others are worsening. The two most notable examples of the latter category are the rapid expansion of the area affected by acid deposition in southwestern China (caused by burning high-sulfur coals in the region's rainy climate), and rising concentrations of tropospheric ozone from more frequent, and more concentrated, episodes of photochemical smog (resulting from the rising intensity of car traffic in the rich coastal provinces).

And, as anywhere else, rising rates of fertilization bring lower yield responses, and higher demand for urban and industrial water uses is now competing with limited supplies for irrigation. Declining response to fertilization is particularly obvious in coastal provinces, where nitrogen applications in triple-cropped fields average 300 kg N/ha and in many localities surpass 500 kg N/ha, levels comparable only to those in the Netherlands.

To some observers, combined effects of these changes were already demonstrable by the post-1984 stagnation of China's grain output (Brown 1995a). That year's record grain harvest of 407 Mt was followed by five years of stagnation, and although a new record, 446 Mt, was set in 1990, it was followed, again, by two years of lower harvests (figure 9.3). Continuation and intensification of these trends could result in annual grain supply deficits amounting to tens of millions of tonnes

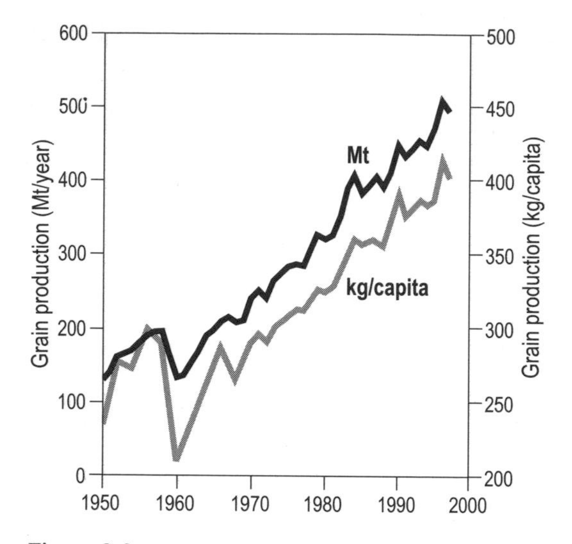

Figure 9.3
China's total and per capita grain production, 1950–1998; plotted and calculated from data in Crook (1988) and FAO (1999).

during the coming decade, and surpassing 100, possibly even 200, Mt before the year 2025.

Climate change could make China's quest for higher yields even more difficult, and its dependence on imports even greater. There are indications that the twentieth century has already brought greater aridity to eastern and northwestern parts of the country, and higher temperatures in the north. Chinese studies of long-term effect of global warming suggest possibilities of lower rice yields in the South, lower corn yields in the East, and lower soybean harvests everywhere except in the Northeast (Smit and Cai 1996).

Implications of China's inability to feed itself would be global. Even if a relatively rich China could afford to buy increasing quantities of cereals on the world market, such purchases would not just lead to price rises in a handful of remaining countries exporting food, they would also gravely reduce, or virtually remove, the access of many poorer nations in Africa and Asia to grain deliveries from the four producers with virtually assured long-term export potential—the United States, Canada,

Australia, and Argentina (the contribution of Russia's and Ukraine's exports, potentially quite large, remains highly uncertain).

Pressure on international grain prices would rise as the country's unmet demand moves substantially above the level of recent Japanese purchases (at almost 30 Mt a year now largest in the world). Annual imports of 100 Mt would be equivalent to half of the recent global grain shipments, and given the fact that future import demand in other populous Asian, African, and Latin American countries will almost certainly grow, it is difficult to see that the country could actually secure such a large share. And if its shortfall amounted to more than 200 Mt a year it could not hope to make it up by imports at any price: doubling of global grain sales during the next generation is extremely unlikely.

Available Resources and Existing Inefficiencies

Realistic appraisals of agricultural possibilities must rest on reliable information. Unfortunately, not a few analyses of China's agricultural prospects have relied on inaccurate figures or interpreted undeniable realities in misleading ways. To begin with, China's population may stay well below the 1.6 billion people assumed by Brown for the year 2030. In fact, the medium variant of the latest UN revision of long-term forecasts foresees a stabilization around 1.5 billion after the year 2030, and the low variant does not even reach 1.4 billion before the totals levels off (United Nations 1998). In addition to a further significant slowdown of China's population growth, overall demand for food might be appreciably lowered by a combination of aging (with low fertilities in place since the early 1980s, China will experience one of the world's fastest demographic transitions), more sedentary lifestyles, and concern about healthy diets (traditionally strong in China).

Farmland, Irrigation, and Fertilizers

As in many countries around the world, China's official figures have been substantially underestimating the country's arable land, which means that the official yields must be adjusted downward, and that there are greater unrealized possibilities to increase future harvests. China's farmland scarcity is thus nowhere near the level observed in South Korea, Taiwan, and Japan, neighbors that have increasingly relied on imports of grain, edible oil, and meat. During much of the 1990s China's *Statistical Yearbook* has contained a note warning that "figures for the cultivated land are under-estimated and must be further verified." This comes as no news: many students of Chinese affairs have known since the early 1980s that the official total of China's farmland—95 Mha listed by the State Statistical Bureau, putting China's per capita mean of arable land below the Bangladeshi average—is wrong.

Even the earliest remote sensing studies based on imagery with inadequate resolution (LANDSAT Multiple Spectral Scanner, with resolution of 80 m) indicated that figures used by the State Statistical Bureau, and hence by virtually all misinformed foreigners, were too low. They came up with total as high as 150 Mha, and detailed sample surveys of the late 1980s came up with the range of 133–140 Mha (Smil 1993). More recently, Wu and Guo (1994) put the total at 136.4 Mha plus additional 7.4 Mha devoted to horticulture. Heilig's (1997) application of land survey data for 1985 to correct the official claims of total cultivated area ended up with 137.1 Mha for the year 1995.

The most extensive, and the most accurate, remote sensing evaluation, used stratified, multistage area estimation approach: samples of the higher resolution classified imagery from the Keyhole (KH) series of intelligence satellites (whose latest models return images with resolution of 15 cm or better) were used as surrogates for *in situ* data to correct estimates derived from much coarser commercially available images, including Advanced Very High Resolution Radiometer and LANDSAT. This analysis yielded the total of 143.3 Mha for the year 1992; with variation of 5.6 percent at the 0.95 confidence interval, the actual area could have been as low as 135.4 and as high as 151.4 Mha (figure 9.4) (MEDEA 1997). This estimate was subsequently revised to the range of 133–147 Mha for the year 1997: failure of the initial appraisal to account properly for fallowed land and intervening conversions of farmland to nonagricultural uses were the main reasons for the reduction.

I believe that even MEDEA's total may be too low inasmuch as it does not include aquacultural ponds and orchards, the two intensive land uses that make very significant contributions to the country's balanced

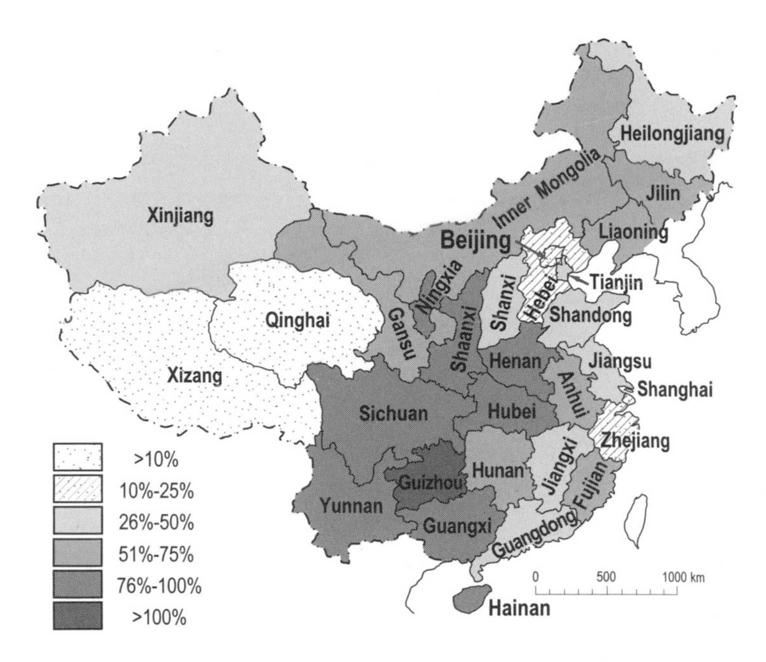

Figure 9.4
Provincial differences between the availability of agricultural land according to the official statistics and to the MEDEA study; based on data in State Statistical Bureau (1995) and MEDEA (1997).

diet. Moreover, a large share of today's orchards and ponds has been converted from crop fields after 1980: land use has changed, but the land has not only remained devoted to food production as it keeps supplying high-quality protein and desirable micronutrients, but its new uses also provide a variety of environmental benefits. Even when assuming an annual yield of no more than four tonnes of fish, a hectare of carp pond will yield about 800 kg of protein, twice as much as the average grain harvest (including the 1.5 multicropping rate) from the same area—and, unlike cereals, carp has an adequate amount of all amino acids. And a hectare of citrus orchard will produce 50 percent more vitamin C and twice as much food energy as the same area planted to cabbages.

With the inclusion of ponds and orchards China's farmland is thus anywhere between 140 and 160 Mha, and the country's 1998 per capita mean is at least 1,100 m^2, compared to 490 m^2 in South Korea,

430 m^2 in Taiwan, and 420 m^2 in Japan. China's per capita farmland availability is thus at least two and almost three times larger than that of its East Asian neighbors. And unlike in those three countries, there are still appreciable opportunities for reclamation of farmland in China. According to official estimates, the country has at least 33 Mha of uncultivated but reclaimable land categorized as wasteland whose eventual development (perhaps one-third of that land could be converted to fairly productive fields, the rest is suitable mostly for planting trees) would result in appreciable food production gains and environmental benefits.

Conversions to appropriate food production uses could be speeded up by large-scale auctioning of rights to long-term (fifty to a hundred years) private use of such land. Hanstad and Li (1997) describe keen interest shown by peasants bidding for the rights to wasteland and relatively high investment of labor and cash they subsequently undertake to use the land for planting trees for fruits, nuts, fuelwood, and timber. The fruit and nut production potential of large-scale wasteland cultivation is obviously substantial, but environmental benefits—above all, reduced soil erosion on previously barren slope lands—may be of equal, or even greater, benefit for the country's agroecosystems.

In addition, much of the farmland in South China can be cropped continuously, yielding three harvests of staple crops, or up to five harvests of vegetables a year. The cold climate in northeastern China allows for only a single crop, while in northern provinces winter wheat commonly follows a summer crop. China's overall multicropping ratio (sown/cultivated area) in the mid-1990s was 1.56, compared to Japan's 1.03 and Korea's 1.14. With proper rotations, further intensification of China's crop cultivation, raising the multicropping ratio to 1.6–1.65, is possible without damaging the affected agroecosystems.

Underreporting of farmland means that the official figures on China's average grain yields are exaggerated. The difference is smallest for rice (less than 10 percent) and largest for corn (in some provinces more than 40 percent). Contrary to Brown's assertion, Chinese yields are not exceptionally high by advanced world standards and could be increased substantially by higher inputs and better agronomic management. Corn, the principal feed grain, averaged officially 4.9 t/ha in 1995, but the actual mean is just below 4 t/ha, or no more than half of the average U.S. yield in a good year. Even the average official rice yield of about 6 t/ha is more than 10 percent behind the Japanese mean. This means that the country has more room to improve crop yields by using additional inputs, better agronomic practices, and price incentives.

Another important consideration is that China's undoubtedly substantial farmland losses have been exaggerated. Since 1980 net farmland loss in China has fluctuated between less than 100,000 ha (in 1989–1991) and 1 Mha (in 1985), with the period of fastest declines between 1992 and 1994. Mean annual loss has been nearly 500,000 ha a year since 1980, but this does not mean that the land devoted to food production simply shrank by that much every year. A large part of the reported loss, over 50 percent in some years, has been due to the restoration of land converted to fields during the years of extremist policies of the Maoist era back to their original, and environmentally much more appropriate, uses as orchards, grasslands, and fish ponds (Smil 1999b). This change has clearly helped China's nutritional balance, and it has enhanced agroecosystemic diversity. Consequently, it is misleading to treat this changed land use as a loss of food production capacity.

Chinese authorities recognize the necessity to protect the country's farmland. New rules to control illegal land use changes came into effect on March 1, 1996, designed to improve the enforcement of stricter regulations. Although it would be naive to expect easy compliance in many regions, it is not unrealistic to foresee an appreciable moderation of annual losses. Ke (1996) put the net reduction of farmland between 1978 and 1994 at 4.5 Mha, and he projects a similar loss (4–5 Mha) during the fifteen years between 1995 and 2010. Even if China were to lose 0.5 Mha a year for the next 25 years, it would still have more than 800 m^2 of farmland per capita in the year 2020 (when its population will be about 1.45 billion people), or twice as much as Japan has today.

A realistic appraisal of China's water availability is also more complex than conveyed by basic, and in this case quite reliable, precipitation and runoff statistics, and less certain data about water stored in aquifers. To begin with, all nationwide means concerning water are not very meaningful as the country's monsoonal regime results in pronounced annual and seasonal precipitation and evapotranspiration differences progressing along the southeast-northwest gradient. The 500-mm isohyet (running from the central Heilongjiang in the northeast to the Sino-Bhutanese border in the southwest) forms an approximate divide between the dry northern and western interior and wet coastal east and inland south (Domros and Peng 1988).

China's water supply is thus determined by a strong seasonality of monsoonal precipitation, by high frequency of droughts north of the great divide, and by large fluctuations in the distribution of annual and seasonal moisture. Densely inhabited parts of northern China, covering about one-third of the country's territory, have about two-fifths of the total population and grow the same share of staple grains—but they receive only about one-quarter of the country's precipitation, and because of high summer evapotranspiration they can access less than 10 percent of the nationwide stream runoff. Not surprisingly, these northern provinces rely heavily on underground water reserves, yet they possess no more than 30 percent of all water in China's aquifers (Smil 1993).

A higher frequency of dry years since the mid-1980s has undoubtedly contributed to northern China's water shortages. At the same time, the extent of this natural precipitation shortfall is not unprecedented. Official statistics on areas affected by drought (where yields are reduced by at least 30 percent in comparison with years of normal precipitation) show large fluctuations of between 1 and 18 Mha a year (figure 9.5) (State Statistical Bureau 1980–1997). Risks of flooding are also considerable inasmuch as about one-tenth of China's territory, inhabited by nearly two-thirds of the population and producing roughly 70 percent of all agricultural output, is below the flood level of major rivers. Flooding serious enough to reduce crop yields by at least 30 percent has been recently affecting 4–9 Mha of farmland annually (figure 9.6).

Any serious cutbacks in China's irrigation would have major repercussions for the country's food production. In 1950 China irrigated no more than 16 percent of its farmland, but now the officially quoted share is 46 percent (State Statistical Bureau 1997). Assuming that the figures on irrigated land are fairly accurate, the actual share (with 140 rather

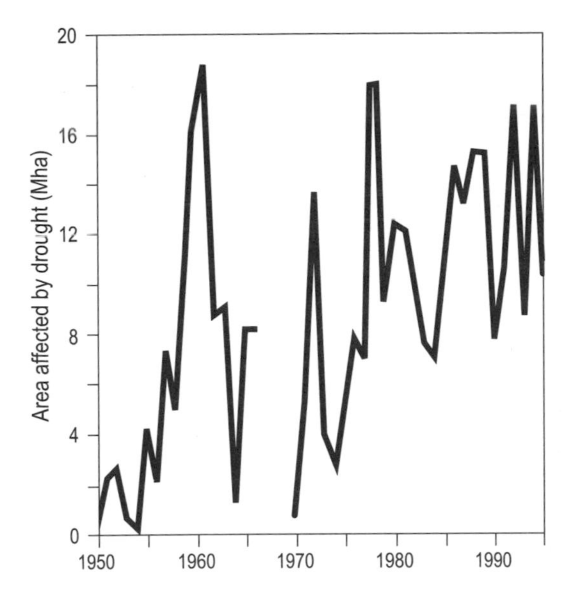

Figure 9.5
China's farmland affected by drought, 1950–1995; plotted from data in State Statistical Bureau (1980–1998).

than just 95 Mha of arable land) would be just above 30 percent. This would make China no more dependent on irrigation than India, which irrigates nearly 30 percent of its farmland.

In spite of so many obvious signs of water shortages, existing Chinese practices do not reflect the growing scarcity of the resource. Inexpensive water prices are primarily responsible for unsustainable and wasteful irrigation, which would be greatly curtailed with the introduction of realistic water fees. Perhaps the best illustration of this giveaway is a comparison of Beijing's water prices with those I have to pay in Winnipeg.

This city of some 700,000 people has no heavy industrial production. It gets its water from Lake of the Woods, one of large glacial lakes left behind by the last Ice Age; this water requires hardly any cleaning, and no pumping is needed as the water flows to the city by gravity. Yet with sewerage rates included, Winnipegers are charged about US\$1.3/m^3. Until the State Council approved higher prices in April 1996, water in

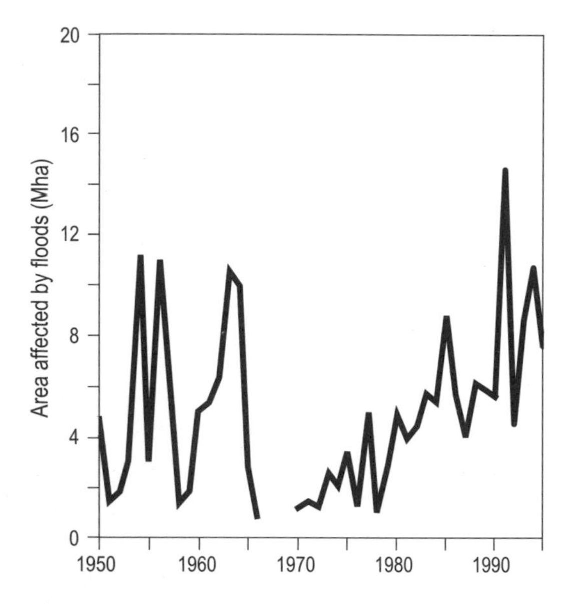

Figure 9.6
China's farmland affected by floods, 1950–1995; plotted from data in State Statistical Bureau (1980–1998).

Beijing—a city of eleven million people where half of all water comes from expensive underground pumping and where Stalinist planners located many water-guzzling heavy industrial enterprises—cost 0.3 yuan/m^3.

This amounts to mere US$0.035/m^3 when converted at the official exchange rate, and still only around US$0.15/m^3 when using a very liberal purchasing parity rate. What is even more remarkable is that in comparison with Winnipeg a cubic meter of Beijing water cost less even as a share of respective average disposable family incomes! The approved increase boosted the new rate to about US$0.25/m^3, still only a fifth of the cost in a city enjoying one of the most abundant, and least costly water supplies in the world (Smil 1997b). The high potential for water savings illustrated by this comparison is the main reason why Nickum (1998) concluded that China's "water crisis" is localized, and is economic and institutional rather than a matter of a disappearing resource.

China is now the world's largest producer and user of nitrogen fertilizers—but it still has a large potential to increase its fertilizer inputs.

Assuming the country has at least 140 Mha of farmland, its mid-1990s nitrogen applications were around 170 kg N/ha, a rate higher than the declining Japanese average (about 120 kg N/ha in 1997). But with average multicropping ratio of 1.5 this prorates to about 110 kg N/ha per crop, a rate lower than the mean applications to single high-yielding crops of U.S. corn or European winter wheat. And while average nitrogen applications in China's coastal rice-growing provinces are just about the highest in the world, large parts of the interior receive considerably less of the nutrient than is the nationwide mean. Clearly, China still has considerable room to increase its average nitrogen applications—and, as I explained in chapter 4, even greater opportunities to combine them with appropriate quantities of P and K.

Postharvest Losses

A key argument of this book—that nearly all assessments of long-term food prospects have been preoccupied with exploring the possibilities of increased supply instead of reducing waste along the whole food chain—is persuasively illustrated by China's enormous postharvest losses resulting from improper storage of crops, low efficiency of animal feeding and very high waste of coked food. As noted in chapter 6, China's antiquated storage methods cost the country roughly one-seventh of its cereal harvest every year (Liang et al. 1993). Better storage could make a huge difference because of China's extraordinarily large amount of grain held in state and private reserves. The total, long considered a state secret, was claimed to be 458 Mt in 1994, more than the harvest of all cereals and tubers, and more than five times as large as standard expectations for setting aside slightly less than one-fifth of annual grain consumption (Crook 1996).

Inefficient feeding of animals, table waste in hundreds of thousands of labor-unit eateries, and wasteful fermenting to alcohol almost doubles total grain loss to more than 50 Mt of staple grain equivalent a year. Current Chinese feeding rates are anywhere between 10 and 50 percent above the norms prevailing in Western countries. The overwhelming majority of China's pigs (pork accounts for no less than 90 percent of the country's meat output) is still not fed well-balanced mixtures but just about any available edible matter and hence it is commonly deficient in protein. Not surprisingly, an average Chinese pig takes at least twice as long to reach slaughter weight than a typical North American animal (twelve to fourteen months rather than just six months)—and its carcass is still lighter and more fatty (Simpson et al. 1994). And hundreds of millions of chickens roaming the country's farmyards take three times as long to reach again lower slaughter weight than North American broilers.

Production of alcoholic beverages is a particularly fine example of waste that could be sharply reduced by relatively simple technical improvements. Brewing beer and fermenting a variety of Chinese liquors (usually rice- or sorghum-based) consumed almost 20 Mt of grain a year in the early 1990s, and the demand has been going up by about 20 percent a year. But drinking more has been only a partial explanation for this huge total: most of China's forty thousand distilleries and breweries are small, inefficient enterprises whose grain consumption is typically 40 percent higher than in the state-of-the-art factories (Liang et al. 1993).

And considering China's still far from abundant average food supply, losses during consumption—particularly in labor unit eating halls and during now ubiquitous banqueting at public expense—are staggering. On December 13, 1994, an official China News Agency report put the annual total of wasted grain at almost 83 Mt, and attributed about three-fifths of this total to losses during consumption. Articles in the Chinese press have repeatedly noted large quantities of leftovers disposed of by restaurants, hotels, and canteens every day (Wu 1996).

Finally, a closer look at the causes of post-1984 stagnation of grain output reveals that a relative neglect of agriculture, rather than a combination of inexorably degradative changes, has been the most important reason for that trend. As the country, and the world, became mesmerized by high rates of China's industrial growth, proportion of state investment in agriculture had been declining, falling by a third between 1991 and 1994 (State Statistical Bureau 1992–1996). Rises in procurement prices paid to farmers for grain delivered under the compulsory quota system lagged behind the high rate of inflation, and often the farmers were not even paid but issued IOUs. Not surprisingly, peasants responded by planting less grain and more cash crops.

One of Brown's main conclusions, that the country faces the prospect of continuously falling grain harvests, was thus quickly disproved. In 1994, when Brown widely publicized his alarming appraisal, China's harvest fell by 2.5 percent compared to the record output of 1993—but it set yet another record in 1995 by reaching almost 467 Mt. And in 1996—in spite of the fact that large parts of the Yangzi valley (particularly Hunan province normally producing about 13 percent of China's rice) experienced some of the worst flooding recorded in modern China—the country's grain harvest reached yet another record of 485 Mt (4 percent above the 1995 level), an increase far ahead the country's rate of population growth.

Realistic Solutions

As I have tried to make clear in preceding chapters, no single efficiency improvement or no single policy change—even if carried to its technical and economic limits or even if representing radical departure from old, irrational ways—has the potential to alter fundamentally a country's long-term food production outlook. Only a combination of such changes, and sustained attention to their diffusion and performance, will make an appreciable difference. Fortunately, in China's case, as in every other large agricultural system, there are many opportunities for addressing the key twin inefficiencies of water and fertilizer use, for improving the management of agroecosystems, for using pricing to promote efficiency, and for investing in research.

And although the country's rapidly unfolding dietary transition—above all the rising consumption of animal foodstuffs—results in substantially higher demand for natural resources, long-term adjustments of China's dominant nutritional pattern could make a substantial contribution toward reconciling the demand for better eating with the capacity of agroecosystems to provide the requisite environmental goods and services.

Raising Efficiencies and Changing Diets

Investment in more efficient forms of irrigation, and more realistic prices for delivered water could yield surprisingly large water savings. Water prices paid by Chinese peasants on the drought-prone North China Plain are about as realistic as those enjoyed by California farmers growing alfalfa and rice in the semidesert climate of the Central Valley (O'Mara 1988). During the late 1980s, a decade of extensive drought and chronic urban water shortages, the typical price of China's irrigation water was mostly between 5 and 20 percent of the actual cost. Higher prices should bring better matching of crops with available moisture and introduce more efficient irrigation.

But higher prices alone may not be sufficient: some regions will also need changes in basic water allocation arrangements. Total distribution, seepage, and evaporation losses in China's traditional ridge-and-furrow irrigation amount commonly to 50–60 percent of carried water. Using an appropriate mixture of water conservation techniques outlined earlier in this chapter—such as irrigating every other furrow, carefully scheduling water applications, or replacing corn, an increasingly popular crop in arid North China, with sorghum—could result in additional and nearly cost-free supply gains.

Fertilizer applications offer equally impressive examples of great efficiency opportunities. Two important approaches—popularization of "fertilizing by prescription" and optimization of applications based on soil analyses and cropping practices—are becoming increasingly popular among China's cost-conscious farmers. Besides the variety of universally applicable approaches aiming at higher fertilizer use efficiency, two measures would greatly improve China's use of nitrogen fertilizers. Major efficiency gains would result from gradual dismantling of small fertilizer factories making ammonium bicarbonate, and from adjusting the nutrient ratio.

Ammonium bicarbonate still accounted for about a third of China's total output of synthetic nitrogenous fertilizers in the mid-1990s, but its high volatility combined with shoddy packaging means that a large share of the nutrient is lost even before it is applied to fields. This, and the underestimated farmland, means that actual applications of nitrogen are much lower than implied by official statistics. Hence, the potential to raise yields by higher fertilization is commensurably higher.

Getting the N:P:K ratios right is a long-overdue goal of Chinese fertilizer applications. Whereas the worldwide mean is now about

100:18:22 (and the United States average is roughly 100:16:35), Chinese applications have been chronically deficient in both P and K, with the nationwide ratio of 100:14:8, and with much higher imbalances in many intensively cultivated regions (FAO 1999). This chronic excess of nitrogen diminishes the efficiency of nitrogen applications and promotes unnecessarily high losses of the nutrient resulting in higher nitrate burden in China's waters, and in higher denitrification rates producing more N_2O (China is already the world's largest emitter of N_2O from farming).

Raising efficiencies of meat production is another area of potentially large rewards. Widespread availability of mixed feeds and better breeds could lower today's feed/meat ratios not only for pigs but also for poultry and for carps and other freshwater fish. This should not mean giving up the feeding of traditionally great variety of waste organic matter. These feeds are more important in China than in any large livestock-producing country. They range from plain and treated cereal straws and root crops to aquatic plants, leucaena leaves, and poultry litter, and they may have contributed as much as 35 percent of all feed energy used in China's animal food production in the early 1990s (Simpson et al. 1994). This share is expected to decline, but it may be still around 30 percent a generation from now as properly prepared and upgraded (urea-treated straw is already common in China) waste feeds will keep reducing the demand for high-quality concentrates.

By far the most important way in which long-term dietary transitions could contribute to higher efficiency of food system would be by producing animal foodstuffs requiring less feed per unit of final product. Animal foods in the average Chinese diet of the late 1990s—dominated by pork (accounting for almost half of energy content), with rising fish, poultry, and egg consumption, and negligible intake of dairy products—take about 3.2 units of grain feed per unit of live weight output.

When animal food shares are split equally among pork, fish, poultry, and eggs, and when typical feeding rates are lowered by just 10–15 percent, this mixture of animal foods requires no more than 2.5 units of grain feed per unit of live weight output, a 20 percent improvement compared to the current state. A diet consisting of equal parts of pork, poultry, eggs, fish, and milk and produced with feeding efficiencies another 10 percent higher (this would still leave them well behind the best Western levels of today) would require just 2.0–2.1 units of concentrate feed. As we have already seen, the Japanese example shows that dairy products, traditionally absent in East Asia, can eventually become a relatively large source of food energy and dietary protein.

The level at which China's meat consumption will eventually saturate is yet another critical variable. Mechanical transfer of Taiwan's experience to China is definitely inappropriate. Taiwan's combination of very high average per capita meat intake (more than 70 kg) and very low direct cereal consumption (less than 110 kg) is exceptional in Asia, and the island's mean per capita cereal intake is even below the OECD's mean of some 130 kg! Differences of scale between the two countries (1.2 billion vs. some 20 million of people) are another obvious matter to consider.

Recent high forecasts of China's meat consumption have been also undoubtedly influenced by erroneous official statistics of average meat supply. FAO's food balance sheets, based on China's official output statistics, put the average per capita meat consumption at 38.1 kg in 1995. *China's Statistical Yearbook* puts per capita purchases of urban households at 23.7 kg in 1995, and a two-year national nutrition survey conducted between 1992 and 1994 found average daily meat consumption of 58 g, or 21.2 kg a year (Cui 1995). This means that eventual doubling of average nationwide per capita meat consumption would result in a rate only marginally higher than the current value claimed by official statistics!

If China's harvest and postharvest grain losses were lowered to a rate still somewhat higher than is common in Western countries—say to no more than 8–10 percent—the country would gain more than 30 Mt of grain a year, a total 1.5 times higher than its exceptionally high cereal imports in 1995, and enough to provide the adequate diet to 75 million people! Building modern grain storages capable of handling China's fluctuating harvests will be the key to solving this problem: during the years of bumper harvests millions of tonnes of grain are left out in the open.

Encouraging Perspectives

Impacts of possibly rapid climate change must be seen in a proper perspective. Given the size of China's territory and the variety of crops grown, global warming would not bring only risks of lower yields but also possibilities for increased harvests. Where an adequate moisture will be provided, Chinese studies forecast higher winter wheat yield throughout the North, better corn yields on the North China Plain, better soybean harvests in the Northeast (the crop's main producing area), and benefits for tea and citrus fruits (Smit and Cai 1996). Other benefits might include northward and westward expansion of wheat area, and northward expansion of corn growing. In addition, a long history leads Chinese researchers to recognize such shifts not only as threatening changes, but also as useful stimuli for adjustments in farming and for spatial shifts in cropping.

Although it is unrealistic to expect that China of the coming generation could appreciably lower the extent and the intensity of its environmental problems, it could substantially reduce the rate of new impacts, and even to turn around some degradative trends. Encouragingly, China's investment in environmental protection is now relatively higher than in any rich nation during a comparable stage of its economic development—and by the year 2000 it should rise to at least 1 percent of total GDP (Smil 1997b). Government spending on environmental protection in rich countries began to make a difference only after their average per capita GDPs passed US$10,000, about five times as high as the Chinese GDP mean today.

Most of these changes will require a greater commitment to agricultural research whose findings are necessary to sustain a variety of technical and managerial innovations. Importance of these innovations for China's agriculture has been quite large. Research by Huang and Rozelle (1996) showed that these advances were at least as important in raising food output even during the early 1980s when most observers interpreted the sharply higher production as the result of newly privatized farming—and during the latter half of the 1980s and in the early 1990s they accounted for almost all of the growth in agricultural productivity.

Because there is a significant time lag in the application of research findings, China should be spending increasing amounts now to enjoy the benefits during the coming decades. Unfortunately, as the country became rapidly richer the real annual expenditures on agricultural research fell between 1985 and 1990, and they surpassed the peak 1985 level only by 1994 (Fan and Pardey 1992; Huang and Rozelle 1996). Currently they amount to less than 0.2 percent of the total gross output value of Chinese agriculture; in contrast, U.S. federal funding alone has been equal to about 1 percent of the value contributed by agriculture to the country's GDP.

Finally, the purchasing power of Chinese consumers is still limited. The International Monetary Fund's exaggerated estimates purchasing power parity (PPP) of average annual per capita GDP at nearly US$3,000 in the early 1990s were recently scaled down by the World Bank to about US$1,800 (1992), or a mere three-fifths of the Indonesian mean (World Bank 1996). Consequently, the rate of dietary transition will not be as rapid in poor counties in China's interior as it has been in the country's large coastal cities.

Considering this evidence of potential capacity for improving harvests, reducing losses and managing demand, it is not surprising that virtually all researchers who spent long time studying China's agriculture agree that the country can do it: that it can feed itself during the coming generations, and that its grain, oil, sugar, and meat imports will not destabilize the global food market. The tenor of these conclusions is remarkably similar.

Alexandratos (1996) uses a wide range of revealing international production and consumption comparisons to make a persuasive case that China's growing grain imports will remain only a fraction of panicky scenarios offered by Brown. He also notes that East Asia's decline of cereal food consumption reflected above all drastic falls in rice consumption—but as a smaller share of China's population consumes a mainly rice-based diet (rice dominates grain output only in fourteen of China's thirty provinces) this trend cannot be duplicated in China (Alexandratos 1997). China is also still much poorer than its smaller neighbors, and (as noted in chapter 8) populations living in poverty will increase their grain intake in early stages of their modernization.

And so it is much more likely that a generation from now China's direct annual grain consumption will be still closer to 200 kg than to 100 kg per capita.

Frederick Crook (1994) expects "Chinese farmers to feed their own population, supplemented by modest quantities of imported grain." Scott Rozelle and his colleagues believe that "China will neither starve the world nor become a major grain exporter. It does seem likely, however, that China will become a much bigger importer in the coming decades" (Rozelle et al. 1996). The president of China's new Agricultural University has an unequivocal answer buttressed by detailed technical explanations: "China should and can feed itself today and in the future" (Ke 1996).

Hence I conclude this appraisal in the same fashion as I summed up my previous assessments of China's ability to feed itself (Smil 1995, 1996b). There do not seem to be any insurmountable biophysical reasons why China should not continue feeding itself during the next two generations. Were this not to happen, it would not be because meeting this challenge requires reliance on as yet unproven bioengineering advances or on unprecedented social adjustments. A combination of well-proven economic and technical fixes, environmental protection measures, and dietary adjustments can extract enough additional food from China's agroecosystems to provide decent nutrition during the coming generations without a further weakening of the country's environmental foundations.

The previous sentence—with "the world's" substituted for "China's" and "the country's"—is a fitting summary of this encouragingly Malthusian book.

References

Aber, J., et al. 1998. Nitrogen saturation in temperate forest ecosystems. *BioScience* 48:921–934.

Agassi, M., ed. 1995. *Soil Erosion, Conservation and Rehabilitation.* New York: Dekker.

Aiello, L. C., and P. Wheeler. 1995. The expensive-tissue hypothesis. *Current Anthropology* 36:199–221.

Alexandratos, N., ed. 1995. *World Agriculture: Towards 2010.* Chichester: John Wiley.

Alexandratos, N. 1996. China's projected cereals deficits in a world context. *Agricultural Economics* 15:1–16.

Alexandratos, N. 1997. China's consumption of cereals and the capacity of the rest of the world to increase exports. *Food Policy* 22:253–267.

Ali, M., and S. C. S. Tsou. 1997. Combating micronutrient deficiencies through vegetables—a neglected food frontier in Asia. *Food Policy* 22:17–38.

Almroth, S., and T. Greiner. 1979. *The Economic Value of Breastfeeding.* Rome: FAO.

American Dietetic Association. 1991. *Survey of American Dietary Habits.* Chicago: ADA.

Apple, R. D. 1996. *Vitamania: Vitamins in American Culture.* New Brunswick, N.J.: Rutgers University Press.

Appleby, M. C., et al. 1992. *Poultry Production Systems.* Wallingford, England: CAB International.

Arce-Diaz, E., et al. 1993. Substitutability of fertilizer and rainfall for erosion in spring wheat production. *Journal of Production Agriculture* 6:72–76.

Archer, J. R., and M. J. Marks. 1997. *Control of Nutrient Loss to Water from Agriculture in Europe.* York: The Fertiliser Society.

Askins, R. A. 1995. Hostile landscapes and the decline of migratory songbirds. *Science* 267:1956–1957.

Assibey, E. O. A. 1974. Wildlife as a source of protein in Africa south of the Sahara. *Biological Conservation* 6:32–39.

Augustin, J., and B. P. Klein. 1989. Nutrient composition of raw, cooked, canned, and sprouted legumes. In *Legumes*, R. H. Matthews, ed., pp. 197–217. New York: Marcel Dekker.

Avery, D. T. 1997. Saving nature's legacy through better farming. *Issues in Science and Technology* Fall 1997:59–64.

Bailey, R. C., et al. 1989. Hunting and gathering in tropical rain forest: is it possible? *American Anthropologist* 91:59–82.

Bailey, S. M. 1982. Absolute and relative sex differences in body composition. In *Sexual Dimorphism in Homo sapiens*, R. L. Hall, ed., pp. 363–390. New York: Praeger.

Baker, L. W., et al. 1996. Ambient air concentrations of pesticides in California. *Environmental Science & Technology* 30:1365–1368.

Baldani, J. I., et al. 1986. Characterization of *Herbaspirillum seropedicae* gen. nov., sp. nov., a root-associated nitrogen-fixing bacterium. *International Journal of Systematic Bacteriology* 36:86–93.

Barnabe, G., ed. 1994. *Aquaculture: Biology and Ecology of Cultured Species.* London: E. Horwood.

Batchelor, C., et al. 1996. Simple microirrigation techniques for improving irrigation efficiency on vegetable gardens. *Agricultural Water Management* 32:37–48.

Bazzaz, F., and W. Sombroek, eds. 1996. *Global Climate Change and Agricultural Production.* New York: John Wiley.

Bekaert, G. 1991. Caloric consumption in industrializing Belgium. *Journal of Economic History* 51:633–655.

Belfiore, F., et al., eds. 1991. *Obesity: Basic Concepts and Clinical Aspects.* Basel: S. Karger.

Bender, W. H. 1994. An end use analysis of global food requirements. *Food Policy* 19:381–395.

Benefice, E., et al. 1984. Nutritional situation and seasonal variations for pastoralist populations of the Sahel (Senegalese Ferlo). *Ecology of Food and Nutrition* 14:229–247.

Bent, M., ed. 1993. *Livestock Productivity Enhancers: An Economic Assessment.* Wallingford, England: CAB International.

Berentsen, P. B. M., and G. W. J. Giesen. 1994. Economic and environmental consequences of different governmental policies to reduce N losses on dairy farms. *Netherlands Journal of Agricultural Science* 42:11–19.

Bhagat, R. M., et al. 1996. Water, tillage and weed interactions in lowland tropical rice: a review. *Agricultural Water Management* 31:165–184.

Bhattacharyya, A. K. 1986. Protein-energy malnutrition (kwashiorkor-marasmus syndrome): terminology, classification and evolution. *World Review Of Nutrition and Dietetics* 47:80–133.

Biggs, R. H., and M. E. Joyner, eds. 1994. *Stratospheric Ozone Depletion—UV-B Radiation in the Biosphere*. Berlin: Springer Verlag.

Bintrim, S. B., et al. 1997. Molecular phylogeny of Archaea from soil. *Proceedings of the National Academy of Sciences* 94:277–282.

Birch, G. G., and K. J. Parker. 1983. *Dietary Fibre*. London: Applied Science Publishers.

Birmingham, C. L., et al. 1999. The cost of obesity in Canada. *Canadian Medical Association Journal* 160:483–488.

Black, A. E., et al. 1991. Critical evaluation of energy intake data using fundamental principles of energy physiology: 2. Evaluating the results of published surveys. *European Journal of Clinical Nutrition* 45:583–599.

Black, A. E., et al. 1996. Human energy expenditure in affluent societies: an analysis of 574 doubly-labelled water measurements. *European Journal of Clinical Nutrition* 50:72–92.

Black, A. L., and A. Bauer. 1990. Stubble height effect on winter wheat in northern Great Plains. II. Plant population and yield relations. *Agronomy Journal* 82:103–109.

Blakstad, F. 1995. Outlook for aquaculture industry: profiting from improved quality control in aquaculture. Paper presented at the Eurofish Report Trade Conference, Brussels.

Bleken, M. A., and L. R. Bakken. 1997. The nitrogen cost of food production: Norwegian society. *Ambio* 26:134–142.

Block, G. 1982. A review of validations of dietary assessment methods. *American Journal of Epidemiology* 115:492–505.

Bloodworth, H., and J. L. Berc. 1998. *Cropland Acreage, Soil Erosion, and Installation of Conservation Buffer Strips: Preliminary Estimates of the 1997 National Resources Inventory*. Washington, DC: USDA.

Bloomgarden, Z. T. 1994. Examining the French Paradox: Does wine prevent coronary heart disease? *Primary Cardiology* 20:6–9.

Blum, L., et al., eds. 1997. *Community Assessment of Natural Food Sources of Vitamin A*. Ottawa: IDRC.

Borrini, G., and S. Margen. 1985. *Human Energetics*. Ottawa: IDRC.

Bouwman, A. F. 1996. Direct emissions of nitrous oxide from agricultural soils. *Nutrient Cycling in Agroecosystems* 46:53–70.

Branca, F., et al. 1993. The nutritional impact of seasonality in children and adults of rural Ethiopia. *European Journal of Clinical Nutrition* 47:840–850.

Bray, G. A. 1997. Recognition of obesity as a chronic disease. *The Canadian Journal of Diagnosis* Supplement 1997:9–11.

Bray, G. A., and D. S. Gray. 1988. Obesity: part I—pathogenesis. *Western Journal of Medicine* 149:429–441.

British Medical Association. 1950. *Report of the Committee on Nutrition.* London: BMA.

British Petroleum. 1999. *BP Statistical Review of Energy 1999*. London: BP.

Brown, L. R. 1974. Global food insecurity. *The Futurist* 8(2):56.

Brown, L. R. 1981. World population growth, soil erosion, and food security. *Science* 214:995–1002.

Brown, L. R. 1989. Feeding six billion. *World Watch* September/October:32.

Brown, L. R. 1995a. *Who Will Feed China?* New York: W. W. Norton.

Brown, L. R. 1995b. Facing food scarcity. *World Watch* November/December:10.

Brown, L. R., and E. C. Wolf. 1984. *Soil Erosion: Quiet Crisis in the World Economy*. Washington, D.C.: Worldwatch Institute.

Bruce, J. P., et al., eds. 1996. *Climate Change 1995: Economic and Social Dimensions of Climate Change*. Cambridge: Cambridge University Press.

Brussaard, L., et al. 1997. Biodiversity and ecosystem functioning in soil. *Ambio* 26:563–570.

Bugbee, B., and O. Monje. 1992. The limits of crop productivity. *BioScience* 42:494–502.

Buller, H. A., et al. 1990. Lactose intolerance. *Annual Review of Medicine* 41:141–148.

Bullock, D. G. 1992. Crop rotation. *Critical Reviews of Plant Sciences* 11:309–326.

Buringh, P. 1977. Food production potential of the world. *World Development* 5:477–485.

Burken, J. G., and J. L. Schnoor. 1997. Uptake and metabolism of atrazine by poplar trees. *Environmental Science & Technology* 31:1399–1406.

Burkitt, D. P., et al. 1972. Effect of dietary fibre on stools and transit times, and its role in the causation of disease. *Lancet* II:1408–1412.

Burns, R. E. 1993. Irrigated rice culture in monsoon Asia: the search for an effective water control technology. *World Development* 21:771–789.

Bussink, D. W., et al. 1994. Ammonia volatilization from nitric-acid-treated cattle slurry surface applied to grassland. *Netherlands Journal of Agricultural Science* 42:293–309.

Buyer, J. S., and D. D. Kaufman. 1996. Microbial diversity in the rhizosphere of corn grown under conventional and low-input systems. *Applied Soil Ecology* 5:21–27.

Campbell-Platt, G. 1980. African locust bean (*Parkia* species) and its West African fermented product, dawadawa. *Ecology of Food and Nutrition* 9:123–132.

Carroll, K. K. 1991. Dietary fat and cancer. *Canadian Medical Association Journal* 144:572.

Carson, R. L. 1962. *Silent Spring*. Boston: Houghton Mifflin.

Carter, T. R., et al. 1991. Climatic warming and crop potential in Europe. *Global Environmental Change* 1:291–312.

Carvalho, M., and G. Basch. 1996. Optimisation of nitrogen fertilisation. *Fertilizer Research* 43:127–130.

Cassell, D. K., 1994. *Encyclopedia of Obesity and Eating Disorders*. New York: Facts on File.

Cassman, K. G., 1999. Ecological intensification of cereal production systems: yield potential, soil quality, and precision farming. *Proceedings of the National Academy of Sciences of the USA* 96:5952–5959.

Cassman, K. G., and R. R. Harwood. 1995. The nature of agricultural systems: food security and environmental balance. *Food Policy* 20:439–454.

Cassman, K. G., and P. L. Pingali. 1995. Intensification of irrigated rice systems: learning from the past to meet future needs. *GeoJournal* 35:299–305.

Cassman, K. G., et al. 1993. Nitrogen use efficiency of rice reconsidered: what are the key constraints? *Plant and Soil* 155/156:359–362.

Cassman, K. G., et al. 1996a. Nitrogen-use efficiency in tropical lowland rice systems: contributions from indigenous and applied nitrogen. *Field Crops Research* 47:1–12.

Cassman, K. G., et al. 1996b. Long-term comparison of the agronomic efficiency and residual benefits of organic and inorganic nitrogen sources for tropical lowland rice. *Experimental Agriculture* 32:927–944.

Chamberlain, G. W. 1993. Aquaculture trends and feed projections. *World Aquaculture* 24(1):19–29.

Chameides, W. 1994. Growth of continental-scale metro-agro-plexes, regional ozone pollution, and world food production. *Science* 264:74–77.

Chameides, W. L., et al. 1997. Ozone pollution in the rural United States and the new NAAQS. *Science* 276:916.

Chang, S. T., and P. G. Miles. 1989. *Edible Mushrooms and Their Cultivation*. Bryn Mawr, Pa: Franklin Publishers.

Chen, C. M. 1986. The national nutrition survey in China, 1982: summary results. *Food and Nutrition* 12(1):59–60.

Chen, J., et al. 1990. *Diet, Lifestyle and Mortality in China: A Study of the Characteristics of 65 Chinese Counties*. Oxford: Oxford University Press.

Chepil, W. S., and N. P. Woodruff. 1963. The physics of wind erosion and its control. *Advances in Agronomy* 15:211–302.

Chevassus-Agnes, S. 1994. Disponibilités des lipides alimentaires dans le monde. *Food, Nutrition and Agriculture* 11:15–22.

Chhabra, R. 1996. *Soil Salinity and Water Quality*. Rotterdam: A. A. Balkema.

China News Digest. 1997. Over 50 million Chinese lack adequate food and clothing. *China News Digest* (http://www.cnd.org), December 12, 1997.

Chittenden, R. H. 1904. *Physiological Economy in Nutrition*. New York: F. A. Stokes.

Choudhury, M., et al. 1996. Review of the use of swine manure in crop production. *Waste Management & Research* 14:581–595.

Choudhury, P. C. 1995. Integrated rice-fish culture in Asia with special reference to deepwater rice. *FAO Aquaculture Newsletter* 10:9–16.

Cihacek, L. J., and J. B. Swan. 1994. Effects of erosion on soil chemical properties in the north central region of the United States. *Journal of Soil and Water Conservation* 49(3):259–265.

Clarke, J. M. 1989. Drying rate and harvest losses of windrowed versus direct combined barley. *Canadian Journal of Plant Science* 69:713–720.

Clawson, D. L. 1984. Harvest security and intraspecific diversity in traditional tropical agriculture. *Economic Botany* 39:56–67.

Clutton-Brock, J. 1989. *A Natural History of Domesticated Animals from Early Times*. Austin, Tex.: University of Texas Press.

Clutton-Brock, T. H., and P. H. Harvey. 1979. Home range size, population density and phylogeny in primates. In *Primate Ecology and Human Origins*, I. S. Bernstein and E. O. Smith, eds., pp. 201–214. New York: Garland STPM Press.

Cock, J. 1982. Cassava: a basic energy source in the tropics. *Science* 218:755–762.

Cockrill, W. R., ed. 1974. *The Husbandry and Health of the Domestic Buffalo*. Rome: FAO.

Cohen, J. E. 1995. *How Many People Can the Earth Support?* New York: W. W. Norton.

Committee on Medical Aspects of Food Policy. 1991. *Dietary Reference Values for Food Energy and Nutrients for the United Kingdom*. London: COMA.

Conway, G., and J. Pretty. 1991. *Unwelcome Harvest: Agriculture and Pollution*. London: Earthscan.

Cook, R. M., et al. 1997. Potential collapse of North Sea cod stocks. *Nature* 385:521–522.

Cooke, G. W. 1988. *Fertilizing for Maximum Yield*. Woodstock, N.Y.: Beekman Publishing.

Cote, J. A. 1984. Use of household refuse as a measure of usual and periodic-specific consumption. *American Behavioral Scientist* 28:129–138.

Cowey, C. B., et al., eds. 1985. *Nutrition and Feeding in Fish*. London: Academic Press.

Crockett, S. J., and D. L. Stuber. 1992. Prestige value of foods: changes over time. *Ecology of Food and Nutrition* 27:51–64.

Crook, F. W. 1988. *Agricultural Statistics of the People's Republic of China*. Washington, D.C.: USDA.

Crook, F. W. 1994. *Could China Starve the World?* Washington, D.C.: USDA.

Crook, F. W. 1996. China's grain stocks: Background and analytical issues. In USDA, *China Situation and Outlook Series*, pp. 35–39. Washington, D.C.: USDA.

Crosson, P., and J. R., Anderson. 1992. *Resources and Global Food Prospects: Supply and Demand for Cereals to 2030*. Washington, D.C.: World Bank.

Crosson, P. 1997. Will erosion threaten agricultural productivity? *Environment* 39:4–9, 29–31.

Cui, L. 1995. Third national nutrition survey. *Beijing Review* 38(4):31.

Cure, J. D., and B. Acock. 1986. Crop response to carbon dioxide doubling: a literature survey. *Agricultural and Forest Meteorology* 38:127–145.

Czech Statistical Office. 1995. *Statisticka Rocenka Ceske Republiky*. Prague: State Statistical Office.

Daily, G., et al. 1998. Food production, population growth, and the environment. *Science* 281:1291–1292.

Darwin, R., et al. 1995. *World Agriculture and Climate Change*. Washington, D.C.: USDA.

Dattilo, A. M. 1992. Effects of omega-3 fatty acids on cardiovascular health. *Journal of Cardiopulmonary Rehabilitation* 12:288–294.

Dausch, J. G. 1992. The problem of obesity: fundamental concepts of energy metabolism gone awry. *Critical Reviews in Food Science and Nutrition* 31:271–298.

Davidek, J., et al., eds. 1990. *Chemical Changes During Food Processing*. Amsterdam: Elsevier.

Davis, J. G. 1994. Managing plant nutrients for optimum water use efficiency and water conservation. *Advances in Agronomy* 53:85–120.

Dawber, T. R. 1980. *The Framingham Study*. Cambridge, Mass.: Harvard University Press.

De, R. 1988. *Efficient Fertilizer Use in Summer Rainfed Areas*. Rome: FAO.

de Boer, J. O., et al. 1988. Energy metabolism and requirements in different ethnic groups. *European Journal of Clinical Nutrition* 42:983–997.

de Datta, S. K. 1995. Plant nutrient balance sheets in lowland rice-based cropping systems. In *Integrated Plant Nutrition Systems*, R. Dudal, and R. N. Roy, eds., pp. 369–394. Rome: FAO.

de Datta, S. K., and R. J. Buresh. 1989. Integrated nitrogen management in irrigated rice. *Advances in Soil Science* 10:143–169.

de Garine, I., and G. A. Harrison, eds. 1988. *Coping with Uncertainty in Food Supply*. Oxford: Clarendon Press.

Dekker, J., and S. O. Duke. 1995. Herbicide-resistant field crops. *Advances in Agronomy* 54:69–116.

DeLany, J. P. 1997. Doubly labeled water for energy expenditure. In *Emerging Technologies for Nutrition Research*, S. J. Carlson-Newberry and R. B. Costell, eds., pp. 281–296. Washington, D.C.: National Academy Press.

den Hartog, A. P. 1992. Dietary change and industrialization: the making of the modern Dutch diet (1850–1985). *Ecology of Food and Nutrition* 27:307–318.

den Hartog, A. P. 1972. Unequal distribution of food within the household. *Nutrition Newsletter* 10(4):8–17.

de Vos, A. 1977. Game as food. *Unasylva* 29(111): 2–12.

Dewey, K. G., et al. 1996. Protein requirements of infants and children. *European Journal of Clinical Nutrition* 50(Supplement):S119–S150.

Diemont, W. H., et al. 1991. Re-thinking erosion on Java. *Netherlands Journal of Agricultural Science* 39:213–224.

Diener, P., and E. E. Robkin. 1978. Ecology, evolution, and the search for cultural origins: the question of Islamic pig prohibition. *Current Anthropology* 19:493–540.

Doberman, A. 1996. Review of H. Lindert; Soil degradation and agricultural change in two developing countries. Archived at sustainable@thecity.sfsu.edu.

Doll, R., and R. Peto. 1981. The causes of cancer: quantitative estimates of avoidable risks of cancer in the United States today. *Journal of the National Cancer Institute* 66:1191–1308.

Domros, M., and G. Peng. 1988. *The Climate of China*. Berlin: Springer Verlag.

Donald, C. M., and J. Hamblin. 1976. The biological yield and harvest index of cereals as agronomic and plant breeding criteria. *Advances in Agronomy* 28:361–405.

Douglas, A. E. 1995. The ecology of symbiotic micro-organisms. *Advances in Ecological Research* 26:69–99.

Draper, S. B. 1996. Breast-feeding as a sustainable resource system. *American Anthropologist* 98:258–265.

Dregne, H., and N. Chou. 1992. Global desertification dimensions and costs. In *Degradation and Restoration of Arid Lands*, H. Dregne, ed., pp. 249–282. Lubbock: Texas Tech University.

Dudal, R., and R. N. Roy, eds. 1995. *Integrated Plant Nutrition Systems*. Rome: FAO.

Dupin, H., et al. 1984. Evolution of the French diet: nutritional aspects. *World Review of Nutrition and Dietetics* 44:57–84.

Durand, J. D. 1974. *Historical Estimates of World Population: An Evaluation*. Philadelphia: University of Pennsylvania Press.

Durnin, J. V. G. A., and R. Passmore. 1967. *Energy, Work and Leisure*. London: Heinemann Educational Books.

Dyson, T. 1996. *Population and Food: Global Trends and Future Prospects*. London: Routledge.

Eastin, J. D., and R. D. Munson, eds. 1971. *Moving off the Yield Plateau*. Madison, Wisc: American Society of Agronomy.

Eaton, S. B., et al. 1997. Paleolithic nutrition revisited: a twelve-year retrospective on its nature and implications. *European Journal of Clinical Nutrition* 51:207–216.

Eck, L. H., et al. 1989. Recall of a child's intake from one meal: are parents accurate? *Journal of American Dietetic Association* 89:784–789.

Edmundson, W. C., and P. V. Sukhatme. 1990. Food and work: poverty and hunger? *Economic Development and Cultural Change* 38:263–280.

Egeland, G. M., and J. P. Middaugh. 1997. Balancing fish consumption benefits with mercury exposure. *Science* 278:1904–1905.

Ehrlich, P. 1968. *The Population Bomb*. New York: Ballantine.

Ehrlich, P. 1969. Eco-Catastrophe! *Ramparts* 8(3):24–28.

Ehrlich, P. R., and A. H. Ehrlich. 1990. *The Population Explosion*. New York: Simon & Schuster.

Ehrlich, P., et al. 1993. Food security, population, and environment. *Population and Development Review* 19:27.

Eisenberg, J. F. 1981. *The Mammalian Radiations*. Chicago: University of Chicago Press.

El-Fouly, M., and A. F. A. Fawzi. 1996. Higher and better yields with less environmental pollution in Egypt through balanced fertilizer use. *Fertilizer Research* 43:1–4.

Engelman, R., and P. LeRoy. 1993. *Sustaining Water*. Washington, D.C.: Population Action International.

Enstrom, J. E. 1993. Vitamin C and mortality. *Nutrition Today* 28:39–42.

Erwin, T. L. 1986. The tropical forest canopy. In *Biodiversity*, E. O. Wilson, ed., pp. 123–129. Washington, D.C.: NAS Press.

Evans, L. T. 1980. The natural history of crop yield. *American Scientists* 68:388–397.

Evenson, R., and M. W. Rosegrant. 1995. *Developing Productivity (Non-price Yield and Area) Projections for Commodity Market Modeling*. Washington, D.C.: IFPRI.

Fageria, N. K. 1992. *Maximizing Crop Yields*. New York: Marcel Dekker.

Fagi, A. M. 1996. *Efficient Water Movement*. Jakarta: Indonesian Society of Agronomy.

Fahey, G. C. 1996. Environmentally friendly methods to process crop residues to enhance fiber digestion. In *Nutrient Management of Food Animals to Enhance and Protect Environment*, E. T. Korngay, ed., pp. 177–198. Boca Raton, Fla.: Lewis Publishers.

Falkenmark, M. 1989. The massive water scarcity now threatening Africa—why isn't it being addressed? *Ambio* 18:112–118.

Fallert, R., et al. 1987. *BST and the Dairy Industry*. Washington, D.C.: USDA.

Fan, S., and P. G. Pardy. 1992. *Agricultural Research in China*. The Hague: International Service for National Agricultural Research.

Fantel, R. J., et al. 1989. World phosphate supply. *Natural Resources Forum* 13:178–190.

FAO. 1950. *Calorie Requirements*. Rome: FAO.

FAO. 1957a. *Calorie Requirements*. Rome: FAO.

FAO. 1957b. *Protein Requirements*. Rome: FAO.

FAO. 1965. *Protein Requirements*. Rome: FAO.

FAO. 1971–1981. *FAO-UNESCO Soil Map of the World*. Paris: UNESCO.

FAO. 1979. *Review of Food Consumption Surveys 1977*. Rome: FAO.

FAO. 1980a. *Maximizing the Efficiency of Fertilizer Use by Grain Crops*. Rome: FAO.

FAO. 1980b. *Assessment and Collection of Data on Post-harvest Foodgrain Losses*. Rome: FAO.

FAO. 1981. *Agriculture: Toward 2000*. Rome: FAO.

FAO. 1983a. *A Comparative Study of Food Consumption Data from Food Balance Sheets and Household Surveys*. Rome: FAO.

FAO. 1983b. *Review of Food Consumption Surveys—1981*. Rome: FAO.

FAO. 1984. *Post-harvest Losses in Quality of Food Grains*. Rome: FAO.

FAO. 1986. *Review of Food Consumption Surveys—1985*. Rome: FAO.

FAO. 1988. *Review of Food Consumption Surveys—1988*. Rome: FAO.

FAO. 1989. *Prevention of Post-harvest Food Losses: Fruits, Vegetables and Root Crops*. Rome: FAO.

FAO. 1993. *World Soil Resources: An Explanatory Note on the FAO World Soil Resources Map at 1:25,000,000 Scale*. Rome: FAO.

FAO. 1994. *Water Harvesting for Improved Agricultural Production*. Rome: FAO.

FAO. 1995a. *Review of the State of World Fishery Resources: Aquaculture*. Rome: FAO.

FAO. 1995b. *Concepts and Definitions of Supply/Utilization Accounts*. Rome: FAO.

FAO. 1995c. *Food Balance Sheets—History, Sources, Concepts and Definitions*. Rome: FAO.

FAO. 1996. *The Sixth World Food Survey*. Rome: FAO.

FAO. 1997. *The State of World Fisheries and Aquaculture*. Rome: FAO.

FAO. 1999. *FAOSTAT Statistics Database*. http://apps.fao.org.

FAO/WHO. 1973. *Energy and Protein Requirements*. Rome: FAO.

FAO/WHO. 1991. *Protein Quality Evaluation*. Rome: FAO.

FAO/WHO/UNU. 1985. *Energy and Protein Requirements*. Geneva: WHO.

Faridi, H., and J. M. Faubion, eds. 1995. *Wheat End Uses Around the World*. St. Paul, Minn.: American Association of Cereal Chemists.

Farquhar, G. D. 1997. Carbon dioxide and vegetation. *Science* 278:1411.

Ferrando, R. 1981. *Traditional and Non-traditional Foods*. Rome: FAO.

Finch, C. E., and R. E. Tanzi. 1997. Genetics of aging. *Science* 278:407–411.

Finkel, H., ed. 1982. *Handbook of Irrigation Technology*. Boca Raton, Fla.: CRC.

Finlayson-Pitts, B. J., and J. N. Pitts. 1997. Tropospheric air pollution: ozone, airborne toxics, polycyclic aromatic hydrocarbons, and particles. *Science* 276: 1045–1051.

Fischer, G., and G. H. Heilig. 1997. Population momentum and the demand on land and water resources. *Philosophical Transactions of the Royal Society B* 352:869–889.

Flegal, K. M. 1996. Trends in body weight and overweight in the U.S. population. *Nutrition Reviews* 54:S97–S100.

Fleming, G. R., and R. van Grandelle. 1994. The primary steps of photosynthesis. *Physics Today* 47(2):48–55.

Fleuret, A. 1979. The role of wild foliage plants in the diet: a case study from Lushoto, Tanzania. *Ecology of Food and Nutrition* 8:87–93.

Fogel, R. W. 1991. The conquest of high mortality and hunger in Europe and America: timing and mechanisms. In *Favorites of Fortune*, P. Higgonet et al. eds., pp. 33–71. Cambridge, Mass.: Harvard University Press.

Foley, R. A., and P. C. Lee. 1990. Ecology and energetics of encephalization in hominid evolution. *Philosophical Transaction of the Royal Society London B* 334:223–232.

Follett, R. F. 1993. Global climate change, U.S. agriculture, and carbon dioxide. *Journal of Production Agriculture* 6:181–190.

Fontenot, J. P., et al. 1996. Potential for recycling animal wastes by feeding to reduce environmental contamination. In *Nutrient Management of Food Animals to Enhance and Protect Environment*, E. T. Kornegay, ed., 199–217. Boca Raton, Fla.: Lewis.

Foster, H. D. 1992. *Health, Disease & the Environment*. London: Belhaven.

Foster, P. 1992. *The World Food Problem*. Boulder: Lynne Rienner.

Fragoso, M. A., ed. 1993. *Optimization of Plant Nutrition*. Amsterdam: Kluwer.

Franzen, D. W., and T. R. Peck. 1995. Field soil sampling density for variable rate fertilization. *Journal of Production Agriculture* 8:568–674.

Frederick, K. D. 1988. Irrigation under stress. *Resources* Spring 1988:1–4.

Friedl, J. 1981. Lactase deficiency: distribution, associated problems, and implications for nutritional policy. *Ecology of Food and Nutrition* 11:37–48.

Frigerio, C., et al. 1991. Is human lactation a particularly efficient process? *European Journal of Clinical Nutrition* 45:459–462.

Frissel, M. J., ed. 1978. *Cycling of Mineral Nutrients in Agricultural Ecosystems*. Amsterdam: Elsevier.

Frissel, M. J., and G. J. Kolenbrander. 1978. The nutrient balances: summarizing graphs and tables. In *Cycling of Mineral Nutrients in Agricultural Ecosystems*, M. J. Frissel, ed. pp. 273–292. Amsterdam: Elsevier.

Garby, L. 1990. Metabolic adaptation to decreases in energy intake due to changes in the energy cost of low energy expenditure regimen. *World Review of Nutrition and Dietetics* 61:173–208.

Garrow, J. S. 1988. *Obesity and Related Disorders*. New York: Churchill.

Gasser, C. S., and R. T. Fraley. 1992. Transgenic crops. *Scientific American* 266(6):62–69.

Ge, K., et al. 1991. Food consumption and nutritional status in China. *Food Nutrition and Agriculture* 1:54–61.

Gebhardt, M. R., et al. 1985. Conservation tillage. *Science* 230:625–630.

General Accounting Office. 1995. *Global Warming: Limitations of General Circulation Models and Costs of Modeling Efforts*. Washington, D.C.: GAO.

General Accounting Office. 1997. *Drinking Water*. Washington, D.C.: GAO.

Gey, K. F., et al. 1991. Inverse correlation between plasma vitamin E and mortality from ischemic heart disease in cross-cultural epidemiology. *American Journal of Clinical Nutrition* 53:326S–334S.

Gifford, R. M., and L. T. Evans. 1981. Photosynthesis, carbon partitioning, and yield. *Annual Review of Plant Physiology* 32:485–509.

Gill, G. J. 1993. *O.K., The Data's Lousy, But It's All We've Got (Being a Critique of Conventional Methods)*. London: International Institute for Environment and Development.

Gilland, B. 1985. Cereal yields in theory and practice. *Outlook on Agriculture* 14:56–60.

Gillis, M., et al. 1991. Taxonomic relationships between (*Pseudomonas*) *rubrisulbalbicans*, some clinical isolates (EF group 1), *Herbaspirillum seropedi-*

cae and (*Aquaspirillum*) *autotrophicum*. In *Nitrogen Fixation*, M. Polsinelli et al., eds., pp. 292–294. Dordrecht: Kluwer Academic.

Glanz, K., et al. 1997. Are awareness of dietary fat intake and actual fat consumption associated? A Dutch-American comparison. *European Journal of Clinical Nutrition* 51:542–547.

Gleick, P. H., ed. 1993. *Water in Crisis: A Guide to the World's Fresh Water Resources*. New York: Oxford University Press.

Glenn, E. P., et al. 1998. Irrigating crops with seawater. *Scientific American* 279(2):76–81.

Glinsmann, W. H., et al. 1996. Dietary guidelines for infants: a timely reminder. *Nutrition Reviews* 54:50–57.

Goedmakers, A. 1989. Ecological perspectives of changing agricultural land use in the European community. *Agriculture, Ecosystems and Environment* 27:99–106.

Goldman, A. S. 1993. The immune system of human milk: antimicrobial, anti-inflammatory and immunomodulating properties. *Pediatric Infectious Disease Journal* 12:664–671.

Gollany, H. T., et al. 1992. Topsoil depth and desurfacing effects on properties and productivity of a Typic Argiustoll. *Soil Science Society of America Journal* 56:220–225.

Gonsolus, J. L. 1990. Non-chemical weed control in corn and soybeans. In *Extending Sustainable Systems*, pp. 331–343. Minneapolis: Minnesota Department of Agriculture.

Goolsby, D. A., et al. 1997. Herbicides and their metabolites in rainfall: origin, transport, and deposition patterns across the Midwestern and Northeastern United States, 1990–1991. *Environmental Science & Technology* 31:1325–1333.

Grabau, L. J., and T. W. Pfeiffer. 1990. Management effects on harvest losses and yield of double-crop soybean. *Agronomy Journal* 82:715–718.

Graves, J., and D. Reavey. 1996. *Global Environmental Change: Plants, Animals and Communities*. London: Longman.

Guerrero, P. F. 1992. *30 Years Since Silent Spring—Many Long-standing Concerns Remain*. Washington, D.C.: GAO.

Gujja, B., and A. Finger-Stich. 1996. What price prawn? *Environment* 38(7): 12–15, 33–39.

Gupta, M. C., B. M. Gandhi, and B. N. Tan. 1974. An unconventional legume—Prosopis cineraria. *American Journal of Clinical Nutrition* 27:1035–1036.

Halver, J. E. 1989. *Fish Nutrition*. San Diego: Academic Press.

Hamdi, Y. A. 1995. Potential and assessment of BNF and its direct contribution in selected cropping systems and ecological conditions. In *Integrated Plant Nutrition Systems*, R. Dudal and R. N. Roy, eds., pp. 201–222. Rome: FAO.

Hammond, K. A., and J. Diamond. 1997. Maximal sustained energy budgets in humans and animals. *Nature* 386:457–462.

Hanley, M. L. 1991. After the harvest. *World Development* 4(1):25–27.

Hanstad, T., and Li Ping. 1997. Land reform in the People's Republic of China: auctioning rights to wasteland. *International & Comparative Law Journal* 19:545–583.

Hargrove, W. L. 1988. Soil, environmental, and management factors influencing ammonia volatilization under field conditions. In *Ammonia Volatilization from Urea Fertilizers*, B. R. Bock and D. E. Kissel, eds., pp. 17–36. Muscle Shoals, Ala.: National Fertilizer Development Center.

Harper, A. E. 1996. Dietary guidelines in perspective. *The Journal of Nutrition* 126:1042S–1048S.

Harris, D. R., ed. 1980. *Human Ecology in Savanna Environments*. New York: Academic Press.

Harris, J. A., and F. G. Benedict. 1919. *A Biometric Study of Basal Metabolism in Men*. Washington, D. C.: Carnegie Institution.

Hassan, N., and K. U. Ahmad. 1992. Studies on food and nutrient intake by urban population of Bangladesh: comparison between intakes of 1962–64 and 1985–86. *Ecology of Food and Nutrition* 28:131–148.

Havlin, J. L., et al., eds. 1994. *Soil Testing: Prospects for Improving Nutrient Recommendations*. Madison, Wisc.: Soil Science Society of America.

Hefner, S. G., and P. W. Tracy. 1995. Corn production using alternate furrow nitrogen fertilization and irrigation. *Journal of Production Agriculture* 8:66–69.

Heichel, G. H. 1987. Legume nitrogen: symbiotic fixation and recovery by subsequent crops. In *Energy in Plant Nutrition and Pest Control*, Z. Helsel, ed., pp. 62–80. Amsterdam: Elsevier.

Heilig, G. K. 1997. Anthropogenic factors in land-use change in China. *Population and Development Review* 23:139–168.

Heini, A. F., et al. 1996. Free-living energy expenditure assessed by two different methods in rural Gambian men. *European Journal of Clinical Nutrition* 50:284–289.

Helsing, E. 1993. Trends in fat consumption in Europe and their influence on the Mediterranean diet. *European Journal of Clinical Nutrition* 47(Supplement):S4–S15.

Hendrey, G. R., ed. 1992. *FACE Free-Air CO_2 Enrichment for Plant Research in the Field*. Boca Raton, Fla.: CRC Press.

Henry, C. K. J., and D. G. Rees. 1991. New predictive equations for the estimation for basal metabolic rates in tropical peoples. *European Journal of Clinical Nutrition* 45:177–185.

Hepher, B. 1988. *Nutrition of Pond Fishes*. Cambridge: Cambridge University Press.

Hera, C. 1995. Contribution of nuclear techniques to the assessment of nutrient availability for crops. In *Integrated Plant Nutrition Systems*, R. Dudal and R. N. Roy, eds., pp. 307–332. Rome: FAO.

Herbert, V. 1993. Does mega-C do more good than harm, or more harm than good? *Nutrition Today* 28:28–32.

Hicks, R. J., et al. 1990. Review and evaluation of the effects of xenobiotic chemicals on microorganisms in soil. *Advances in Applied Microbiology* 35: 195–253.

Higgitt, D. L. 1991. Soil erosion and soil problems. *Progress in Physical Geography* 15:91–100.

Higgs, R. L., et al. 1990. Crop rotations: sustainable and profitable. *Journal of Soil and Water Conservation* 45:68–70.

Hillel, D. 1994. *Rivers of Eden*. Oxford: Oxford University Press.

Hillel, D. 1997. *Small-scale Irrigation for Arid Zones*. Rome: FAO.

Hinman, C. W. 1986. Potential new crops. *Scientific American* 255(1):32–37.

Hollingsworth, D. F. 1983. Rationing and economic constraints on food consumption in Britain since the Second World War. *World Review of Nutrition and Dietetics* 42:191–218.

Hollis, G. R. 1993. *Growth of the Pig*. Wallingford, England: CAB International.

Hoogerbrugge, I. D., and L. O. Fresco. 1993. *Homegarden Systems: Agricultural Characteristics and Challenges*. London: IIED.

Hooper, D. U., and P. M. Vitousek. 1997. The effects of plant composition and diversity on ecosystem processes. *Science* 277:1302–1305.

Hopkins, P. N., and R. R. Williams. 1981. A survey of 246 suggested coronary risk factors. *Atherosclerosis* 40:1–52.

van Horn, H. H., et al. 1994. Components of dairy manure management systems. *Journal of Dairy Science* 77:2008–2030.

Horwith, B. 1985. A role for intercropping in modern agriculture. *BioScience* 35:286–291.

Houghton, J. T., et al., eds. 1996. *Climate Change 1995: The Science of Climate Change*. Cambridge: Cambridge University Press.

Huang, J., and S. Rozelle. 1995. Environmental stress and grain yields in China. *American Journal of Agricultural Economics* 77:853–864.

Huang, J., and S. Rozelle. 1996. Technological change: rediscovering the engine of productivity growth in China's agricultural economy. *Journal of Development Economics* 49:337–369.

Huang, P. C., and C. P. Lin. 1981. Protein requirements of young Chinese male adults. In *Protein-Energy Requirements of Developing Countries: Evaluation of New Data*, B. Torun et al., eds., pp. 63–70. Tokyo: United Nations University.

Hui, Y. H., ed. 1993. *Dairy Science and Technology Handbook: 2: Product Manufacturing*. New York: VCH.

Hulme, M., et al. 1999. Relative impacts of human-induced climate change and natural climate variability. *Nature* 397:688–691.

Hulse, J. H., and O. Pearson. 1980. The nutritional status of the population of the semi-arid tropical countries. In *Nutritional Status of the Rural Population of the Sahel*, pp. 86–92. Ottawa: IDRC.

Humphries, C. J., et al. 1995. Measuring biodiversity value for conservation. *Annual Review of Ecology and Systematics* 26:93–111.

Huston, M. 1993. Biological diversity, soils, and economics. *Science* 262: 1676–1679.

Huston, M. 1994. *Biological Diversity: The Coexistence of Species on Changing Landscapes*. Cambridge: Cambridge University Press.

Hynes, R. K., et al. 1995. Inoculants/Additives. *Journal of Production Agriculture* 8:547–552.

Hytten, F. E., and I. Leitch. 1964. *The Physiology of Human Pregnancy*. Edinburgh: Blackwell.

Inness, R. 1989. Quinoa's comeback. *IDRC Reports* April 1989:22–23.

Intergovernmental Panel on Climate Change. 1996: *Climate Change 1995*. New York: Cambridge University Press.

Jaffe, G. M. 1984. Vitamin C. In *Handbook of Vitamins*, L. J. Machlin, ed., pp. 199–220. New York: Marcel Dekker.

James, D. G. 1986. The prospects for fish for the malnourished. *Food and Nutrition* 12(2):20–30.

James, W. P. T., and E. C. Schofield. 1990. *Human Energy Requirements: A Manual for Planners and Nutritionists*. Oxford: Oxford University Press.

Jankun, J., et al. 1997. Why drinking green tea could prevent cancer. *Nature* 387:561.

Jarnagin, S. K., and M. A. Smith. 1993. *Soil Erosion and Effects on Crop Productivity*. Paper prepared for Project 2050.

Jarrige, R., and C. Beranger. 1992. *Beef Cattle Production*. Amsterdam: Elsevier.

Jarvis, L. S. 1996. *The Potential Effect of Two New Biotechnologies on the World Dairy Industry*. Boulder: Westview.

Jenkinson, D. S. 1982. The nitrogen cycle in long-term experiments. *Philosophical Transaction of the Royal Society London* B 296:563–571.

Jensen, M. 1993. Soil conditions, vegetation structure and biomass of a Javanese homegarden. *Agroforestry Systems* 34:171–186.

Johansson, I., et al. 1996. The Norwegian diet during the last hundred years in relation to coronary heart disease. *European Journal of Clinical Nutrition* 50:277–283.

Johnston, A. E. 1997. The value of long-term field experiments in agricultural, ecological, and environmental research. *Advances in Agronomy* 59: 291–333.

Jones, J. M. 1992. *Food Safety*. St. Paul, Minn.: Eagan Press.

Jongbloed, A. W., and C. H. Henkens. 1996. Environmental concerns of using animal manure—the Dutch case. In *Nutrient Management of Food Animals to Enhance and Protect Environment*, E. T. Kornegay, ed., pp. 315–332. Boca Raton, Fla.: Lewis.

Juma, N. G., et al. 1993. Crop yield and soil organic matter over 60 years in a Typic Cryoboralf at Breton, Alberta. In *The Breton Plots*, pp. 31–46. Department of Soil Science, University of Alberta, Edmonton.

Kaihura, F. B., et al. 1996. Topsoil thickness effect on soil properties and maize (*Zea mays*) yield in three ecoregions of Tanzania. *Journal of Sustainable Agriculture* 9:1–15.

Kaiser, J., and R. Gallagher. 1997. How humans and nature influence ecosystems. *Science* 277:1203–1205.

Karl, T. R., et al. 1997. The coming climate. *Scientific American* 276(5):78–83.

Karlen, D. L., et al. 1994. Crop rotations for the 21st century. *Advances in Agronomy* 53:1–45.

Karlovsky, J. 1981. Cycling of nutrients and their utilisation by plants in agricultural ecosystems. *Agro-Ecosystems* 7:127–144.

Ke, B. 1996. *Grain Production in China*. Beijing: China Agricultural University.

Kearney, M., et al. 1997. Perceived need to alter eating habits among representative samples of adults from all member states of the European Union. *European Journal of Clinical Nutrition* 51(Supplement 2):S30–S35.

Kelley, H. W. 1983. *Keeping the Land Alive: Soil Erosion—Its Causes and Cures*. Rome: FAO.

Kennedy, E., and T. Reardon. 1994. Shift to non-traditional grains in the diets of East and West Africa: role of women's opportunity cost of time. *Food Policy* 19:45–56.

Kerr, J., and N. K. Sanghi. 1992. *Indigenous Soil and Water Conservation in India's Semi-arid Tropics*. London: International Institute for Environment and Development.

Keys, A. 1948. Nutrition in relation to etiology and course of degenerative diseases. *Journal of the American Dietetic Association* 24:281–285.

Keys, A. 1952. Human atherosclerosis and diet. *Circulation* 5:115–118.

Keys, A. 1980. *Seven Countries: A Multivariate Analysis of Death and Coronary Heart Disease*. Cambridge, Mass.: Harvard University Press.

Keys, A., and M. Keys. 1975. *How to Eat Well and Stay Well the Mediterranean Way*. New York: Doubleday.

Kimball, B. A. 1983. Carbon dioxide and agricultural yield: an assembling and analysis of 430 prior observations. *Agronomy Journal* 75:779–788.

Kimball, B. A., et al. 1993. Effects of free-air CO_2 enrichment on energy balance and evapotranspiration of cotton. *Agricultural and Forest Meteorology* 70:259–278.

Klatsky, A. L., et al. 1997. Red wine, white wine, liquor, beer, and risk of coronary artery disease hospitalization. *The American Journal of Cardiology* 80:416–420.

Kleges, R. C., et al. 1987. Validation of the 24-hour dietary recall in preschool children. *Journal of American Dietetic Association* 87:1383–1385.

Kleiber, M. 1961. *The Fire of Life*. New York: John Wiley.

Kolpin, D. W., et al. 1991. *Herbicides and Nitrates in Near-surface Aquifers in the Midcontinental United States*. Denver: USGS.

Korn, M. 1996. The dike-pond concept: sustainable agriculture and nutrient recycling in China. *Ambio* 25:6–13.

Kramer, E. M., et al. 1997. Intrahousehold allocation of energy intake among children under five years and their parents in rural Bangladesh. *European Journal of Clinical Nutrition* 51:750–756.

Kuczmarski, R. J., et al. 1994. Increasing prevalence of overweight among US adults. *Journal of American Medical Association* 272:205–211.

Kuhnlein, H. V., and G. H. Pelto. 1997. *Culture, Environment, and Food to Prevent Vitamin A Deficiency*. Ottawa: IDRC.

Kushi, L. H., et al. 1995. Health implications of Mediterranean diets in light of contemporary knowledge. 1. Plant foods and dairy products. *American Journal of Clinical Nutrition* 61(Supplement):1407S–1415S.

Kushner, R. F. 1993. Body weight and mortality. *Nutrition Reviews* 51:127–136.

Lal, R. 1995. Erosion-crop productivity relationships for soils in Africa. *Soil Science Society of America Journal* 59:661–667.

Lal, R., and B. A. Stewart. 1990. *Soil Degradation*. Berlin: Springer Verlag.

Lamberg, B. A. 1993. Iodine deficiency disorders and endemic goitre. *European Journal of Clinical Nutrition* 47:1–8.

Lansing, J. S. 1991. *Priests and Programmers: Technologies of Power in the Engineered Landscape of Bali*. Princeton, N.J.: Princeton University Press.

Lappalainen, R., et al. 1997. Difficulties in trying to eat healthier: descriptive analysis of perceived barriers for healthy eating. *European Journal of Clinical Nutrition* 51(Supplement 2):S36–S40.

Larson, B. A., and G. B. Frisvold. 1996. Fertilizers to support agricultural development in sub-Saharan Africa: what is needed and why. *Food Policy* 21:509–525.

Larson, W. E., et al. 1983. The threat of soil erosion to long-term crop production. *Science* 219:458–465.

Law, M. R., and J. K. Morris. 1998. By how much does fruit and vegetable consumption reduce the risk of ischaemic heart disease? *European Journal of Clinical Nutrition* 52:549–556.

Lawrence, M., et al. 1988. Between-group differences in basal metabolic rates: an analysis of data collected in Scotland, the Gambia and Thailand. *European Journal of Clinical Nutrition* 42:877–891.

Lee, L. K. 1990. The dynamics of declining soil erosion rates. *Journal of Soil and Water Conservation* 45:622–624.

Lefohn, A. S. 1991. *Surface-Level Ozone Exposures and Their Effects on Vegetation*. Boca Raton, Fla.: Lewis.

Legg, J. O., and J. J. Meisinger. 1984. Soil nitrogen budgets. *Agronomy* 22: 503–566.

Lemnitzer, K-H. 1977. *Ernährungssituation und wirtschaftlichen Entwickelung*. Saarbrücken: Verlag der SSIP-Schriften Breitenbach.

Lemon, P. W. 1996. Is increased dietary protein necessary or beneficial for individuals with a physically active lifestyle? *Nutrition Review* 54:S169–S174.

Levin, S. A. 1995. Scale and sustainability: a population and community perspective. In *Defining and Measuring Sustainability: The Biophysical Foundations*, M. Munasinghe and W. Shearer, eds., pp. 103–116. Washington, D.C.: The United Nations University.

Liang, L., et al. 1993. China's post-harvest grain losses and the means of their reduction and elimination. *Jingji dili* (*Economic Geography*) 1 (March 1993): 92–96.

Lichtenstein, A. H., et al. 1998. Dietary fat consumption and health. *Nutrition Reviews* 56:S3–S28.

Liebig, J. 1840. *Chemistry in Its Application to Agriculture and Physiology*, London: Taylor & Walton.

Lifson, N., et al. 1955. Measurement of total carbon dioxide production by means of D_2O^{18}. *Journal of Applied Physiology* 7:705–710.

Lindert, P. H. 1996. *Soil Degradation and Agricultural Change in Two Developing Countries*. http://thecity.sfsu.edu/~sustain/lindert.

Lindroos, A-K., et al. 1993. Validity and reproducibility of a self-administered dietary questionnaire in obese and non-obese subjects. *European Journal of Clinical Nutrition* 47:461–481.

Lissner, L., et al. 1989. Body composition and energy intake: do overweight women overeat and underreport? *American Journal of Clinical Nutrition* 49:320–325.

Little, C. E. 1987. *Green Fields Forever: The Conservation Tillage Revolution in America*. Washington, D.C.: Island Press.

Livingstone, M. B. E., et al. 1990. Accuracy of weighed dietary records in studies of diet and health. *British Medical Journal* 300:708–712.

Lowenberg-DeBoer, J., and S. M. Swinton. 1995. *Economics of Site-specific Management in Agronomic Crops*. Department of Agricultural Economics, Michigan State University.

Lu, Y., et al. 1997. The current state of precision farming. *Food Reviews International* 13(2):141–162.

Lutz, W., et al. 1997. Doubling of world population unlikely. *Nature* 387: 803–805.

Maberly, G. F., et al. 1994. Programs against micronutrient malnutrition: ending hidden hunger. *Annual Review of Public Health* 15:277–301.

Mader, T. L., et al. 1991. Long-term storage losses of alfalfa stored in loaf stacks. *The Professional Animal Scientist* 7(3):13.

Madsen, E. L. 1995. Impacts of agricultural practices on subsurface microbial ecology. *Advances in Agronomy* 54:1–67.

Magadza, C. H. D. 1994. Climate change: some likely multiple impacts in southern Africa. *Food Policy* 39:165–190.

Mahler, M., ed. 1991. *Science and Cultivation of Edible Fungi*. Rotterdam: A.A. Balkema.

Malthus, T. R. 1798. *An Essay on the Principle of Population, as it Affects the Future Improvement of Society*. London: J. Johnson.

Malthus, T. R. 1803. *An Essay on the Principle of Population; or, A View of Its Past and Present Effect on Human Happiness*. London: J. Johnson.

Mann, C. C. 1999. Crop scientists seek a new revolution. *Science* 283:310–314.

Marx, J. J. M. 1997. Iron deficiency in developed countries: prevalence, influence of lifestyle factors and hazards of prevention. *European Journal of Clinical Nutrition* 51:491–494.

Masood, E. 1997. Aquaculture: a solution, or source of new problems? *Nature* 386:109.

Massee, T. W., and H. O. Waggoner. 1985. Productivity losses from soil erosion in the intermountain area. *Journal of Soil and Water Conservation* 40:447–450.

Matthews, R. H. 1989. *Legumes*. New York: Marcel Dekker.

Mbagwu, J. S. C., et al. 1984. Effects of desurfacing Alfisols and Ultisols in Southern Nigeria: I. Crop performance. *Soil Science Society of America Journal* 48:28–33.

McEvedy, C., and R. Jones. 1979. *Atlas of World Population History*. London: Allen Lane.

McGrady-Steed, J., et al. 1997. Biodiversity regulates ecosystem predictability. *Nature* 390:162–165.

McKee, D. J. 1993. *Tropospheric Ozone: Human Health and Agricultural Impacts*. Boca Raton, Fla.: Lewis.

McKnight, T. L. 1990. Irrigation technology: a photo-essay. *Focus* 40(2):1–6.

McLachlan, M. 1996. Bioaccumulation of hydrophobic chemicals in agricultural food chains. *Environmental Science & Technology* 30:252–259.

McLarney, M. J., et al. 1996. Pattern of amino acid requirements in humans: an interspecies comparison using published amino acid requirement recommendations. *The Journal of Nutrition* 126:1871–1882.

Mead, R., et al. 1986. Stability comparison of intercropping and monocropping systems. *Biometrics* 42:253–266.

Meadows, D. H., et al. 1972. *Limits to Growth*. New York: Universe Books.

Meadows, D. H., et al. 1992. *Beyond the Limits*. London: Earthscan.

Mearns, L. O., et al. 1996. The effect of changes in daily and interannual climatic variability on CERES-Wheat: a sensitivity study. *Climatic Change* 32: 257–292.

MEDEA. 1997. *China Agriculture: Cultivated Land Area, Grain Projections, and Implications*. Washington, D.C.: National Intelligence Council.

Meershoek, S. 1993. The economic value of breastfeeding. *Breastfeeding Review* 2:354–357.

Melville, H. 1851. *Moby-Dick; or, The Whale*. New York: Harper & Brothers.

Mench, J. A., and A. van Tienhoven. 1986. Farm animal welfare. *American Scientist* 74:598–603.

Mepsted, R., et al. 1996. Effects of enhanced UV-B radiation on pea (*Pisum sativum* L.) grown under field conditions in the UK. *Global Change Biology* 2:325–334.

Mielke, L. N., and J. S. Schepers. 1986. Plant response to topsoil thickness on an eroded loess soil. *Journal of Soil and Water Conservation* 41:59–63.

Milio, N. 1991. Food rich and health poor. *Food Policy* 16:311–318.

Miller, E. R., et al., eds. 1991. *Swine Nutrition*. Boston: Butterworth-Heinemann.

Millward, D. J., and J. P. W. Rivers. 1988. The nutritional role of indispensable amino acids and the metabolic basis for their requirements. *European Journal of Clinical Nutrition* 42:367–393.

Ministry of Health and Welfare. 1995. *National Nutrition Survey*. Tokyo: MHW.

Mitchell, D. O., et al. 1997. *The World Food Outlook*. New York: Cambridge University Press.

Montanez, A., et al. 1995. The effect of temperature on nodulation and nitrogen fixation by five *Bradyrhizobium japonicum* strains. *Applied Soil Ecology* 2:165–174.

Morgan, R. P. C. 1986. *Soil Erosion and Conservation*. London: Routledge.

Morris, M. C., et al. 1995. Fish consumption and cardiovascular disease in the Physicians' Health Study: A prospective study. *American Journal of Epidemiology* 142:166–175.

Mortvedt, J. J. 1996. Heavy metal contaminants in inorganic and organic fertilizers. *Fertilizer Research* 43:55–61.

Mosier, A. R., et al. 1998. Assessing and mitigating N_2O emissions from agricultural soils. *Climatic Change* 40:7–38.

Mount, T., and Z. Li. 1994. *Estimating the Effects of Climate Change on Grain Yield and Production in the U.S.* Washington, DC.: USDA.

Muehlbauer, F. J., et al. 1995. Production and breeding of lentil. *Advances in Agronomy* 54:283–332.

Munson, R. D., and C. F. Runge. 1990. *Improving Fertilizer and Chemical Efficiency Through "High Precision Farming."* Center for International Food and Agricultural Policy, University of Minnesota.

Naeem, S., and S. Li. 1997. Biodiversity enhances ecosystem reliability. *Nature* 390:507–509.

Nakayama, F. S., and D. A. Bucks. 1986. *Trickle Irrigation for Crop Production.* Amsterdam: Elsevier.

National Center for Health Statistics. 1979. *Food Consumption Profiles of White and Black Persons Aged 1–74 Years: United States 1971–1974.* Hyatsville, Md.: US DHEW.

National Research Council. 1943. *Recommended Dietary Allowances.* Washington, D.C.: NRC.

National Research Council. 1982. *Diet, Nutrition, and Cancer.* Washington, D.C.: National Academy Press.

National Research Council. 1987. *Predicting Feed Intake of Food-Producing Animals.* Washington, D.C.: National Academy Press.

National Research Council. 1988a. *Nutrient Requirements of Dairy Cattle.* Washington, D.C., National Academy Press.

National Research Council. 1988b. *Nutrient Requirements of Swine.* Washington, D.C.: National Academy Press.

National Research Council. 1989. *Recommended Dietary Allowances.* Washington, D.C.: National Academy Press.

National Research Council. 1994. *Nutrient Requirements of Poultry.* Washington, D.C.: National Academy Press.

National Research Council. 1996a. *Lost Crops of Africa.* Washington, D.C.: National Academy Press.

National Research Council. 1996b. *Nutrient Requirements of Beef Cattle.* Washington, D.C.: National Academy Press.

Naylor, R., et al. 1997. Variability and growth in grain yields, 1950–1994: does the record point to greater instability? *Population and Development Review* 23:41–58.

Nestle, M. 1995. Mediterranean diets: historical and research overview. *American Journal of Clinical Nutrition* 61(Supplement):1313S–1320S.

Nevison, C. D., et al. 1996. A global model of changing N_2O emissions from natural and perturbed soils. *Climatic Change* 32:327–378.

Nichols, N. 1997. Increased Australian wheat yield due to recent climate trends. *Nature* 387:484–485.

Nickum, J. E. 1995. *Dam Lies and Other Statistics: Taking the Measure of Irrigation in China, 1931–91.* Honolulu: East-West Center.

Nickum, J. E. 1998. Is China living on the water margin? *The China Quarterly* 156:880–898.

Njiforti, H. L. 1996. Preferences and present demand for bushmeat in north Cameroon: some implications for wildlife conservation. *Environmental Conservation* 23:149–155.

Nolan, B. T., et al. 1997. Risk of nitrate in groundwaters of the United States—a national perspective. *Environmental Science & Technology* 31:2229–2236.

Norse, D. 1994. Multiple threats to regional food production: environment, economy, population? *Food Policy* 19:133–148.

Nova Scotia Heart Health Program. 1993. *Report of the Nova Scotia Nutrition Survey.* Halifax: Nova Scotia Department of Health.

Oldeman, I. R., et al. 1990. *World Map of the Status of Human-Induced Soil Degradation: An Explanatory Note.* International Soil Information and Reference Center, Wageningen, and UNEP, Nairobi.

Olinto, M. T. A., et al. 1995. Twenty-four-hour recall overestimates the dietary intake of malnourished children. *Journal of Nutrition* 125:880–884.

Olk, D. C., et al. 1996. Changes in chemical properties of organic matter with intensified rice cropping in tropical lowland soils. *European Journal of Soil Science* 47:293–303.

Olson, K. R., et al. 1994. Evaluation of methods to study soil erosion-productivity relationships. *Journal of Soil and Water Conservation* 49(6):586–590.

O'Mara, G. T., ed. 1988. *Efficiency in Irrigation.* Washington, D.C.: World Bank.

Orskov, E. R. 1990. *Energy Nutrition in Ruminants.* New York: Elsevier.

Osler, M., et al. 1997. Ten year trends in the dietary habits of Danish men and women: cohort and cross-sectional data. *European Journal of Clinical Nutrition* 51:535–541.

Pace, N. R. 1997. A molecular view of microbial diversity and the biosphere. *Science* 276:734–740.

Paoletti, M. G., and D. Pimentel. 1996. Genetic engineering in agriculture and the environment. *BioScience* 46:665–673.

Paoletti, M. G., et al. 1992. Agroecosystem biodiversity: matching production and conservation biology. *Agriculture, Ecosystems and Environment* 40:3–23.

Parker, R. O. 1995. *Aquaculture Science.* Albany, N.Y.: Delmar.

Patton, J. F. 1997. Measurement of oxygen uptake with portable equipment. In *Emerging Technologies for Nutrition Research*, S. J. Carlson-Newberry and R. B. Costell, eds., pp. 297–314. Washington, D.C.: National Academy Press.

Pauling, L. 1986. *How to Live Longer and Feel Better.* New York: W. H. Freeman.

Pellett, P. L. 1983. Changing concepts on world malnutrition. *Ecology of Food and Nutrition* 13:115–125.

Pellett, P. L. 1990a. Food energy requirements in humans. *American Journal of Clinical Nutrition* 51:711–722.

Pellett, P. L. 1990b. Protein requirements in humans. *American Journal of Clinical Nutrition* 51:723–737.

Penning de Vries, F. T. W., et al. 1995. Natural resources and limits of food production in 2040. In *Eco-regional Approaches for Sustainable Land Use*, J. Bouma et al., eds., pp. 65–87. Dordrecht: Kluwer Academic.

Penning de Vries, F. T. W., et al. 1996. The role of soil science in estimating global food security in 2040. In *The Role of Soil Science in Interdisciplinary Research*, pp. 17–33. Madison, Wisc.: American Society of Agronomy and Soil Science.

Pennington, J. A. T. 1996. Intakes of minerals from diets and foods: is there a need for concern? *The Journal of Nutrition* 126:2304S–2308S.

Perren, R. 1985. The retail and wholesale meat trade 1880–1939. In *Diet and Health in Modern Britain*, D. J. Oddy and D. S. Miller, eds., pp. 46–65. London: Croom Helm.

Pierce, F. J., et al. 1984. Productivity of soils in the Cornbelt: an assessment of the long-term impact of erosion. *Journal of Soil and Water Conservation* 39:131–136.

Piers, L. S., and P. S. Shetty. 1993. Basal metabolic rates of Indian women. *European Journal of Clinical Nutrition* 47:586–591.

Piers, L. S., et al. 1997. The validity of predicting the basal metabolic rate of young Australian men and women. *European Journal of Clinical Nutrition* 51:333–337.

Pimentel, D., ed. 1993. *World Soil Erosion & Conservation*. New York: Cambridge University Press.

Pimentel, D., et al. 1995. Environmental and economic costs of soil erosion and conservation benefits. *Science* 267:1117–1122.

Pimm, S. L. 1997. In search of perennial solutions. *Nature* 389:126–127.

Pingali, P. L., et al. 1997. *Asian Rice Bowls: The Returning Crisis?* Wallingford, England: CAB International.

Plotkin, M. J. 1986. The outlook for new agricultural and industrial products from the tropics. In *Biodiversity*, E. O. Wilson, ed., pp. 106–115. Washington, D.C.: National Academy of Sciences.

Poleman, T. T. 1993. Food values. *The Economist* 326(7795):8.

Poleman, T. T. 1996. Global hunger: the methodologies underlying the official estimates. *Population and Environment* 17:545–568.

Poleman, T. T., and L. T. Thomas. 1995. Income and dietary change. *Food Policy* 20:149–157.

Pollitt, E., ed. 1995. The relationship between undernutrition and behavioral development in children. *Journal of Nutrition* 125(Supplement): 2211S–2284S.

Pollitt, E., and A. Amante. 1984. *Energy Intake and Activity*. New York: Alan R. Liss.

Pond, W. G., et al. 1991. *Pork Production Systems*. New York: Van Nostrand Reinhold.

Popkin, B. M. 1993. Nutritional patterns and transitions. *Population and Development Review* 19:138–157.

Popkin, B. M., et al. 1993. The nutrition transition in China: a cross-sectional analysis. *European Journal of Clinical Nutrition* 47:333–346.

Popkin, B. M., and C. M. Doak. 1998. The obesity epidemic is a worldwide phenomenon. *Nutrition Reviews* 56:106–114.

Poppendieck, J. 1997. The USA: hunger in the land of plenty. In *First World Hunger*, G. Riches, ed., pp. 134–164. New York: St. Martin Press.

Population Action International. 1995. *Catching the Limit: Population and the Decline of Fisheries*. Washington, D.C.: Population Action International.

Postel, S. 1999. *Pillar of Sand Can the Irrigation Miracle Last?* New York: W. W. Norton.

Postel, S. L., et al. 1996. Human appropriation of renewable fresh water. *Science* 271:785–788.

Power, J. F., and R. F. Follett. 1987. Monoculture. *Scientific American* 256(3):78–86.

Prasad, R., and J. F. Power. 1995. Nitrification inhibitors for agriculture, health, and the environment. *Advances in Agronomy* 54:233–281.

Prasad, R., and J. F. Power. 1997. *Soil Fertility Management for Sustainable Agriculture*. Boca Raton, Fla.: Lewis.

Pratt, N. D., et al. 1997. Estimating areas of land under small-scale irrigation using satellite imagery and ground data for a study area in N.E. Nigeria. *The Geographical Journal* 163:65–77.

Prentice, A. M. 1984. Adaptations to long-term low energy intake. In *Energy Intake and Activity*, E. Pollitt and P. Amante, eds., pp. 3–31. New York: Alan R. Liss.

Prentice, A. M., et al. 1985. Unexpectedly low levels of energy expenditures in healthy women. *Lancet* 1(1985):1419–1422.

Prentice, A. M., et al. 1986. High levels of energy expenditure in obese women. *British Medical Journal* 292:983–987.

President's Science Advisory Committee. 1967. *The World Food Problem*. Washington, D.C.: The White House.

Preston, D. A. 1989. Too busy to farm: under-utilisation of farm land in Central Java. *The Journal of Development Studies* 26:43–57.

Price, L. L. 1997. Wild plant food in agricultural environments: a study of occurrence, management, and gathering rights in Northeast Thailand. *Human Organization* 56:209–221.

Prochaska, H. J., et al. 1992. Rapid detection of inducers of enzymes that protect against carcinogens. *Proceedings of the National Academy of Science USA* 89: 2394–2398.

Prudencio, Y. C., and R. Al-Hassan. 1994. The food security stabilization role of cassava in Africa. *Food Policy* 19:57–64.

Putman, J., et al. 1988. Using the erosion-productivity impact calculator (EPIC) model to estimate the impact of soil erosion for the 1985 RCA appraisal. *Journal of Soil and Water Conservation* 43(4):321–326.

Rao, M. K., and R. Nagarcenkar. 1977. Potentialities of the buffalo. *World Review of Animal Production* 13(3):53–62.

Raskin, P. 1997. *Water Futures: Assessment of Long-Range Patterns and Problems*. Boston: Stockholm Environment Institute.

Rathje, W. L. 1994. Beyond the pail. Less fat? Aw, baloney. *Nutrition Quarterly* 18:30–32.

Rathje, W. L., and C. Murphy. 1992. *Rubbish! The Archaeology of Garbage*. New York: HarperCollins.

Ravindran, V. 1990. Bananas. In *Nontraditional Feed Sources for Use in Swine Production*, P. A. Thacker and R. N. Kirkwood, pp. 13–20. Boston: Butterworth.

Rees, M. K. 1983. Cholesterol and the heart: will we act too late to lead? *Modern Medicine of Canada* 38(6):21.

Rees, R. M., et al. 1997. The effects of fertilizer placement on nitrogen uptake and yield of wheat and maize in Chinese loess soils. *Nutrient Cycling in Agroecosystems* 47:81–91.

Reij, C., et al., eds. 1996. *Sustaining the Soil*. London: Earthscan.

Reineccius, G. A. 1989. Flavor and nutritional concerns relating to the quality of refrigerated foods. *Food Technology* 43(1):84–89.

Renaud, S., et al. 1995. Cretan Mediterranean diet for prevention of coronary heart disease. *American Journal of Clinical Nutrition* 61(Supplement): 1360S–1367S.

Repetto, R., and S. S. Baliga. 1996. *Pesticides and the Immune System: The Public Health Risks*. Washington, D.C.: World Resources Institute.

Revelle, R. 1976. The resources available for agriculture. *Scientific American* 235(3):164–178.

Revelle, R., and H. E. Suess. 1957. Carbon dioxide exchange between atmosphere and ocean and the question of an increase of atmospheric CO_2 during the past decades. *Tellus* 9:18–27.

Rhoades, J. D., et al. 1992. *The Use of Saline Waters for Crop Production*. Rome: FAO.

Richards, J. F. 1990. Land transformation. In *The Earth as Transformed by Human Action*, B. L. Turner II et al., eds., pp. 163–178. New York: Cambridge University Press.

Richardson, G., et al. 1989. Gypsum blocks "tell a water tale." *Journal of Soil and Water Conservation* 44:192–195.

Richardson, T., and J. W. Finley, eds. 1985. *Chemical Changes in Food During Processing*. Westport, Ct.: AVI.

Riches, G., ed. 1997. *First World Hunger*. New York: St. Martin's Press.

Richter, D. D., and D. Markewitz. 1995. How deep is soil? *BioScience* 45: 600–609.

Riha, S. J., et al. 1996. Impact of temperature and precipitation variability on crop model predictions. *Climatic Change* 32:293–311.

Rinehart, K. E. 1996. Environmental challenges as related to animal agriculture—poultry. In *Nutrient Management of Food Animals to Enhance and Protect the Environment*, E. T. Kornegay, ed., pp. 21–28. Boca Ration, Fla.: Lewis.

Ripsin, C. M., and J. M. Keenan. 1992. The effects of dietary oat products on blood cholesterol. *Trends in Food Science and Technology* 3(6):137–141.

Robertson, G. P., et al. 1997. Soil resources, microbial activity, and primary production across an agricultural ecosystem. *Ecological Applications* 7:158–170.

Robles, M. D., and I. C. Burke. 1997. Legume, grass, and soil conservation reserve program effects on soil organic matter recovery. *Ecological Applications* 7:345–357.

Romanoff, S., and J. Lynam. 1992. Cassava and African food security: some ethnographic examples. *Ecology of Food and Nutrition* 27:29–41.

Rose, A. H., ed. 1982. *Fermented Foods*. London: Academic Press.

Rosegrant, M. W. 1997. *Water Resources in the Twenty-first Century: Challenges and Implications for Action*. Washington, D.C.: International Food Policy Research Institute.

Rosenberg, N. J. 1982. The increasing CO_2 concentration in the atmosphere and its implications on agricultural production. II. Effects through CO_2-induced climatic change. *Climatic Change* 4:239–254.

Rosenzweig, C., and M. L. Parry. 1994. Potential impact of climate change on world food supply. *Nature* 367:133–138.

Rosenzweig, C., et al. 1994. The effects of potential climate change on simulated grain crops in the United States. In *Implications of Climate Change for International Agriculture: Crop Modelling Study*, C. Rosenzweig and A. Iglesias, eds., pp. 1–24. Washington, D.C.: USEPA.

Rosenzweig, C., and D. Hillel. 1998. *Climate Change and the Global Harvest*. New York: Oxford University Press.

Rozanov, B. G., et al. 1990. Soils. In *The Earth as Transformed by Human Action*, B. L. Turner II et al., eds., pp. 203–214, New York: Cambridge University Press.

Rozelle, S., et al. 1996. Why China will not starve the world. *Choices* 1996(1):18–25.

Rubner, M. 1909. *Kraft und Stoff im Haushalte der Natur.* Leipzig: Akademische Verlagsgesellschaft.

Ruddle, K., and G. Zhong. 1988. *Integrated Agriculture-Aquaculture in South China: The Dike-Pond System of Zhujiang Delta*. Cambridge: Cambridge University Press.

Runge, C. F., et al. 1990. *Agricultural Competitiveness, Farm Fertilizer and Chemical Use, and Environmental Quality: A Descriptive Analysis.* Minneapolis: Center for International Food and Agricultural Policy.

Safina, C. 1995. The world's imperiled fish. *Scientific American* 273(5):46–53.

Sasaki, S., and H. Kesteloot. 1992. Value of Food and Agriculture Organization data on food-balance sheets as a data source for dietary fat intake in epidemiologic studies. *American Journal of Clinical Nutrition* 56:716–723.

Sathaye, J., and Gadgil A. 1992. Aggressive cost-effective electricity conservation. *Energy Policy* 20:163–172.

Sauvy, A. 1949. Le 'faux probleme' de la population mondiale. English translation in *Population and Development Review* 16:759–774.

Sawyer, J. E. 1994. Concepts of variable rate technology with considerations for fertilizer application. *Journal of Production Agriculture* 7:195–201.

Scharf, P. C., and M. M. Alley. 1994. Residual soil nitrogen in humid region wheat production. *Journal of Production Agriculture* 7:81–85.

Scharf, P. C., and M. M. Alley. 1995. Nitrogen loss inhibitors evaluated for humid-region wheat production. *Journal of Production Agriculture* 8:269–275.

Schertz, D. L., et al. 1989. Effect of past soil erosion on crop productivity in Indiana. *Journal of Soil and Water Conservation* 44:604–608.

Schimmelpfennig, D., et al. 1996. *Agricultural Adaptation to Climate Change.* Washington, D.C.: USDA.

Schipper, L., and S. Meyers. 1992. *Energy Efficiency and Human Activity.* Cambridge: Cambridge University Press.

Schlegel, A. J., and J. L. Havlin. 1995. Corn response to long-term nitrogen and phosphorus fertilization. *Journal of Production Agriculture* 8:181–185.

Schoeller, D. A. 1990. How accurate is self-reported dietary energy intake? *Nutrition Reviews* 48:373–379.

Schofield, W. N., et al. 1985. Predicting basal metabolic rate: new standards and review of previous work. *Human Nutrition and Clinical Nutrition* 39C(Supplement 1):5–41.

Seckler, D., and M. Rock. 1995. *World Population Growth and Food Demand to 2050.* Winrock International Institute for Agricultural Development.

Semenov, M. A., and J. R. Porter. 1995. Climatic variability and the modelling of crop yields. *Agriculture and Forest Meteorology* 72:265–283.

Sen, A. 1984. Family and food: sex bias in poverty. In *Resources, Values and Development,* pp. 346–348. Cambridge: Harvard University Press.

Sere, C., and H. Steinfeld. 1996. *World Livestock Production Systems.* Rome: FAO.

Serra-Majem, L., et al. 1995. How could changes in diet explain changes in coronary heart disease mortality in Spain? The Spanish paradox. *American Journal of Clinical Nutrition* 61(Supplement):1351S–1359S.

Service, R. F. 1997. Microbiologists explore life's rich, hidden kingdoms. *Science* 275:1740–1742.

Shaffer, M. J., et al., 1994. Long-term effects of erosion and climate interactions on corn yield. *Journal of Soil and Water Conservation* 49(3):272–275.

Sharma, P. K., and S. R. D. Datta. 1994. Rainwater utilization efficiency in rain-fed lowland rice. *Advances in Agronomy* 52:85–130.

Sharma, S. N., and R. Prasad. 1996. Use of nitrification inhibitors (neem and DCD) to increase N efficiency in maize-wheat cropping system. *Fertilizer Research* 44:169–175.

Shepard, M. 1991. How to improve energy efficiency. *Issues in Science and Technology* 7:85–91.

Shiklomanov, I. A. 1993. World fresh water resources. In *Gleick, ed Water in Crisis: A Guide to the World's Fresh Water Resources*, P. H. Gleick, ed., pp. 13–24. New York: Oxford University Press.

Shorland, F. B. 1988. Is our knowledge of human nutrition soundly based? *World Review of Nutrition and Dietetics* 57:126–213.

Sillence, M. N. 1996. Evaluation of new technologies for the improvement of nitrogen utilization in ruminants. In *Nutrient Management of Food Animals to Enhance and Protect Environment*, E. T. Kornegay, ed., pp. 105–133. Boca Raton, Fla.: Lewis.

Simon, J. L. 1981. *The Ultimate Resource.* Princeton, N.J.: Princeton University Press.

Simon, J. L. 1996. *The Ultimate Resource 2.* Princeton, N.J.: Princeton University Press.

Simoons, F. J. 1978. The geographic hypothesis and lactose malabsorption: a weighing of evidence. *The American Journal of Digestive Diseases* 23:963–980.

Simpson, J. R., et al. 1994. *China's Livestock and Related Agriculture: Projections to 2025.* Wallingford, England: CAB International.

Sims, J. T., and D. C. Wolf. 1994. Poultry waste management: agricultural and environmental issues. *Advances in Agronomy* 52:1–63.

Sinclair, T. R., et al. 1984. Water-use efficiency in crop production. *BioScience* 34:36–40.

Sinclair, T. R., and T. Horie. 1989. Leaf nitrogen, photosynthesis and crop radiation use efficiency: a review. *Crop Science* 29:90–98.

Skirvin, D. J., et al. 1997. The effect of climate change on an aphid-coccinellid interaction. *Global Change Biology* 3:2–11.

Smale, M. 1997. The Green Revolution and wheat genetic diversity: some unfounded assumptions. *World Development* 25:1257–1269.

Smil, V. 1985. China's Food. *Scientific American* 253(6):116–124.

Smil, V. 1987. *Energy Food Environment.* Oxford: Oxford University Press.

Smil, V. 1989. Coronary heart disease, diet, and Western mortality. *Population and Development Review* 15:399–424.

Smil, V. 1991. *General Energetics*. New York: John Wiley.

Smil, V. 1993. *China's Environmental Crisis*. Armonk, N.Y.: M. E. Sharpe.

Smil, V. 1994a. How many people can the Earth feed? *Population and Development Review* 20:255–292.

Smil, V. 1994b. *Energy in World History*. Boulder: Westview.

Smil, V. 1995. Who will feed China? *The China Quarterly* 143: 801–813.

Smil, V. 1996a. *Environmental Problems in China: Estimates of Economic Costs*. Honolulu: East-West Center.

Smil, V. 1996b. Who Will Feed China? Fourth Annual Hopper Lecture, University of Guelph.

Smil, V. 1997a. *Cycles of Life*. New York: Scientific American Library.

Smil, V. 1997b. China's environment and security: simple myths and complex realities. *SAIS Review* 17:107–126.

Smil, V. 1998. China's energy and resource uses: continuity and change. *The China Quarterly* 156:935–951.

Smil, V. 1999a. Crop residues: agriculture's largest harvest. *BioScience* 49:299–308.

Smil, V. 1999b. China's agricultural land. *The China Quarterly* 158:130–145.

Smil, V. 2000. *Transforming the World: Synthesis of Ammonia and Its Consequences*. Cambridge, Mass.: MIT Press.

Smil, V., and Y. Mao, eds. 1998. *The Economic Costs of China's Environmental Degradation*. Cambridge, Mass.: American Academy of Arts and Sciences.

Smit, B., and Cai, Y. 1996. Climate change and agriculture in China. *Global Environmental Change* 6:205–214.

Smit, B. et al. 1996. Agricultural adaptation to climatic variation. *Climatic Change* 33:7–29.

Smith, R. 1997. Pork producers starting to look at a concept of wean-to-finish production. *Feedstuffs* July 28:9.

Snow, A. A., and P. M. Palma. 1997. Commercialization of transgenic plants: potential ecological risks. *BioScience* 47:86–96.

Soane, B. D., and C. van Ouwerkerk. 1994. *Soil Compaction in Crop Production*. Amsterdam: Elsevier.

Sohal, R. S., and R. Weindruch. 1996. Oxidative stress, caloric restriction, and aging. *Science* 273:59–67.

Song, S., et al. 1997. Comprehensive sustainable development of dryland agriculture in Northwest China. *Journal of Sustainable Agriculture* 9:67–83.

Southgate, D. A. T. 1991. Nature and variability of human food consumption. *Philosophical Transactions of the Royal Society B* 334:281–288.

Splinter, W. E. 1976. Center-pivot irrigation. *Scientific American* 234:89–97.

Spurgeon, D. 1997. End agreed for ozone-destroying pesticide. *Nature* 389:310.

Stanford, C. B. 1996. The hunting ecology of wild chimpanzees: implications for the evolutionary ecology of Pliocene hominids. *American Anthropologist* 98: 96–113.

Stanford, C. B. 1998. *Chimpanzee and Red Colobus*. Cambridge, Mass.: Harvard University Press.

Stanhill, G. 1985. The water resource for agriculture. *Philosophical Transactions of the Royal Society London* B310:161–173.

Stanhill, G. 1986. Water use efficiency. *Advances in Agronomy* 39:53–85.

Stanhill, G. 1990. *Irrigation in Israel*. Bet Dagan, Israel: Agricultural Research Organization.

Staples, C. R. 1992. Forage selection, harvesting, storing, and feeding. In *Large Dairy Herd Management*, H. H. Van Horn and C. J. Wilcox, eds., pp. 347–357. Champaign, Ill.: American Dairy Science Association.

State Statistical Bureau. 1980–1998. *China Statistical Yearbook*. Beijing: SSB.

Statistics Bureau. 1950–1998. *Japan Statistical Yearbook*. Tokyo: Statistics Bureau.

Steffens, W. 1989. *Principles of Fish Nutrition*. Chichester, England: Ellis Horwood.

Steinmetz, K. A. 1996. Vegetables, fruit, and cancer prevention: a review. *Journal of the American Dietetic Association* 96:1027–1039.

Steverink, M. H. A., et al. 1994. The influence of restricting nitrogen losses of dairy farms on dairy cattle breeding goals. *Netherlands Journal of Agricultural Science* 42:21–27.

Suarez, F. L., and D. A. Savaiano. 1997. Diet, genetics, and lactose intolerance. *Food Technology* 51(3):74–76.

Subcommittee on Biological Energy. 1981. *Nutritional Energetics of Domestic Animals*. Washington, D.C.: NAS.

Sukhatme, V. P. 1970. Size and nature of the protein gap. *Nutrition Reviews* 28:223–226.

Sutton, M. A., et al. 1993. The exchange of ammonia between the atmosphere and plant communities. *Advances in Ecological Research* 24:301–390.

Tacon, A. 1994. Dependence of intensive aquaculture systems on fishmeal and other fishery resources. *FAO Aquaculture Newsletter* 6:10–16.

Tacon, A. G. J. 1995. Aquaculture feeds and feeding in the next millennium: major challenges and issues. *FAO Aquaculture Newsletter* 1195(10):2–8.

Tacon, A. G. J. 1996. *Feeding Tomorrow's Fish: The Asian Perspective*. Rome: FAO.

Tangney, C. C. 1997. Vitamin E and cardiovascular disease. *Nutrition Today* 32:13–22.

Terborgh, J. 1992. Why American songbirds are vanishing. *Scientific American* 266(5):98–104.

Thacker, P. A., and R. N. Kirkwood. 1990. *Nontraditional Feed Sources for Use in Swine Production.* Boston: Butterworths.

Thomas, D. S. G. 1993. Sandstorm in a teacup? Understanding desertification. *The Geographical Journal* 159:318–331.

Thomas, D. S. G., and N. J. Middleton. 1994. *Desertification: Exploding the Myth.* Chichester, England: John Wiley & Sons.

Thrupp, L. A. 1997. *Linking Biodiversity and Agriculture.* Washington, D.C.: World Resources Institute.

Tiffen, M., et al. 1994. *More People, Less Erosion: Environmental Recovery in Kenya.* Chichester, England: John Wiley & Sons.

Tilman, D., et al. 1997. The influence of functional diversity and composition on ecosystem processes. *Science* 277:1300–1302.

Toutain, J-C. 1971. La consommation alimentaire en France de 1789 a 1964. *Économies et Sociétés* 5:1909–2049.

Trichopoulou, A., et al. 1993. The traditional Greek diet. *European Journal of Clinical Nutrition* 47(Supplement):S76-S81.

Trichopoulou, A., et al. 1995. Diet and survival of elderly Greeks: a link to the past. *American Journal of Clinical Nutrition* 61(Supplement):1346S-1350S.

Ulijaszek, S. J. 1991. Human dietary change. *Philosophical Transactions of the Royal Society B* 334:271–279.

UN Advisory Committee on the Application of Science and Technology to Development. 1968. *International Action to Avert the Impending Protein Crisis.* New York: UN.

Unger, P. W., and T. M. McCalla. 1980. Conservation tillage systems. *Advances in Agronomy* 33:1–58.

UNICEF. 1995. *The State of the World's Children 1995.* New York: Oxford University Press.

United Nations. 1998. *World Population Prospects: The 1998 Revision.* New York: UN.

U.S. Department of Agriculture. 1910–1999. *Agricultural Yearbook.* Washington, D.C.: USDA.

U.S. Department of Agriculture. 1972. *How to Control Wind Erosion.* Washington, D.C.: USDA.

U.S. Department of Agriculture. 1997a. The Continuing Survey of Food Intakes by Individuals (CSFII) and the Diet and Health Knowledge Survey (DHKS), 1994–96. http://sun.ars-grin.gov/ars/Beltsville

U.S. Department of Agriculture. 1997b. More than one-fourth of U.S. food wasted, USDA study finds. http://www.usda.gov/fcs.

U.S. Department of Health and Human Services. 1988. *The Surgeon General's Report on Nutrition and Health*. Washington, D.C.: USDHHS.

U.S. Office of Technology Assessment. 1992. *A New Technological Era for American Agriculture*. Washington, D.C.: OTA.

Valiela, I., et al. 1997. Nitrogen loading from coastal watersheds to receiving estuaries: new method and application. *Ecological Applications* 7:358–380.

van Itallie, T. B. 1997. Epidemiology of obesity in North America. *The Canadian Journal of Diagnosis* Supplement 1997:3–5.

van der Voet, E., et al. 1996. Nitrogen pollution in the European Union—origins and proposed solutions. *Ambio* 23:120–132.

van Dijk, J. A., and M. H. Ahmed. 1993. *Opportunities for Expanding Water Harvesting in Sub-Saharan Africa: The Case of the Teras of Kassala*. London: IIED.

van Noordwijk, M. 1996. Decomposition: driven by nature or nurture? *Applied Soil Ecology* 4:1–3.

van Steenbergen, W., et al. 1984. Food consumption of different household members in Machakos, Kenya. *Ecology of Food and Nutrition* 14: 1–19.

Veenstra, J., et al. 1993. Alcohol consumption in relation to food intake and smoking habits in the Dutch National Food Consumption Survey. *European Journal of Clinical Nutrition* 47:482–489.

Verplancke, H. J., et al., eds. 1992. *Water Saving Techniques for Plant Growth*. Amsterdam: Kluwer Academic.

Vollenweider, R. A., et al., eds. 1992. *Marine Coastal Eutrophication*. Amsterdam: Elsevier.

Waggoner, P. E., et al. 1996. Lightening the tread of population on the land: American examples. *Population and Development Review* 22:531–545.

Walker, A. R. P., and I. Segal. 1996. Fibre and colorectal cancer. *Lancet* 348: 957.

Walker, A. R. P., and B. F. Walker. 1995. Nutrition-related diseases in Southern Africa: with special reference to urban populations in transition. *Nutrition Research* 15:1053–1094.

Wang, H. L., et al. 1979. *Soybeans as Human Food—Unprocessed and Simply Processed*. Washington, D.C.: USDA.

Wardle, D. A., et al. 1997. The influence of island area on ecosystem properties. *Science* 277:1296–1299.

Wargo, J. 1996. *Our Children's Toxic Legacy*. New Haven: Yale University Press.

Wasan, H. S., and R. A. Goodland. 1996. Fibre-supplemented foods may damage your health. *Lancet* 348:319–320.

Watson, R. T., et al. 1990. Greenhouse gases and aerosols. In *Climate Change: The IPCC Scientific Assessment*, J. T. Houghton et al., eds., pp. 1–40. Cambridge: Cambridge University Press.

Watson, R. T., et al., eds. 1996. *Climate Change 1995: Impacts, Adaptations and Mitigation of Climate Change: Scientific Analysis.* Cambridge: Cambridge University Press.

Webb, A. L. J., et al. 1997. Urea as a nitrogen fertilizer for cereals. *Journal of Agricultural Science* 128:263–271.

Webb, G. P. 1994. A survey of 50 years of dietary standards, 1943–1993. *Journal of Biological Education* 28:39–46.

Weesies, G. A., et al. 1994. Effect of soil erosion on crop yield in Indiana: results of a 10 year study. *Journal of Soil and Water Conservation* 49:597–600.

Weindruch, R. 1996. Caloric restriction and aging. *Scientific American* 274(1):46–52.

Westerterp, K. R., et al. 1986. Use of the doubly labeled water technique in humans during hearvy sustained exercise. *Journal of Applied Physiology* 61:2162–2167.

Westerterp-Plantega, M. S., et al. 1994. *Food Intake and Energy Expenditure.* Boca Raton, Fla.: CRC Press.

Whitt, F. R., and D. G. Wilson. 1993. *Bicycling Science.* Cambridge, Mass.: MIT Press.

Whittemore, C. T. 1993. *The Science and Practice of Pork Production.* Essex, England: Longman.

Wibawa, W. D., et al. 1993. Variable fertilizer application based on yield goal, soil fertility, and soil map unit. *Journal of Production Agriculture* 6:255–261.

Widdowson, E. M. 1947. *A Study of Individual Children's Diets.* London: Medical Research Council.

Willett, W. 1990. *Nutritional Epidemiology.* New York: Oxford University Press.

Williams, D. M., and T. M. Embley. 1996. Microbial diversity: domains and kingdoms. *Annual Review of Ecology and Systematics* 27:569- 595.

Williams, M. 1996. *The Transition in the Contribution of Living Aquatic Resources to Food Security.* Washington, D.C.: IFPRI.

Wilson, C. S. 1985. Nutritionally beneficial cultural practices. *World Review of Nutrition and Dietetics* 45:68–96.

Wilson, E. O., ed. 1986. *Biodiversity.* Washington, D.C.: National Academy of Sciences.

Wittwer, S. H. 1974. Maximum production capacity of food crops. *BioScience* 24:216–224.

Woods, S. C., et al. 1998. Signals that regulate food intake and energy homeostasis. *Science* 289:1378–1383.

World Bank. 1996. *Poverty in China: What Do the Numbers Say?* Washington, D.C.: World Bank.

World Health Organization. 1993. *Global Prevalence of Iodine Deficiency Disorders.* Geneva: WHO.

World Resources Institute. 1996. *World Resources 1996–97*. New York: Oxford University Press.

Worrest, R. C., and M. M. Caldwell, eds. 1986. *Stratospheric Ozone Reduction, Solar Ultraviolet Radiation and Plant Life*. Berlin: Springer Verlag.

Wu, G., and H. Guo, eds. 1994. *Land Use of China*. Beijing: Beijing Science Press.

Wu, H. 1996. Wining and dining at public expense in post-Mao China from the perspective of sayings. *East Asia Forum* 5(Fall 1996):1–37.

Wuest, P. J., et al., eds. 1987. *Cultivating Edible Fungi*. Amsterdam: Elsevier.

Wuest, S. B., and K. G. Cassman. 1992. Fertilizer-nitrogen use efficiency of irrigated wheat: I. Uptake efficiency of preplant versus late-season application. *Agronomy Journal* 84:682–688.

Yang, C. S. 1997. Inhibition of carcinogenesis by tea. *Nature* 389:134–135.

Yao, H., et al. 1996. Estimation of methane emissions from rice paddies in mainland China. *Global Biogeochemical Cycles* 10:641–649.

Yates, S. R., et al. 1997. Methyl bromide emissions from agricultural fields: bare-soil, deep injection. *Environmental Science & Technology* 41:1136–1143.

Young, V. R., et al. 1989. A theoretical basis for increasing current estimates of the amino acid requirements in adult man, with experimental support. *American Journal of Clinical Nutrition* 50:80–92.

Young, V. R., and P. L. Pellett. 1990. Current concepts concerning indispensable amino acid needs in adults and their implications for international nutrition planning. *Food and Nutrition Bulletin* 12:289–300.

Zelitch, I. 1982. The close relationship between net photosynthesis and crop yield. *BioScience* 32:797–802.

Zeuner, F. E. 1963. *A History of Domesticated Animals*. London: Hutchinson.

Zizza, C. 1997. The nutrient content of the Italian food supply 1961–1992. *European Journal of Clinical Nutrition* 51:259–265.

Index

71

122

166

248

249 A true reformer . . .

267 I'll bet promotion is irrelevant